Remote Sensing of Glaciers

Remote Sensing of Glaciers

Techniques for Topographic, Spatial and Thematic Mapping of Glaciers

Petri Pellikka

Department of Geosciences and Geography
University of Helsinki, Finland

W. Gareth Rees

Scott Polar Research Institute, University of Cambridge
England

CRC Press
Taylor & Francis Group
Boca Raton London New York

CRC Press is an imprint of the
Taylor & Francis Group, an **informa** business

A BALKEMA BOOK

Cover information:

Heinrich Schatz and Hans Hess enjoying a smoke while surveying the confluence of Hintereisferner and Kesselwandferner in 1929.

Svartisen ice caps in Norway in Landsat ETM+ satellite image of September, 7, 1999. (Heiskanen, J., K. Kajuutti, M. Jackson, H. Elvehøy & P. Pellikka, 2003. Assessment of glaciological parameters using Landsat satellite data in Svartisen, Northern Norway. EARSeL eProceedings 2 1/2003, 34-42. Observing our Cryosphere from Space).

CRC Press
Taylor & Francis Group
6000 Broken Sound Parkway NW, Suite 300
Boca Raton, FL 33487-2742

© 2010 by Taylor & Francis Group, LLC
CRC Press is an imprint of Taylor & Francis Group, an Informa business

Typeset by Vikatan Publishing Solutions (P) Ltd, Chennai, India

No claim to original U.S. Government works

ISBN-13: 978-0-415-40166-1 (hbk)
ISBN-13: 978-0-367-38464-7 (pbk)

Library of Congress Cataloging-in-Publication Data

Remote sensing of glaciers : techniques for topographic, spatial, and
thematic mapping of glaciers / editors, Petri K.E. Pellikka, W. Gareth
Rees Petri K.E. Pellikka.
    p. cm.
    Includes bibliographical references and index.
    ISBN 978-0-415-40166-1 (hardcover : alk. paper) – ISBN 978-0-203-85130-2 (e-book)
    1. Glaciers–Remote sensing. 2. Glaciers–Measurement. 3. Climatic
changes. I. Pellikka, Petri K.E. II. Rees, Gareth, 1959-III. Title.

GB2401.72.R42R45 2010
551.31'20223–dc22

                                   2009047393

Visit the Taylor & Francis Web site at
http://www.taylorandfrancis.com

and the CRC Press Web site at
http://www.crcpress.com

Contents

Foreword

Glaciers are an important part of the global cryosphere. They account for less than 1% of the total volume of ice locked up on land as glaciers and ice sheets, but are at present the dominant cryospheric contributor to global sea-level rise. The question of whether glaciers are growing or decaying, and at what rate, is therefore not only an academic one but is also of direct interest to governments and policymakers. Glaciers, and the larger ice caps beyond the margins of the great ice sheets of Antarctica and Greenland, cover a total of several hundred thousand square kilometres. This very large spatial scale implies that remote sensing instruments, whether deployed from the air or from satellite platforms, are essential to the scientific investigation of glaciers and their mass balance or state of health.

In fact, most glaciers, whether in the Arctic or in the mountain regions of the World, are shrinking; that is to say, they are retreating and thinning. A clear task for remote sensing is to provide quantitative datasets that can be used to measure the rates at which such changes are taking place. These data need to be acquired over as long a time series as possible, and with wide geographical coverage, so that any acceleration in the rate of mass loss from glaciers in our warming World can be detected.

This book, edited by Petri Pellikka and Gareth Rees, brings together a number of experts in the field of remote sensing of glaciers, to provide a set of authoritative chapters that range from early terrestrial photogrammetry to the latest airborne laser and satellite radar methods for investigating glacier change. Chapters on the mass balance and flow of glaciers, and on the physical principles that underlie remote sensing at various wavelengths, provide important background to those on instruments and techniques of measurement. These chapters, taken together, provide a clearly focused view of glacier measurement and monitoring that is an important contribution to investigations of glacier mass balance, with its important implications for global sea-level change over the coming century and beyond.

Julian Dowdeswell
Director
Scott Polar Research Institute
University of Cambridge

Acknowledgments

This book is the final outcome of the OMEGA project, which was funded by the 5th Framework Programme of the European Commission under the research programme Energy, Environment and Sustainable Development, thematic priority Global Change, Climate and Biodiversity and research area Development of new long-term observation capacity. The OMEGA (Development of Operational Monitoring System for European Glacial Areas – synthesis of Earth observation data of the present, past and future) project was funded from 2001 to 2004. I am grateful to the co-proposers of the project, Prof. Henrik Haggrén, Mr. Kari Kajuutti and Ms. Kukka-Maaria Luukkonen, to our collaborators at the University of Turku, Helsinki University of Technology, Novosat Ltd., the University of Innsbruck, Joanneum Research, Norut Tromsø, the Norwegian Water and Energy Directorate (NVE), SITO Ltd. and the Bavarian Academy of Sciences and Humanities as well as to the European Commission for its support. Herewith, I also acknowledge the companies who carried out the airborne flight campaigns over the glaciers West Svartisen (Norway) and Hintereisferner (Austria), Topscan GmbH and DLR, the German Aerospace Centre.

It was a very interesting personal experience to coordinate OMEGA, in which partners from three different countries and more than 35 people of ten different nationalities were involved. I would like to thank all the partners and individuals of the OMEGA consortium for their participation in the project, which was a great pleasure for me to coordinate. I hope that the partnership and friendship created will last for the decades to come.

My personal work for this book was carried out at the Department of Geography of the University of Helsinki and at the Scott Polar Research Institute at the University of Cambridge, with a grant from the Academy of Finland. Herewith, I have the honour to express my gratitude to SPRI for the inspiring atmosphere which the staff, institute and Cambridge itself provided. Magnus Magnusson from the International Glaciological Society provided valuable help for finding authors for the last missing chapter.

This book would not have been possible without the specialists writing chapters and without the reviewers commenting on the manuscripts. Gratitude is expressed by presentation of a small biography of each author and by including the name of the reviewers at the end of the book. It can be observed from this list that specialists from many institutes were involved. In the end, one third of the book was written by colleagues outside the OMEGA consortium based on the collaborations OMEGA had during its operation. Some of the figures were provided by other scientists and journals, which are listed at the end of the book. I would also like to thank my co-editor, Dr. Gareth Rees from SPRI, for helping me with the book. Being experienced in producing books about remote sensing and physics and as a native English speaker, Gareth was both a great help and a friend.

Finally, I would like to thank my wife Kristiina for being patient, also with this work, and last but not least my three daughters, Reetta, Venla and Noora.

Petri Pellikka

Author biography

Petri Pellikka obtained his B.Sc. and M.Sc. degrees at the University of Helsinki, Finland, specializing in Physical geography and Development geography. He carried out his Ph.D. studies at the University of Munich, Germany and at the University of Oulu, Finland, on airborne video camera sensor calibration and its application in studies of forest phenology in the German Alps. He did his post-doctoral research at Carleton University, Canada (1998–1999) on the remote sensing of forest damage caused by freezing rain. Dr. Pellikka has been a professor of geoinformatics at the Faculty of Science of the University of Helsinki since 2002. He leads the GeoInformatics Research Group of 10 researchers and Ph.D. students at the Department of Geosciences and Geography. In 2007–2008 he spent a sabbatical year at the University of Ghent, Belgium and at the Scott Polar Research Institute of the University of Cambridge, England. Dr. Pellikka is a Member of the Commission for Glaciology of the Bavarian Academy of Sciences and Humanities and a lecturer of physical geography, especially its remote sensing applications, at the University of Turku. His current research activities are concentrated in remote sensing of forest, land cover and land use changes in East Africa, particularly in Kenya (http://www.helsinki.fi/science/taita). Dr. Pellikka was the coordinator of the OMEGA project (http://omega.utu.fi), and participated in the field work activities, both in the Hintereisferner and Svartisen glaciers. He has supervised various Ph.D. and M.Sc. students at the University of Turku and University of Helsinki who worked on subjects related to the OMEGA. He has published over 110 scientific papers and over 40 popular articles.

Petri Pellikka, Department of Geosciences and Geography, University of Helsinki, P.O. Box. 64, FIN-00014 Helsinki, Finland. petri.pellikka@helsinki.fi

 **Gareth Rees** studied Natural Sciences as an undergraduate, specialising in Physics, and Radio Astronomy for his Ph.D., at the University of Cambridge. Since 1985 he has been a member of the academic staff at the Scott Polar Research Institute (SPRI), specialising in the application of spaceborne and airborne remote sensing techniques to the investigation of polar environments. He directs the Polar Landscape and Remote Sensing Group at SPRI. The work of this group includes field-based investigation of glaciers and ice caps in Svalbard and Iceland and laboratory-based studies of snow and terrestrial ice in other parts of the world including Siberia and the Caucasus Mountains. The group also has a major research focus on the dynamics of high-latitude vegetation (tundra and the northern part of the boreal forest) and its response to global climate change. Dr. Rees is also the co-director of a core project of the International Polar Year in which the Arctic treeline region is investigated. Dr. Rees conducts frequent fieldwork in Svalbard and in northern Russia. He has published seven books and over 70 scientific papers.

W. Gareth Rees, Scott Polar Research Institute, University of Cambridge, Lensfield Road Cambridge CB2 1ER, England. wgr2@cam.ac.uk

Preface:
Remote sensing of glaciers – glaciological research using remote sensing

The acronym OMEGA stands for Development of an Operational Monitoring System for European Glacial Areas – synthesis of Earth observation data of the present, past and future. The aim of the project was to develop methodologies using every potential remote sensing data source and method for constructing elevation models and areal delineations of glaciers. The data types used were terrestrial photography, aerial photography, digital airborne imagery, airborne laser scanner data, airborne radar data, optical high resolution satellite data (Landsat TM, ETM+), very high resolution optical satellite data (Ikonos, EROS), multiangular satellite data (MISR) and radar satellite data (ERS-1, Radarsat). In addition, old cartographic products dating back to late 1800s, were used. The time span of the remote sensing data ranged from 1907 (old terrestrial photography) to 2003.

A second aim was to develop an operational monitoring system for glacial areas. This system would compile the data developed in the project and could be updated periodically. In the first thoughts semi-automatic or automatic processing chains for satellite imagery were planned to be part of the system, but finally the system was modified to be a data management system rather than a monitoring system. In the system, the partners maintained their data among themselves, but shared the data using a common node set up at Norut in Tromsø, Norway. A few years after the project ended it became clear that the OMEGA system did not stand alone as planned as there was no sustainability and at the same time other glacier monitoring instruments, such as the World Glacier Monitoring System (WGMS) and Global Land Ice Measurements from Space (GLIMS), were functioning.

The glaciers Hintereisferner located in Öztal, Austria and Engabreen (part of the Svartisen ice cap, Norway) were chosen as study areas. The glaciers were different as Hintereisferner is a narrow retreating valley glacier and Engabreen is part of an ice cap and has advanced slightly during the 1990s. Common to both glaciers is the fact that they have been intensively studied: the University of Innsbruck has been studying Hintereisferner since the 1950s, and the Norwegian Water and Energy Directorate (NVE) has been studying the Engabreen outlet, which enabled us to use existing remote sensing data and glaciological measurements. In addition, there were old glacier maps of Hintereisferner and surrounding glaciers. The background knowledge was also very helpful when planning the field work activities.

During the project we faced some problems with the changing weather conditions, typical of mountainous areas, and also some technical and logistical problems related to airborne remote sensing. First of all, the time window for optical remote sensing data acquisition is narrow. The data should be acquired at the end of the glacier year, which means August-September in Europe and there should not be new snow on the glacier surface, since we also wanted to study the elevation of snow and firn lines at the end of the glacier's year. Sometimes this time window is too narrow to use airborne remote sensing, as well as for acquiring up-to-date satellite imagery.

However, we were able to acquire all the remote sensing data planned and also to analyse the data and produce the results. For example, laser scanner data were acquired ten times over Hintereisferner glacier in order to study the winter and summer balance changes. The D-day of the OMEGA project was August 12, 2003, when during one day, aerial photography, digital camera imagery, laser scanner data, Ikonos data and terrestrial photography were acquired. The field staff was also simultaneously in the field with their field measurement devices. In addition, two days later there was a MISR satellite overpass. These data allowed us to evaluate the accuracy of various elevation models over melting glacier surfaces, without temporal differences in various data types. More information about the OMEGA project can be found at http://omega.utu.fi.

This book presents the outcomes of the first aim of the OMEGA project, which was to develop methodologies using every potential remote sensing data and a method for constructing elevation models and areal delineations of glaciers. The target readers of this book are M.Sc. and Ph.D. students and professionals in natural sciences, especially in glaciology and remote sensing. It consists of chapters presenting principles of remote sensing, the background to glacier formation, and the main data types and their use in the remote sensing of glaciers.

In Chapter 1, remote sensing principles are introduced by Dr. Gareth Rees and by me, since basic knowledge of the physical background and data types of remote sensing is needed in the later chapters. In Chapter 2, Prof. Michael Kuhn presents the formation and dynamics of mountain valley glaciers, such as the Hintereisferner. In Chapter 3, the parameters that can be derived by remote sensing of glaciers are presented in order to highlight the connection between glaciology and remote sensing. In Chapter 4, Dr. Christoph Meyer presents the early history of remote sensing of glaciers, which is much older than aerial photography or satellite imagery. In Chapter 5, professors Matti Leppäranta and Hardy B. Granberg present the physics of the remote sensing of glaciers. In the following seven chapters, each data type and its applications are presented. In Chapter 6, Mr. Kari Kajuutti and his co-authors present the principles of terrestrial photogrammetry and in Chapter 7, Prof. Andreas Kääb presents aerial photogrammetry in glacier studies. The use of optical high resolution satellite imagery is presented in Chapters 8 and 12 by Dr. Frank Paul and Mr. Johan Hendriks. In Chapter 9, the principles of SAR imaging in glaciology are highlighted especially for Arctic glaciers by Dr. Kjell Arild Høgda and his co-authors. In Chapter 10, the most promising new data type, airborne laser scanner data, is presented by Dr. Thomas Geist and Prof. Hans Stötter. In Chapter 11, the use of a ground penetrating radar is presented for analyzing the internal characteristics and thickness of glaciers by Prof. Francisco Navarro and Dr. Olaf Eisen. In Chapter 13, the accuracies of the elevation models created from the data acquired on August 12, 2003, are presented by Dr. Olli Jokinen, and in Chapter 14, the accuracy aspects of topographical change detection are discussed by Jokinen and Geist. In the last Chapter 15, perspectives for world-wide glacier monitoring are provided by Prof. Andreas Kääb.

I hope that this book will be of interest to scientists specialized in glaciology, remote sensing and environmental issues in mountainous areas.

Petri Pellikka
Jalasjärvi, Finland
October 24, 2009

Abbreviations

3D	3 dimensional
AAR	accumulation area ratio
ADS-40	airborne digital sensor 40
AGC	automatic gain control
ALOS	Advanced Land Observing Satellite
ALS	airborne laser scanner
ALTM	Airborne Laser Terrain Mapper
ASAR	Advanced Synthetic Aperture Radar
ASIRAS	Airborne SAR/Interferometric Radar Altimeter System
ASTER	Advanced Spaceborne Thermal Emission and Reflection Radiometer
AVHRR	Advanced Very High Resolution Radiometer
AWI	Alfred Wegener Institute
BEW	Bundesamt für Eich- und Vermessungswesen (Austrian Mapping Agency)
BIE	bounds for the interpolation error
BRDF	bidirectional reflectance distribution function
BV	brightness value
CCD	charge coupled device
CIAS	Correlation Image Analysis (software)
CMP	common-midpoint method
CRT	conventional radiative transfer
CSC	Finnish IT Centre for Science
CTS	cold-temperate transition surface
DEM	digital elevation model
DInSAR	differential interferometric synthetic aperture radar
DLR	Deutsches Zentrum für Luft- und Raumfahrt (German Aerospace Center)
DMRT	dense medium radiative transfer
DN	digital number
DOS	dark object subtraction
DSM	digital surface model

DSS	drainage system side
DTM	digital terrain model
EL(A)	equilibrium line (altitude)
EROS	Earth Resources Observation and Science (of the United States Geological Survey)
ERS	European Remote Sensing Satellite
ESA	European Space Agency
E-SAR	Experimental Synthetic Aperture Radar
ETM+	Enhanced Thematic Mapper
FFT	fast Fourier transform
FM	frequency modulated
FMCW	frequency modulated continuous wave
FP5	Framework Programme 5 (of the European Commission)
GCM	global climate model
GCOS	Global Climate Observing System
GCP	ground control point
GeoTIFF	geographic tagged image file format
GHOST	Global Hierarchical Observing Strategy
GIFOV	ground-projected instantaneous field of view
GIS	Geographic Information System
GLIMS	Global Land Ice Measurements from Space
GPR	ground penetrating radar
GPS	Global Positioning System
GRACE	gravity recovery and climate experiment
GTOS	Global Terrestrial Observing System
HH	horizontal-horizontal (polarization)
HRSC-A	High Resolution Stereo Camera-Airborne
HRV	High Resolution Visible (SPOT)
HV	horizontal-vertical (polarization)
ICESat	Ice, Cloud and land Elevation Satellite
IGS	International Glaciological Society
IGY	International Geophysical Year
IMU	inertial measurement unit
INS	inertial navigation system
InSAR	interferometric SAR
IPY	International Polar Year
IR A	near infrared (NIR)
IR B	short wavelength infrared (SWIR)
IR C	mid wavelength infrared, long wavelength infrared (TIR)
IT	information technology
ITCZ	Inter Tropical Convergence Zone
LiDAR	Light Detection And Ranging
LISS (IRS)	Linear Imaging Self Scanning Sensor (Indian Remote Sensing)
LOS	line of sight
MIR	middle infrared
MISR	Multi-angle Imaging Spectroradiometer
MODIS	Moderate Resolution Imaging Spectroradiometer

NASA (JPL)	National Aeronautics and Space Administration (Jet Propulsion Laboratory)
NDSI	Normalised Difference Snow Index
NDVI	Normalised Difference Vegetation Index
NIR	near infrared
NSIDC	National Snow and Ice data Centre (of the United States)
NVE	Norwegian Water Resources and Energy Directorate
OMEGA	Operational Monitoring of European Glacial Areas
PALSAR	Phased Array type L-band Synthetic Aperture Radar
Pol-InSAR	Polarimetric SAR Interferometry
Radar	radio detection and ranging
RCM	regional climate model
RES	radio-echo sounding
RMS(E)	root mean square (error)
RWV	radio-wave velocity
SAR	synthetic aperture radar
SNR	signal to noise ratio
SPOT	Satellite Pour l'Observation de la Terre
SPRI	Scott Polar Research Institute
SRTM	shuttle radar topography mission
SWIR	short wavelength infrared
TIN	triangulated irregular network
TIR	thermal infrared
TM	Thematic Mapper
TOA	top of atmosphere
TUM	Technische Universität München (Technical University of Munich)
UHF	ultra high frequency
USD	United States dollar
USGS	United States Geological Survey
UTM	Universal Transverse Mercator
VHF	very high frequency
VNIR	visible and near infrared
VV	vertical-vertical (polarization)
VH	vertical-horizontal (polarization)
VHR	very high resolution
WGI	World Glacier Inventory
WGMS	World Glacier Monitoring Service
WGS	World Geodetic System
WiFS (IRS)	Wide Field Sensor (Indian Remote Sensing)
WMO	World Meteorology Organisation

Chapter 1

Principles of remote sensing

W. Gareth Rees
Scott Polar Research Institute, University of Cambridge, England

Petri Pellikka
Department of Geosciences and Geography, University of Helsinki, Finland

1.1 BACKGROUND

Remote sensing is, in general, the collecting of information from an object without making direct physical contact with it. The term is usually used in a more restricted sense in which the observation is made from above the object of interest, from a sensor carried on an airborne or spaceborne platform, and the information is carried by electromagnetic radiation, i.e. visible light, infrared or ultraviolet radiation, or radio waves (Rees 2001, 2006). This radiation can occur naturally, in which case the type of remote sensing is said to be passive, or it can be transmitted from the sensor to the object under investigation, in which case the remote sensing is said to be active. Passive remote sensing developed originally from aerial photography, and can be thought of as an extension of the idea of aerial photography to include other parts of the electromagnetic spectrum, other technologies for detecting the radiation and storing the data, and other platforms to carry the sensor. Active remote sensing grew from the military development of radar during the Second World War.

One of the most significant factors in the increasing applicability of remote sensing to many investigations in the environmental sciences, amongst other disciplines, has been the use of spaceborne platforms. Although remote sensing instruments were carried into orbit around the Earth in the 1960s, the age of satellite remote sensing effectively began in 1972 with the launch of Landsat 1 satellite. Spaceborne remote sensing provides a number of advantages compared with airborne observations. Information can be obtained from huge areas in a short time, and from locations that could be difficult or dangerous to overfly. As important as these, however, is the scope for continuity of data collection. While an individual satellite mission does not normally have a planned lifetime exceeding three to five years, even this is enormously longer than the period of continuous data collection achievable from an airborne platform. However, missions are often designed to provide continuity of consistent data coverage with previous missions, in some cases for several decades. These are all advantages that make spaceborne remote sensing particularly valuable for the study of glaciers, and the glaciological research community was and remains quick to identify and exploit the possibilities offered by satellite data.

1.2   ELECTROMAGNETIC RADIATION

The electromagnetic spectrum differentiates EMR (electromagnetic radiation) according to its wavelength, or equivalently its frequency. Different regions of the spectrum are given conventional names, more or less precisely defined, and while the entire spectrum includes many types of named EMR, only a few of these are important in remote sensing (Figure 1.1). We shall essentially be concerned only with that part of the spectrum having wavelengths between a few tenths of a micrometre (1 μm, a millionth of a metre, often informally called a *micron*) and a few centimetres, i.e. embracing the visible (possibly also the ultraviolet), infrared and microwave regions. Microwaves are a particular kind of radio wave, having wavelengths between 1 mm and 1 m. There are two main reasons for this restriction: firstly, it is in this part of the spectrum that virtually all naturally occurring radiation is found, and secondly, the Earth's atmosphere is more or less opaque to other forms of EMR, severely limiting the possibility of acquiring useful data about the Earth's surface, especially from a spaceborne platform. Within the optical part of EMR, several atmospheric gases, such as N_2O, O_2, O_3, CO_2 and H_2O, absorb or scatter the solar radiation, so that very little radiation reaches the earth's surface. In some parts of the spectrum radiation travels freely to the surface, where it is absorbed, reflected or transmitted. These wavelength regions are called atmospheric windows (Figure 1.1), and remote sensing sensors are designed to operate within these windows.

There are two important natural sources of EMR. The first of these is the Sun. Solar radiation (sunlight) is composed principally of ultraviolet, visible and infrared radiation, although much of the ultraviolet and infrared components are filtered out by the Earth's atmosphere. Passive remote sensing instruments that operate in this range of wavelengths are usually measuring reflected solar radiation. The other major

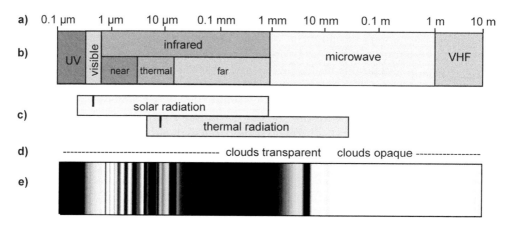

Figure 1.1 Region of the electromagnetic spectrum important in remote sensing. a) wavelength; b) spectral regions; c) approximate ranges of radiation emitted by the Sun and by objects at typical terrestrial temperatures (peak emissions are shown by heavier lines); d) transparency of clouds to EMR; e) typical transparency of the clear atmosphere to EMR. Lighter shades denote increasing transparency (atmospheric windows).

natural source of EMR is variously termed thermal radiation, black-body radiation or Planck radiation. This is radiation emitted by all bodies having a temperature above absolute zero (i.e. by all bodies), and the distribution of the radiation with wavelength depends on the temperature of the body. For example, for an object at 0°C (273 K) the dominant wavelength is around 11 μm and most normal terrestrial temperatures generate radiation in this region of the spectrum. For this reason, the part of the infrared spectrum between about 3 μm and 15 μm is often referred to as the *thermal infrared* region (also as mid-wavelength or long wavelength infrared, see Chapter 5). In fact, the distribution of the radiation is such that, while almost none is emitted at wavelengths significantly shorter than the dominant wavelength, there is a long 'tail' of emission at longer wavelengths. The Earth's atmosphere is highly opaque to wavelengths between about 20 μm and a few millimetres, but measurable amounts of thermal radiation can be detected at longer wavelengths, in the microwave region. This forms the basis of *passive microwave* remote sensing.

1.3 WHAT PROPERTIES OF EMR CAN BE MEASURED?

As we noted above, the fundamental idea of remote sensing is that information about the object being investigated is carried by EMR. This involves measuring some physical property of the EMR. In the case of passive systems that detect reflected solar radiation, the basic property is the *radiance* of the radiation reaching the sensor. Radiance is a measure of the intensity of the radiation in a particular direction, often specified as a function of wavelength, and in a typical remote sensing instrument the radiance is measured for a number of different wavebands across the visible and near-infrared spectrum. The radiance is often used to calculate a derived quantity, the *reflectance* of the surface from which the radiation was reflected. This depends on knowing how much radiation was incident on the Earth's surface in the same waveband and the geometry of the observation. However, a complication is introduced by the presence of the Earth's atmosphere, which both attenuates the radiance of EMR passing through it and also adds to the radiance, as a result of absorption and scattering processes. The reflectance should be calculated from the at-surface radiances, but the remote sensing instrument measures radiances above the atmosphere (or above some of the atmosphere in the case of an airborne observation). It is thus necessary to convert the at-satellite radiance to an at-surface radiance through atmospheric correction algorithms like MODTRAN (Berk et al. 1999) or 6S (Vermote et al. 1997).

The concept of radiance also applies to thermal radiation. However, another useful way of specifying the radiance in this case is as a *brightness temperature*. This is the temperature of a theoretical perfect emitter (a *black body*) that would produce the same radiance. As with measurements of reflected solar radiation, it is necessary to distinguish between at-surface and at-satellite brightness temperatures. A related concept is the *emissivity*. This is the ratio of the actual (at-surface) radiance to the radiance of a perfect emitter at the same physical temperature. Since no real object can emit more radiation than a perfect emitter at the same temperature, the emissivity has a maximum value of 1. Knowing the at-surface brightness temperature of a body whose emissivity is also known allows its physical temperature to be determined.

Table 1.1 The radar wavelengths and frequencies used in active microwave remote sensing.

Band	Wavelength (cm)	Frequency (GHz)	Sensor
K	0.75–2.4	40.0–12.5	
X	2.4–3.75	12.5–8.0	SIR-C/X-SAR
C	3.75–7.5	8.0–4.0	Radarsat, ERS-1/2, Envisat ASAR
S	7.5–15.0	4.0–2.0	
L	15.0–30.0	2.0–1.0	JERS-1, SEASAT
P	30.0–100.0	1.0–0.3	

Active remote sensing systems can provide more flexibility in the choice of measurable quantities, since the characteristics of the illuminating radiation can be controlled. There are two principal types of active instrument: *ranging* instruments, which measure the distance from the instrument to some reflecting interface which may be the surface of the Earth (for example, a glacier surface) but may instead be some sub-surface interface such as that between a glacier and the underlying bedrock, and instruments that measure the surface reflectance. The distinction is not a rigid one, and instruments may combine both aspects. There are ranging instruments using both the visible (in fact more usually the near-infrared) and the radio parts of the spectrum, but in all cases the primary variable to be measured is simply the two-way travel time for a short pulse of EMR transmitted from, and subsequently received back at, the instrument. Active instruments that measure the surface reflectance operate almost entirely in the microwave part of the spectrum, and are hence all broadly speaking *radars* (RAdio Detection And Ranging) of one kind or another. Although the concept is very similar to that of surface reflectance, the variable that is usually measured by these instruments is the *backscattering coefficient* ($\sigma°$) measured as decibels (dB). The backscattering coefficient often depends on the geometry of the observation, as well as on the wavelength of the radiation. The *polarization* of the transmitted and received radiation can also be significant in determining the backscattering coefficient. Here the polarization refers to radiation transmitted by the sensor and received by it. The signal can be transmitted and received in horizontal (H) or vertical (V) polarization. In the other words, there are four possible polarization combinations: HH, VV, HV, VH (Table 9.1). The *phase* of the returned signal can also be important, in the case of radar interferometry. The wavelengths of the EMR used in radar are much longer than in the visible and infrared parts of the spectrum, typically centimetres (Figure 1.1), and they are sometimes named using various letters of the alphabet (K, X, C, S, L, P). Alternatively, the region of the electromagnetic spectrum occupied by radar radiation can be specified by its frequency, and this can be particularly useful since frequency remains constant while wavelength changes when radiation passes through material of different refractive indices (Jensen 2000). The wavelengths and frequencies of the radar bands are described in Table 1.1. Similarly to optical measurements the radar wavelength is an important variable in the detection of various phenomena on the ground.

1.4 RESOLUTION

After the wavelength and type of measurement, one of the most important systems concepts that can be used to assess the suitability of a particular instrument for its intended application is its resolution, or resolving power. In very general terms, this relates to the ability of the instrument to distinguish between two similar things. It is convenient to identify four aspects to resolution: spatial, spectral, radiometric and temporal, which are described in Table 1.2 for some representative spaceborne optical remote sensing data applied to remote sensing of glaciers.

1.4.1 Spatial resolution

Most remote sensing instruments form two-dimensional images of the Earth's surface. In this context, spatial resolution can be thought of as the ability to distinguish between a point object and a horizontally extended object, or to recognise spatially distributed detail in an object. It is often assumed to be more or less coincident with the *pixel size*, or more precisely the size of the element on the surface that is imaged in a single pixel (picture element). This is something of an oversimplification, but it is often a useful approximation unless the imagery has been significantly oversampled (in which case the resolution will be coarser than the pixel size). In the case of analogue (film) photography, the image is not represented by pixels and it is more usual to specify the resolution by stating the scale of the photograph. However, as aerial photographs are increasingly being digitised using scanners, the pixel size or *ground resolution* is a practical characterisation of its spatial resolution. The effect of ground resolution on our ability to distinguish various features in a digital camera image over Svartisen glacier in Norway is shown in Figure 1.2. The finest resolutions in this figure (0.6 and 2.4 m) are similar to very high resolution (VHR) satellite data, the 12 and 24 m resolution to high resolution data (ASTER and Landsat TM/ETM+), and the coarsest (60 and 120 m) similar to Landsat MSS data. The ability to detect different surface features increases with finer spatial resolution.

Table 1.2 The resolutions of typical optical satellite sensors applied in glaciological research. VNIR = visible − near infrared, SWIR, short wavelength infrared, TIR = thermal infrared.

Resolution	Landsat MSS	Landsat TM	Landsat ETM+	SPOT HRV	SPOT 4	ASTER	MODIS	NOAA AVHRR	Ikonos
Spatial (m)	80 (VNIR) 240 TIR	30 (VNIR, SWIR) 120 TIR	30 (VNIR, SWIR) 60 TIR	20 (VNIR) 10 PAN	20 (VNIR, SWIR) 10 PAN	15 (VNIR) 30 (SWIR) 90 (TIR)	250 (1–2) 500 (3–7) 1000 (8–36)	1100	1 (PAN) 4 (VNIR)
Spectral	5 bands	7 bands	7 bands	3 bands	4 bands	14 bands	36 bands	6 bands	4 bands
Radiometric	6-bits 64 DN	8-bits 256 DN	8-bits 256 DN	8-bits 256 DN	8-bits 256 DN	8-bits 256 DN	12-bits 4096 DN	10-bits 1024 DN	11-bits 2048 DN
Temporal	18 days	16 days	16 days	26 days	26 days	16 days	1–2 days	<1 day	<3 days
Swath width (km)	185	185	185	60	60	60	2330	2700	11

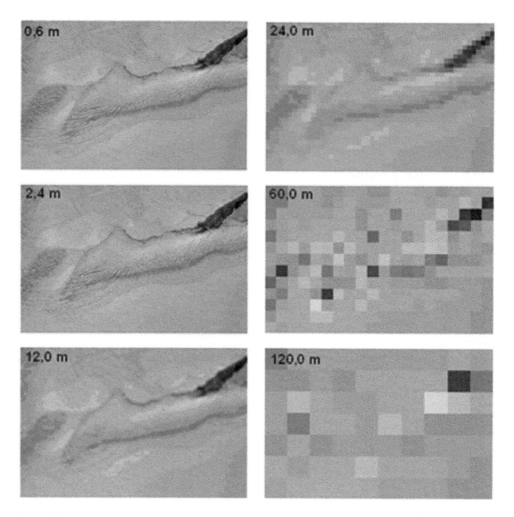

Figure 1.2 The effect of spatial resolution on the detection of features on a glacier surface, Svartisen, September 24, 2001. In the original NIKON D1H digital camera imagery of 0.6 m resolution several features can be detected (rock, ice, snow, crevasses, various firn lines), but at 60 and 120 m resolution only glacier and rock can be distinguished.

The spatial resolution of imaging systems other than synthetic aperture radars can be described as the product of the *angular resolution* with the range (i.e. the height above the surface for a downward-looking instrument). The angular resolution is more or less determined by the wavelength of the radiation and by the diameter of the objective lens (or, in the case of microwave systems, the antenna), short wavelengths and large lenses being needed for the finest angular resolutions. Thus the same instrument, deployed at different heights above the Earth's surface, will yield a spatial resolution proportional to the height. This has profound significance for the choice of an airborne as opposed to a spaceborne system. Aircraft used to acquire remote sensing data generally fly one or a few kilometres above the surface, while satellites used for the same purpose usually orbit at least 700 km above the surface. Airborne systems thus generally offer maximum spatial resolutions that are finer, by a factor of a hundred, than those achievable from spaceborne systems. For example, spaceborne imaging systems that operate in the visible/near infrared part of the spectrum typically give resolutions of 10s or 100s of metres (finer resolutions are possible), while airborne systems can give resolutions of 10s of centimetres. This difference, however, has diminished recently with the advent of very high resolution satellite imagery, such as from Ikonos, QuickBird or EROS, which provides spatial resolutions of the order of one metre or finer (Table 1.2). Not all imaging systems are designed to maximise the spatial resolution, since that may cause limitations elsewhere. This point is discussed under *temporal resolution* in Section 1.4.4. In the case of ranging instruments such as laser scanner (also known as LiDAR, Light Detection And Ranging) and ground-penetrating radar (GPR), the concept of spatial resolution also includes the *range resolution*. This is mainly determined by the duration of the transmitted pulse and hence largely independent of the height.

1.4.2   Spectral resolution

Spectral resolution refers to the ability to distinguish between radiation of different wavelengths. Typically spectral resolution is defined through the number and width of the instrument's wavebands, i.e. spectral ranges of sensitivity (Table 1.2). Narrower bands correspond to higher spectral resolutions. Imaging systems that operate in the visible/near infrared part of the spectrum are usually designed to capture the variation in radiance, and hence reflectance, with wavelength, since this can be characteristic of the surface type or condition. Colour photography achieves this aim to some extent, responding to the intensity of light reaching the film in the red, green and blue regions of the spectrum. These three spectral ranges are defined by filters, within the film in the case of analogue photography or in the camera in the case of digital camera imagery. The spectral resolution is comparatively coarse, each range being roughly 100 nm wide. Most visible/near infrared imagers provide a number of wavebands, typically 50 to 100 nm wide and defined by filters, to achieve what is usually termed *multispectral imaging*. Some instruments give significantly finer spectral resolution, usually to resolve particular features such as absorption lines, and *hyperspectral imagers*, such as AISA (www.specim.fi), provide hundreds or even thousands of narrow contiguous wavebands. Hyperspectral instruments have not yet found significant application to the study of glaciers, but some some research has been done on snow surfaces (e.g. Painter et al. 2003).

1.4.3   Radiometric resolution

Radiometric resolution is the ability to distinguish between two similar but not identical radiances. It can be thought of as having two components: the range of values of radiance to which the instrument can respond without the response being saturated, and the number of levels into which this range is divided. In the case of digital data, the latter is controlled by the number of bits (binary digits) used to represent the value of the radiance. Commonly, digital imaging systems use 8-bit data since this is simply compatible with many computer-based processing systems. This means that the value of the radiance (or whatever other parameter describing the radiometric response, such as the brightness temperature or the backscattering coefficient) is represented by an integer between 0 and 255, and that the range between the minimum and maximum values that can be represented is divided into 255 steps (digital numbers, DN). However, sometimes the number is less than 255, since some numbers may be reserved to indicate particular conditions such as that the data are not reliable for some reason. For example, data from band 1 of the ETM+ sensor carried on Landsat 7 are recorded as 8-bit numbers, which can be used as an approximation of radiometric resolution (Table 1.2). The lowest representable value of the radiance in this band is -6.2 W m^{-2} sr^{-1}μm^{-1} and the highest value is either 293.7 or 191.6 in the same units, depending on whether the band is set to its low or high gain setting respectively. The radiometric resolution of this band is thus either 1.18 or 0.78 W m^{-2} sr^{-1}μm^{-1}, corresponding to a resolution of around 0.2 or 0.1% in the reflectance. Similar considerations apply to thermal infrared channels. Again taking the ETM+ instrument as an example, the band 6 (thermal infrared) data are represented as 8-bit numbers with the minimum possible value corresponding to a radiance of 0 and the maximum value to 17.04 W m^{-2} sr^{-1}μm^{-1} (in low-gain mode). This gives a radiometric resolution of 0.067 W m^{-2} sr^{-1}μm^{-1}, and a corresponding resolution in the at-satellite brightness temperature of about 0.6 K (°C) for brightness temperatures around 273 K (0°C).

1.4.4 Temporal resolution

The concept of temporal resolution is normally applied to situations in which data are acquired repeatedly from the same location, in which case it is essentially the frequency of data collection. It is possible to identify some general principles governing temporal resolution in the case of spaceborne sensors.

Many remote sensing satellites are placed in *exactly-repeating orbits*, in which the sub-satellite track forms a closed curve on the Earth's surface, revisiting precisely the same location after n_1 days and n_2 orbits, n_1 and n_2 being integers. For satellites in low-Earth orbits, the ratio n_2/n_1 is close to 14.3. For example, Landsat 7 is in an exactly-repeating orbit with $n_1 = 16$ and $n_2 = 233$. There is thus one sense in which the temporal resolution of the ETM+ sensor carried by Landsat 7 is 16 days: a given point on the Earth's surface can potentially be imaged with the same geometry once every 16 days. Whether the potential is realised or not depends on the decision by the operating agency whether to collect data during a particular orbit of the satellite, and whether the location is free of cloud.

However, in another sense the temporal resolution of the ETM+ sensor is shorter than 16 days. The spacing between spatially adjacent sub-satellite tracks of the Landsat 7 satellite is $360°/233 = 1.55°$ in longitude, or about 120 km at a latitude of 45° North or South. Since the swath width of the ETM+ instrument is 185 km, a given point on the Earth's surface can potentially be imaged more than once in every 16 days, albeit not with the same geometry (Rees 1992). In fact, we would expect an average interval of roughly $16 \times 120/185 \approx 10$ days between these opportunities. Because of the convergence of lines of longitude, the temporal resolution increases towards the poles. The most important factor controlling the temporal resolution is the swath width of the sensor: wider swaths give more frequent viewing opportunities (Rees et al. 2002). Swath width and spatial resolution cannot be chosen completely independently of one another, and the maximum swath width usually corresponds to not more than about 10000 pixels. This implies that spatial and temporal resolution can also not be chosen independently, and is an important reason why coarser resolution sensors continue to be operated in parallel with finer-resolution sensors. For example, as we have noted, the ETM+ on Landsat 7 has approximately one viewing opportunity every 10 days at a latitude of 45°. On the other hand the MODIS instrument, with a swath width of around 2300 km (and a correspondingly coarser spatial resolution of 250–1000 m), gives more than one viewing opportunity per day (Table 1.2).

1.5 HOW ARE ELECTROMAGNETIC MEASUREMENTS CONVERTED INTO INFORMATION ABOUT GLACIERS?

We noted in Section 1.3 that what is measured in remote sensing is some property or combination of properties of electromagnetic radiation, detected at the sensor. From this, some property of the glacier has to be deduced. This is the purpose of a retrieval algorithm (Campbell 2007), and these are described in later chapters. In some cases there is an obvious link between the sensed property of the radiation and the geophysical parameter of interest. Examples of this situation include converting the range in a sounding or ranging measurement to the position of the reflecting point or surface, or calculating the physical surface of the temperature from the brightness temperature of the detected radiation. Less obvious retrievals have no direct link of this kind. For example, one might wish to delineate the boundary of a snow-covered area from an aerial photograph or a digital image acquired in the visible/near infrared. This delineation might be based on the fact that the reflectance of snow is higher than that of any other materials represented in the image, so that the image can be classified into snow/non-snow areas on this basis. Such classification could be performed manually or automatically, and might require training, in which case areas known (e.g. by field work) to represent snow and non-snow types are used to calibrate the process. More sophisticated forms of classification make use of multi-dimensional datasets, for example multispectral data from a sensor such as the ETM+, SAR backscattering data from different dates, and so on. Many commercial and public-domain computer packages, such as ENVI, Erdas Imagine, ER Mapper, Idrisi, Grass, etc. are available for performing image processing steps and classification.

1.6 PASSIVE REMOTE SENSING SYSTEMS

In Sections 1.6 and 1.7 we classify the principal types of remote sensing system considered in this book, and briefly outline their characteristics. The topics treated in the remainder of this chapter are covered in greater detail by, for example, Rees (2001, 2006).

1.6.1   Aerial photography

Aerial photography is the longest established technique in remote sensing, and as Chapters 3, 4 and 7 demonstrate, it has a long history of application to the study of glaciers in particular. It uses the visible and near-infrared regions of the electromagnetic spectrum, and hence cannot be carried out at night or through clouds. It is now almost entirely an airborne, as opposed to a spaceborne, technique, although there is a huge archive of spaceborne photography from the US military reconnaissance programme Corona[1] in the 1960s and 1970s, and some space photography is still carried out. Since the spatial resolution of an aerial photograph depends on the height from which it was acquired, space photographs generally have much poorer resolution than aerial photographs, although they do offer correspondingly increased spatial coverage.

Despite the inconvenience of the photochemical processing involved in film photography, it retains its value because of its relative simplicity, widespread availability, and perhaps above all its potentially very high geometric fidelity. This last aspect is successfully exploited in cartographic and stereophotogrammetric applications. Although a film photograph is not digital, it can easily be made so by scanning it. Application of digital camera imagery obtained by metric digital camera systems, such as, Leica ADS40, UltraCamD or Intergraph DMC, or by low cost systems, such as EnsoMOSAIC (Holm et al. 1999, Pellikka et al. 2009) are in any case beginning to rival film photography in glacier studies (e.g. Parviainen 2006) as elsewhere. However, one difficulty in three-dimensional mapping of glacier surfaces using either film or digital photographs is that of identifying surface features due to the uniformly bright glacier surface.

Aerial photography or digital camera imagery are still superior for smaller glaciers or for areas for which a laser scanner campaign would be complicated and too expensive. A good example is aerial photography of Mt. Kenya in equatorial Africa over 6000 m a.s.l. (Rostom and Hastenrath 1994). Aerial photography has remained in use for change detection for a long time: until the 1990s it was the only data type that could give a spatial resolution better than 10 m. A particular reason that favours aerial photography over very high resolution satellite data in high latitudes is shadowing (e.g. Dowdeswell 1986). Satellite sensors typically acquire images around 10 a.m., at a time when the solar elevation angle is fairly small causing long shadows in mountainous areas. Airborne remote sensing campaigns offer the flexibility to minimise this problem.

1.6.2   Visible/near infrared scanners

Electro-optical scanners, providing calibrated digital imagery in the visible and near-infrared regions of the electromagnetic spectrum, are the workhorses of remote sensing. They are in many ways similar to aerial photography, covering the same spectral region (and hence being subject to the same limitations of inoperability through clouds or at night), and can be thought of as being similar to digital aerial photography,

---

1 http://eros.usgs.gov/products/satellite/declass1.php

although a typical instrument usually records data in several spectral bands. Apart from their potential for generating multispectral data, scanners offer a number of advantages with respect to photography: the data are usually calibrated radiometrically, meaning that a given output corresponds to a known at-satellite radiance, the data are digital (and hence immediately compatible with computer-based processing), and the data can be transmitted by radio signal from the instrument to a suitable receiving station. This last feature is especially important for spaceborne data collection, and scanners are widely used from satellites although airborne systems are also available.

There is a wide variety of imagery from visible/near infrared scanners. One of the most useful ways to differentiate between them is by the spatial resolution. At the coarsest end of the scale are so called low resolution (LR) satellite imagery, which has a spatial resolution of the order of one kilometre. These include the geostationary imagers such as METEOSAT and GOES (Geostationary Operational Environmental Satellites), whose primary purpose is to collect synoptic meteorological data, and coarse resolution imagers such as MISR (Multiangle Imaging SpectroRadiometer), MODIS (Moderate Resolution Imaging Spectroradiometer) and AVHRR (Advanced Very High Resolution Radiometer) carried on satellites in low Earth orbits. The main advantage of these coarse resolution imagers is that they have correspondingly wide coverage (Figure 1.3), a single swath of data covering a couple of thousand kilometres up to (Table 1.2), in the case of the geostationary imagers, the whole of the Earth's disc. This in turn means that these imagers can collect data from huge areas very

Figure 1.3 An approximately 800 km wide extract of 32-day composite MODIS image between June 10 and July 11, 2005 (band 1) showing the Alps. The white areas are the highest mountain ridges covered still by last winter's snow. The longitudinal alpine valleys and glacier tongue lakes are clearly visible on both the southern side (Italy) and northern side (Switzerland, Germany) of the Alps.

quickly, so in effect they offer very high temporal resolution and the expense of coarse spatial resolution. MISR also provides an interesting multiangular view of the same land surface as its uses nine different sensors looking at the ground (Diner et al. 1999). This new kind of information have been used extensively for vegetation studies (e.g. Heiskanen 2006, Canisius and Chen 2007), but also for glacier studies (Nolin and Payne 2007).

The LR data are also often available almost or absolutely free of cost. However, they are not represented in this book, which focusses much more on the study of comparatively small glaciers and hence requires higher spatial resolution. Optical low resolution data is more suitable for the study of larger glaciers, ice-sheets and ice shelves (e.g. Steffen et al. 1993, Stroeve et al. 1997, Scambos et al. 2004, Stroeve et al. 2007, Nolin and Payne 2007) than for narrow alpine glaciers. For example, using AVHRR data over Hintereisferner approximately 10 pixels would cover the glacier and even then almost all the pixels would be mixed pixels including glacier and its surroundings (land, vegetation, etc.).

The next main class of scanner, on the basis of their spatial resolution, includes systems providing resolutions in the range 10 to 100 m. This category can be named high resolution (HR) satellite imagery. This is the most numerous class of spaceborne scanner, and it is well represented by the instruments carried on board the Landsat series of satellites (Table 1.2). The earlier Landsat satellites, till 1993, carried the Multispectral Scanner (MSS) which had four spectral bands and a spatial resolution of 80 m, while the later satellites (from 1982) have carried the Thematic Mapper (TM), with six spectral bands in the visible, near infrared and SWIR, and a spatial resolution of 30 m. For glaciological studies, the improvement in spatial resolution was a great leap, as 80 m resolution of MSS was too coarse for alpine glaciers. The Enhanced Thematic Mapper (ETM+), operational on Landsat 7 since 1999, has six spectral bands in this part of the spectrum, and a spatial resolution of 15–30 m. The coverage of a single Landsat image is approximately square, with a side length of 185 km. Landsat images have typically cost a few hundred Euros, although the cost model has varied over the years and Landsat data, including the historical archive, are now free of cost. For a variety of reasons, including the good match between the spatial resolution and coverage of scanner images and the requirements for investigating glaciers, data of this kind have found widespread application in glaciology, as discussed in particular in Chapters 8 and 12. Unfortunately, Landsat ETM+ has had reduced functionality since 2003, but on the other hand it has been replaced by ASTER (Advanced Spaceborne Thermal Emission and Reflection Radiometer), which has similar temporal resolution and a spatial resolution of 15 m in the VNIR area and 30 m in the thermal infrared (Table 1.2). Having also short-wave and thermal infrared bands it is useful for many masking operations (Chapter 8). SPOT HRV (Satellite Pour l'Observation de la Terre High Resolution Visible) and SPOT 4 have even finer spatial resolution than Landsat TM/ETM+, but the absence of SWIR bands in SPOT HRV (1, 2, 3) made the data incapable of cloud detection and the creation of masks. The areal coverage of SPOT makes it also somewhat unsuitable, although many glaciers do fit into the 60 by 60 km coverage.

The main disadvantage of Landsat and SPOT imagery for glacier studies is the poor temporal resolution, being 16 days for Landsat and 26 days for SPOT. With the overlapping paths, a 10 day revisit time is possible in the Alps (shorter at higher latitudes), but frequent cloud cover means that the opportunities for collecting imagery of the surface are much less frequent in practice. This can be a serious difficulty if it is necessary to obtain an image at a particular moment, such as when the glacier surface is free of new snow at the end of the glacier's year (see Chapter 3). Marshall et al. (1993) presented an analysis of the limitations to multitemporal visible satellite data imposed by cloud cover in European Arctic and found out that in the Russian High Arctic and Iceland only 7 to 15% of the images are useful and in Svalbard, Norway and Novaja Zemlja (Russia) only 15 to 27%. In East Greenland, the probability of obtaining cloud-free imagery was the highest, between 38 to 54%. Therefore quite often an aerial photograph acquisition is ordered to ensure that there are remote sensing data for the crucial time window.

The third main class of scanner, again on the basis of spatial resolution, provides resolutions of a few metres down to one metre or even finer. This can be termed very high resolution (VHR) satellite imagery. An example of such an instrument is provided by QuickBird, a commercial satellite whose sensor provides digital panchromatic (black and white) imagery with a resolution of 0.6–0.7 m and 4-band multispectral imagery with a resolution of 2.4–2.8 m. The swath width is correspondingly small at 16.5 km. As seen in the Table 1.2, the parameters are similar to another source of VHR data, Ikonos. This kind of imagery is thus an approximate substitute for aerial photography, although generally with poorer spatial resolution. The applications are

Figure 1.4 3D-visualisation of Hintereisferner made using an Ikonos stereopair of August 12, 2003. The peak of Langtauferer Spitze is located on the right at 3529 m a.s.l. while Hintereisferner in the front is at 2700 m a.s.l. Courtesy of OMEGA project (provided by Vesa Roivas, Pöyry Environment).

similar as presented by Sharov and Etzold (2007), who applied Ikonos stereopairs over Engabreen, Norway and Hintereisferner, Austria, in order to construct orthoimagery and elevation models and to map glacier boundaries and margins. VHR imagers can also sometimes be programmed to acquire stereo imagery, thus enabling the construction of digital elevation models (Chapter 13) and 3D visualisations over glaciers (Figure 1.4). The disadvantage of all of the VHR data is fairly high cost, typically a few thousand Euros or more if the image acquisition is ordered by the customer. Archived images are significantly cheaper.

1.6.3   Thermal infrared scanners

Scanners designed to measure the brightness temperature of radiation in the thermal infrared are often combined with visible/near infrared multispectral scanners. For example, the TM and ETM+ carried on Landsat satellites both include a band from 10.4 to 12.5 μm (Figure 1.5). The spatial resolution of thermal infrared bands tends to be a little coarser than that of the associated visible/near infrared bands, primarily because of the significantly longer wavelength and less recordable signal (Table 1.2). Since thermal infrared scanners detect emitted radiation rather than reflected solar radiation they can be operated at night. However, they cannot penetrate through

Figure 1.5 Part of Landsat 7 ETM+ band 6 (thermal infrared) image of July 11, 1999 of the Brøgger Peninsula, Svalbard, processed to show variations in at-satellite brightness temperature. The minimum brightness temperature in this extract is about 269 K (−4°C); the maximum is about 297 K (24°C). The darkest (coldest) areas correspond to glaciers, middle-tone grey areas correspond to the sea surface and the lightest areas are land.

cloud cover. The use of thermal infrared data for the study of glaciers is described in Chapters 3 and 8.

1.7 ACTIVE REMOTE SENSING SYSTEMS

1.7.1 Laser scanner (LiDAR)

Laser scanners, also called LiDAR or laser profilers, transmit short pulses of EMR from a laser. The wavelength range is typically in the near infrared region, e.g. 1064 nm (Chapter 10), but LiDARs have also been operated in green light for bathymetric applications. The laser pulse propagates through the atmosphere at very close to the speed of light, and is reflected from the first solid or liquid interface it encounters, provided that it is not absorbed by it. In the case of the green laser, some of the pulse also penetrates through the shallow water column and is reflected from the lake or sea bed. Measurement of the two-way propagation time gives the range from the instrument to the reflecting surface, and if the position and orientation of the instrument are known, the location of the reflecting point can be determined. Modern airborne laser scanner systems transmit tens or hundreds of thousands of pulses a second, and the direction of view is scanned from side to side by an oscillating mirror, so that a dense two-dimensional pattern of reflecting points can be laid down on the Earth's surface. Their position and orientation is determined using precise GPS methods. Although the data rate of laser scanners is relatively high for an airborne system, it is still comparatively slow. It takes around half an hour to collect the data from a typical small glacier, such as Hintereisferner with an area to be covered of around 10 km^2.

Laser scanners are thus very well suited to the measurement of topography of glaciers, forest canopies, ground or buildings, or even sea beds. An airborne laser scanner can achieve point densities of one or more points per square metre and vertical accuracies as high as 5 cm, which is dense and precise enough to give a number of applications for glaciology (Lutz et al. 2003, Arnold et al. 2006, Rees and Arnold 2007) in addition to the generation of precise digital elevation models (DEMs) (Figure 1.6). Nearly all laser scanner data are collected from airborne sensors, although satellite instruments do exist. The Geoscience Laser Altimeter System (GLAS) is carried on the ICESat satellite, launched in 2003. While it has a similar vertical accuracy to airborne systems, the spatial density of the reflecting points is much lower, with an along-track spacing of around 180 m. This somewhat unsuits it to the study of all but the largest glaciers. Chapter 10 discusses the application of laser scanner data to the study of glaciers, while Chapters 13 and 14 consider DEMs, whether generated by laser scanner data or other data, in general.

1.7.2 Ground-penetrating radar

Ground-penetrating radar is somewhat analogous to laser scanning in the sense that short pulses, in this case of radio-wavelength radiation, usually in the range 1–20 m, are transmitted by the instrument and the return signal is detected and analysed. In the ice, the wavelength is significantly shorter as a result of refraction. Freshwater ice is largely transparent at these wavelengths, provided it is cold enough, and the pulse propagates through the ice and is reflected from the bottom of the glacier. Reflections

Figure 1.6 Visualisation of the surface topography of Midre Lovénbreen, Svalbard, by shaded relief of airborne laser scanner data. Crevasses and surface meltwater channels are clearly visible.

also occur from significant stratification of the ice. Ground-penetrating radar is thus, in a glaciological setting, primarily a means of investigating the interior structure of a glacier. It does not, perhaps, fall entirely within the definition of a remote sensing technique given in Section 1.1 since it is not normally an airborne method and certainly not a spaceborne one. In fact the radar equipment is usually towed across the glacier surface, behind a snow scooter or similar vehicle (Figure 11.3). Since the transparency of the atmosphere is not a relevant consideration, the remarks made in Section 1.2 about the useful wavelengths for remote sensing also do not apply to ground-penetrating radar. Nevertheless, it is usual to describe it as a remote sensing technique. For the principles of ground-penetrating radars and their applications to glaciers, please refer to Chapter 11.

1.7.3 Synthetic Aperture Radar

Synthetic Aperture Radar (SAR) is an active imaging technique operating in the microwave part of the electromagnetic spectrum. As such, it is independent of daylight and operational through cloud cover. The SAR instrument transmits microwave radiation towards the Earth's surface, from an antenna, and the radiation reflected (scattered) from the Earth's surface is used to build up a two-dimensional representation of the variation of backscattering coefficient ($\sigma°$) across the scene. This simplified description would in fact apply to any imaging radar; what distinguishes a SAR from other types is the fact that the forward motion of the antenna, as it is carried on its airborne or spaceborne platform, is used to synthesise the effect of a very long antenna and hence to achieve a high spatial resolution in the forward (azimuth) direction. High resolution in the sideways (range) direction is achieved by transmitting short pulses of radiation and time-resolving the return signal. For this reason, the antenna has to point sideways rather than vertically downwards. The spatial resolution of a SAR is determined to a large extent by the signal processing applied after the data have been collected, and can range, for a typical spaceborne system, from 5 to 1000 m. As with visible/near infrared systems, fine spatial resolution is generally associated with narrow swath widths. For example, the Radarsat instrument can be operated in a number of different modes. The finest spatial resolution is 10 m, with a 45 km swath width, while at the opposite extreme it can be operated in a mode having a spatial resolution of 100 m and swath width of 500 km.

SAR imagery shares some similarities with visible/near infrared imagery, but there are some important differences too. The first generation SAR systems ERS-1/2 and Radarsat 1 (Table 9.1) provided only a single channel of data – a single wavelength and polarization – so the resulting image resembles a black-and-white photograph. However, when compared with a photograph the SAR image is very noisy. This is a consequence of the fact that the transmitted and scattered radiation is *coherent*, i.e. has a well-defined phase, which means that interference occurs between the radiation returned from adjacent parts of the Earth's surface. This phenomenon, which is termed *speckle* (Figure 1.7), reduces the radiometric resolution of the imagery. It can be reduced by averaging, although at the expense of the spatial resolution.

SAR imagery is necessarily acquired from an oblique viewing angle, as noted above. This has several consequences for the appearance of a SAR image if there is any appreciable variation in altitude across the region being imaged. *Foreshortening* is a terrain-induced distortion whereby the image of an elevated point is displaced by a terrain-dependent amount towards the ground-track of the radar. This has the effect of foreshortening the images of slopes that face towards the radar and lengthening those that face away from it. *Layover* is an extreme case of foreshortening, which occurs when the incident angle is smaller than the foreslope. In this case, the mountain has so much relief that the summit backscatters energy toward the antenna before the energy pulse from the antenna reaches the base of the mountain (Jensen 2000). *Highlighting* is the phenomenon whereby radar-facing slopes are brighter (i.e. exhibit higher backscattering coefficient) than the same type of material in a horizontal configuration. Finally, if one part of the terrain intercepts the microwave radiation from the radar and hence prevents it from reaching another part of the terrain, no return signal at all will be received from the latter. This is termed *shadowing*, but unlike optical shadows radar shadows are completely missing land surface information as the radar signal never reached it. Layover, highlighting and shadowing are effects faced especially in mountainous areas, which are typical environments for valley glaciers. Apart from shadowing, these geometrical effects can be corrected if a sufficiently high resolution DEM of the terrain is available.

Figure 1.7 Part of an ERS-1 SAR image (April 21, 2000) of the Brøgger Peninsula, Svalbard, showing the phenomena of speckle, foreshortening and highlighting. Image © ESA 2000.

A major advantage of SAR, not shared with VNIR systems, derives from the fact that it uses coherent radiation. This means that a change in the relative position of the radar and a scattering object of the order of one wavelength can produce a measurable change in the scattered signal from the object, and since the wavelength of a typical SAR system is just a few centimetres this provides a potentially huge precision in geometric measurements. This is the basis of *interferometric SAR* (InSAR). There are essentially two types of measurement that can be made using InSAR. The first of these is the determination of the three-dimensional shape of static objects by acquiring SAR images from two (or more) slightly different locations. This can be used to construct DEMs over glaciers (Chapter 9). The second type of measurement determines the motion of an object by acquiring SAR images from essentially the same location and two different times (Chapter 3). This has been used to measure the flow of glaciers. InSAR is not particularly straightforward to implement. It is often effected by *repeat-orbit interferometry*, in which data are acquired from the same satellite at different times as its orbit carries it to almost the same location. This can lead to some confusion between the effects of shape and of motion which can be resolved in a number of ways.

A typical spaceborne SAR is that carried on the Envisat satellite (Tables 1.1, 9.1). This transmits and receives at a wavelength of 5.6 cm in both HH and VV polarizations, and can be operated in different modes giving spatial resolutions of 30, 100 or 1000 m with swath widths of around 60 to 400 km. The principles and results of SAR imaging and InSAR as applied to glaciers are discussed in Chapter 9.

1.8 HOW ARE DATA OBTAINED? WHAT DO THEY COST?

In general the user is responsible for scheduling the collection of his or her own data from airborne sensors, or retrieving data that have already been collected. The situation with regard to spaceborne data is rather diverse. Some currently operational spaceborne systems collect all possible data, while others collect data only according to some programme which the individual scientists may or may not have the possibility to influence. Some satellite data are available free of charge, either to anyone, or to approved users, while some are strictly commercial (and can in some cases be rather expensive). As a general principle, the coarsest resolution data are collected systematically and available free of charge, while the highest-resolution imagery tends to be commercial and acquired on demand. Radar imagery is not usually free. Most agencies and companies that supply satellite data have searchable online archives and ordering systems.

REFERENCES

Arnold, N.S., W.G. Rees, B.J. Devereux and G.S. Amable (2006). Evaluating the potential of high-resolution airborne LiDAR data in glaciology. *International Journal of Remote Sensing* 27, 1233–1251.

Berk, A., G.P. Anderson, L.S. Bernstein, P.K. Acharya, H. Dothe, M.W. Matthew, S.M. Adler-Golden, J.H. Chetwynd, S.C. Richtsmeier, B. Pukall, C.L. Allred, L.S. Jeong and M.L. Hoke (1999). MODTRAN4 radiative transfer modelling for atmospheric correction. *SPIE Proceedings Series*, Vol. 3756, 348–353. Proceedings of the Optical Spectroscopy Techniques and Instrumentation for Atmospheric and Space Research III, Denver, U.S.A., July 19–21, 1999.

Campbell, J.B. (2007). *Introduction to Remote Sensing*. 4th ed., Guilford Press, New York, 626 p.

Canisius, F. and J.M. Chen (2007). Retrieving forest background reflectance in a boreal region from Multi-angle Imaging SpectroRadiometer (MISR) data. *Remote Sensing of Environment* 107 (1–2), 312–321.

Diner, D.J., G.P. Asner, R. Davies, Y. Knyazikhin, J.P. Muller, A.W. Nolin, B. Pinty, C.B. Schaaf and J. Stroeve (1999). New directions in Earth observing: Scientific applications on multiangle remote sensing. *Bulletin of the American Meteorological Society* 80, 2209–2228.

Dowdeswell, J.A. (1986). Remote sensing of ice cap outlet glacier fluctuations on Nordaustlandet, Svalbard. *Polar Research* 4, 25–32.

Heiskanen, J. (2006). Tree cover and height estimation in the Fennoscandian tundra–taiga transition zone using multiangular MISR data. *Remote Sensing of Environment* 103, 97–114.

Holm, M., A. Lohi, M. Rantasuo, S. Väätäinen, T. Höyhtyä, J. Puumalainen, J. Sarkeala and F. Sedano (1999). Creation of large image mosaics of airborne digital camera imagery. *Proceedings of the 4th International Airborne Remote Sensing Conference and Exhibition*, Ottawa, Canada, June 21–24, 1999, 2, 520–526.

Jensen, J.R. (2000). *Remote Sensing of Environment – an Earth Resource Perspective*. Prentice Hall, Upper Saddle River, 544 p.

Lutz, E., T. Geist and J. Stötter (2003). Investigations of airborne laser scanning signal intensity on glacial surfaces – utilising comprehensive laser geometry modelling and orthophoto surface modelling. *International Archives of Photogrammetry, Remote Sensing and Spatial Information Sciences* 35(3/W13), 143–148.

Marshall, G.J., J.A. Dowdeswell and W.G. Rees (1993). Limitations to multitemporal visible band satellite data from polar regions imposed by cloud cover. *Annals of Glaciology* 17, 113–120.

Nolin, A.W. and M.C. Payne (2007). Classification of glacier zones in western Greenland using albedo and surface roughness from the Multi-angle Imaging SpectroRadiometer (MISR). *Remote Sensing of Environment* 107, 264–275.

Painter, T.H., J. Dozier, D.A. Roberts, R.E. Davis and R.O. Green (2003). Retrieval of sub-pixel snow-covered area and grain size from imaging spectrometer data. *Remote Sensing of Environment* 85, 64–77.

Parviainen, P. (2006). *Detection of glacier facies and parameters using false colour digital camera data – case study Hintereisferner, Austria.* M.Sc. thesis, Department of Geography, University of Helsinki, 74 p. + appendices.

Pellikka, P., M. Lötjönen, M. Siljander and L. Lens (2009). Airborne remote sensing of spatiotemporal change (1955-2004) in indigenous and exotic forest cover in the Taita Hills, Kenya. *International Journal of Applied Earth Observations and Geoinformation* 11(4), 221–232. DOI: 10.1016/j.jag.2009.02.002

Rees, W.G. (1992). Orbital subcycles for Earth remote sensing. *International Journal of Remote Sensing* 13(5), 825–833.

Rees, W.G. (2001). *Physical Principles of Remote Sensing.* 2nd ed., Cambridge University Press, Cambridge, 372 p.

Rees, G., I. Brown, K. Mikkola, T. Virtanen and B. Werkman (2002). How can the dynamics of the tundra-taiga boundary be remotely monitored? *Ambio Special Report* 12, 56–62.

Rees, W.G. (2006). *Remote Sensing of Snow and Ice.* CRC Press, Taylor & Francis Group, Boca Raton, 285 p.

Rees, W.G. and N.S. Arnold (2007). Mass balance and dynamics of a valley glacier measured by high-resolution LiDAR. *Polar Record* 43(227), 311–319.

Rostom, R.S. and S. Hastenrath (1994). Variations of Mount Kenya's glaciers 1987–1993. *Erdkunde* 48, 174–80.

Scambos, T., J. Bohlander, B. Raup and T. Haran (2004). Glaciological characteristics of Institute Ice Stream using remote sensing. *Antarctic Science* 16(2), 205–213.

Sharov, A.I. and S. Etzold (2007). Stereophotogrammetric mapping and cartometric analysis of glacier changes using IKONOS imagery. *Zeitschrift für Gletscherkunde und Glazialgeomorphologie* 41, 107–130.

Steffen, K., R. Bindschadler, G. Casassa, J. Comiso, D. Eppler, F. Fetterer, J. Hawkins, J. Key, D. Rothrock, R. Thomas, R. Weaver and R. Welch (1993). Snow and ice applications of AVHRR in polar regions: report of a workshop held in Boulder, Colorado, 20 May 1992. *Annals of Glaciology* 17, 1–16.

Stroeve, J., A. Nolin and K. Steffen (1997). Comparison of AVHRR derived and in situ surface albedo over the Greenland ice sheet. *Remote Sensing of Environment,* 62(3), 262–276.

Stroeve, J.C., J.E. Box and T. Haran (2006). Evaluation of the MODIS (MOD10A1) daily snow albedo product over the Greenland ice sheet. *Remote Sensing of Environment* 105(2), 155–171.

Vermote, E.F., D. Tanré, J.L. Deuzé, M. Herman and J.-J. Morcrette (1997). Second simulation of the satellite signal in the solar spectrum, 6S: an overview. *IEEE Transactions on Geoscience and Remote Sensing* 35(3), 675–686.

Chapter 2

The formation and dynamics of glaciers

Michael Kuhn
Institute of Meteorology and Geophysics, University of Innsbruck, Austria

2.1 INTRODUCTION: HOW DO GLACIERS FORM?

Glaciers form when in the course of a typical year more snow is deposited than is wasted, that is, when accumulation exceeds ablation. Generally, accumulation is made up of snowfall, wind drift and avalanches, sometimes also of rime and frost or the freezing of rain or melt water in the snow pack. Ablation may be caused by melting, sublimation or wind erosion, loss of ice by avalanches or ice falls, or by the calving of icebergs, i.e. the breaking off of ice into an ocean or lake.

When there is net accumulation at the end of a summer season, snow will densify into ice and will build up until it has become sufficiently thick to deform under its own weight and move downwards. The ice will continue its downward flow until it reaches lower altitudes where temperature and other environmental conditions waste just as much ice as is resupplied annually from higher altitudes. In that case the glacier is said to have reached a steady state and a stationary size and shape.

The processes determining accumulation, ice flow and ablation interact in many ways that make climate and topography the decisive factors for the size and shape of any glacier. As far as decadal or short term variation is concerned, glaciers respond predominantly to climatic changes, but their response to a given change may vary from one glacier to another depending on their respective topographies.

Apart from basic scientific curiosity, recent interest in the study of glaciers has been motivated by their clear reaction to climate changes. As glaciers cover so many climatic regions worldwide, from the Tropics to the poles, their study enables a survey of significant aspects of global climate change. As many of their features are detectable from space, such investigations are well suited for the methods of remote sensing.

2.2 THE CLIMATE OF TODAY'S GLACIER ENVIRONMENT

The remote sensing of glaciers has become a very valuable tool for documenting their fast response to changing climate (Bamber and Kwok 2003, Kuhn 2007, Pellikka 2007, Solomon et al. 2007). Of the climatic forcings of glacier mass balance, snowfall is a basic requirement for glacier growth. Apart from the general circulation of the atmosphere and from regional weather situations, snowfall depends on a critical air temperature. For many mid-latitude locations the following empirical relation between air temperature at 2 m height (T in degrees C) and the fraction of precipitation that falls as snow Q, may be used in the range of -4 to $6°C$.

$$Q = 0.6 - 0.1T \tag{2.1}$$

On a regional scale snow falls predominantly on the windward side of mountains in the prevailing wind direction. For example, the Scandinavian mountains which extend across the prevailing westerly winds have annual precipitation in excess of 2 m at their Atlantic windward side and about 0.5 m on their lee side in the east of the Scandinavian mountains. In the Alps, advective precipitation may arrive from a north-westerly direction from the Atlantic or from a south-westerly direction from the Mediterranean, creating two zones of maximum precipitation (again of the order of 2 m annually) in the northern and southern chains, while the dry interior receives only 0.5 m annually.

The associated distribution of cloudiness affects both solar radiation and temperature. Glaciers on the windward sides of mountains receive more accumulation and less ablation so that they may exist at an altitude of 2500 m in the Alps. Those in the dry, warm, sunny interior of the Alps receive less accumulation and experience more ablation, with the result that there are peaks of 3200 m altitude that are not ice covered in spite of suitable topography.

Figure 2.1 illustrates the interconnection of summer temperature and winter accumulation plus summer precipitation at the equilibrium line altitude (ELA) where accumulation is exactly equal to ablation for a given free atmosphere mean summer temperature (June, July, August in the northern hemisphere or December, January, February in the southern hemisphere) of $4°C$. Apart from temperature and precipitation, radiation fluxes are a third climatic variable that influences this balance: the change from 210 to 240 W m^{-2} is equivalent to a rise in winter accumulation plus summer precipitation of nearly 2000 mm water equivalent.

Another example of the relation between the elevation of snow line (nearly identical with the equilibrium line altitude in Figure 2.1 is given in Figure 2.2 for a meridional transect of the Andes Mountains. The elevation of the $0°C$ summer isotherm is close to the elevation of the snow line in the Tropics from $10°N$ to $10°S$. In the dry Subtropics from $10°S$ to $33°S$ the snow line elevation is far above the $0°C$ summer isotherm on account of the energy spent in sublimation (Kuhn 1981, 1987). In the wet zone south of $33°S$ it drops significantly below the $0°C$ isotherm because of abundant snowfall.

At a local scale, near-surface air temperature decreases with altitude, at a rate generally less than the mid-latitude free atmosphere lapse rate of $0.7°C$ per 100 m. Figure 2.3 shows that these values are reached at times of strong atmospheric convection in April and May in the European Alps while in other months diurnal and seasonal cooling creates temporary inversions that reduce mean lapse rates towards the winter months.

The seasonal course of precipitation and of its change with altitude is determined by seasonal changes of air temperature and humidity on the one side and by the dominance in spring and summer of convective over advective precipitation on the other

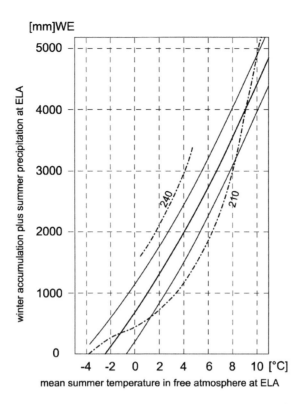

Figure 2.1 Temperature and precipitation at the equilibrium line modified from Ohmura et al. (1992). The solid black line and the grey lines indicate the square regression line and the standard deviation, respectively. The dotted and dashed lines indicate the best-fit curves for the glaciers with summer radiation of 240 and 210 W m^{-2}. ELA = equilibrium line altitude.

side (e.g. Barry 2008). At all altitudes in the European Alps there is two or three times as much precipitation in a summer month than in a winter month due to the prevalence of abundant convective precipitation in summer. Precipitation events are associated with above-average temperatures in winter and with below-average temperatures in summer so that summer snowfall is frequent at altitudes of glaciers (Fuchs 2006).

In alpine climates and in most mid-latitude mountains the increase of precipitation with altitude is stronger in winter (Figure 2.4) as it is, in a way, proportional to the product of water vapour density and wind speed, the latter increasing strongly with altitude in advective precipitation, which prevails in winter in mid-latitudes. The convective precipitation of summer depends more on topography than on absolute elevation.

On a local scale, wind is responsible for the redistribution of snow on the ground, taking snow from the ridges and depositing it in valleys and basins, thus often doubling net accumulation with respect to mere snowfall (Kuhn 1981, 2003).

The so called glacier wind is a local scale characteristic of the baroclinic atmospheric boundary layer over a glacier in summer. Due to the temperature difference

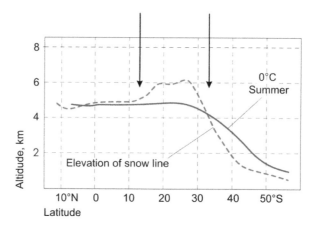

Figure 2.2 A meridional transect of the elevation of snow line (nearly identical with the equilibrium line altitude in Figure 2.1 versus the mean elevation of the 0°C summer isotherm. The snow line is far above the 0°C line in the dry Subtropics, and below in the wet zone south of 33°S. Adapted from Schwerdtfeger (1976).

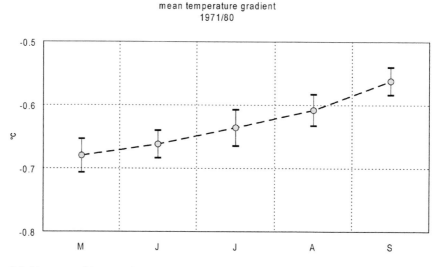

Figure 2.3 Mean monthly vertical temperature gradients at alpine stations, expressed as temperature change per 100 m elevation, and range of mean values at five pairs of stations.

between the air in the valley atmosphere and that in contact with the 0°C ice at the same elevation, a layer of denser air of several tens of metres vertical extent is accelerated downward over the glacier tongue, usually decoupled from the gradient wind and valley wind system.

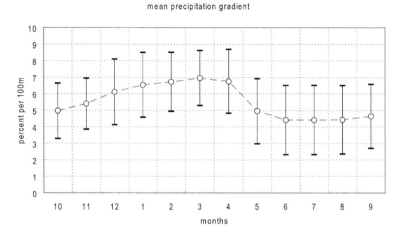

Figure 2.4 Mean monthly precipitation increase in Eastern Alps basins expressed as percent increase per 100 m elevation.

2.3   ACCUMULATION AND THE FORMATION OF ICE

Glacier ice originally comes from the sky. It is nourished by solid precipitation and rime and is redistributed by wind and avalanches (Kuhn 2003). Freshly fallen snow has a bulk density of about 100 kg m^{-3} and is subsequently compacted by several processes. Wind drift densifies it, hardens it and may give it a surface structure of ripples, dunes and cornices. Once the snow is deposited, destructive metamorphism transforms it from stars, dendrites, prisms, plates or needles into nearly spherical snow grains by moving molecules from the tips and points to the concave sites, and hardens it by sintering material into the contact regions between neighbouring grains. When additional snow is deposited at the surface, the lower layers are compressed by the new overburden which causes further densification.

Because snow is a good thermal insulator, strong temperature gradients may form between the relatively warm base of the snow pack and the cold snow surface. Temperature gradients imply gradients of saturation vapour pressure in the pore space and these in turn give rise to vapour diffusion upward which builds new, cup-like crystals, called depth hoar. A survey of crystal shapes and a summary of metamorphism is given by Colbeck et al. (1992). The snow/climate relation was recently summarized by Armstrong and Brun (2008).

Melting usually sets in at the surface of the snow cover, and melt water percolates into the snow, where it refreezes and increases density. When the snow is soaked with water the smallest grains melt first, temporarily increasing its permeability for air and water. As this decreases its mechanical strength, the snow pack settles again and reaches densities between 500 and 600 kg m^{-3}, the values typical for summer snow. It may then take several years before the firn reaches a density of about 830 kg m^{-3} where, from experience, it becomes impermeable and is classified as ice by convention.

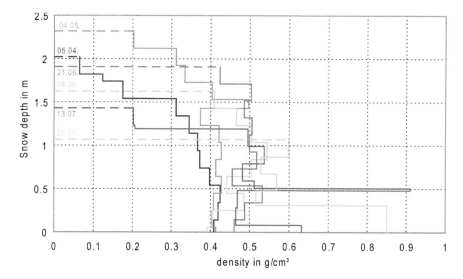

Figure 2.5 Density profiles in the spring to summer snow pack on Hintereisferner (Matzi 2004). (See colour plate section).

In the accumulation area of alpine glaciers solid ice is overlain by up to about 20 m of firn and seasonal snow. Note that glaciologists speak of firn when snow has survived one summer season while alpine skiers speak of firn when several melt-freeze cycles have given the snow a fairly uniform grain size. A long-term series of firn densities was analysed by Matzi (2004). An example of the seasonal development of snow density is given in Figure 2.5.

2.4 ENERGY BALANCE AND ABLATION

In mid-latitude glaciers, melting is the predominant form of ablation, while in polar glaciers iceberg calving, i.e. breaking off of ice into water, is more important. In the subtropics and other dry environments, sublimation may prevail. As these phase transitions consume energy ($L_M = 0.33$ MJ kg^{-1} for melting, $L_E = 2.5$ MJ kg^{-1} for evaporation, $L_S = 2.83$ MJ kg^{-1} for sublimation) ablation is closely connected to the energy balance of snow and ice. It is useful to express the energy balance in terms of energy per time per unit surface area (W m^{-2}), i.e. as energy flux densities, and to postulate that all energy fluxes reaching and leaving a surface must sum to zero, $\Sigma Q_i = 0$. Thus,

$$S\downarrow + S\uparrow + L\downarrow + L\uparrow + H + LS + LM + C = 0 \tag{2.2}$$

where S and L are the respective downward and upward short wave and long wave radiation flux densities, H is the turbulent exchange of sensible heat, LS is the turbulent exchange of latent heat of sublimation, LM refers to melting and C summarizes all energy fluxes penetrating into or coming out of snow and ice. It needs to be pointed out that LS includes all water that is first melted and then evaporated, while LM represents the melt water that actually runs off. The sign convention is chosen such that energy reaching the surface from above or below is positive.

Rather than listing examples of numerical values of these flux densities, suffice it to say that at the tongues of alpine glaciers ice ablation has reached values of 10 m in the hot summer of 2003, and that 10 cm of ice may be ablated on a single day. For further information the reader is referred to Armstrong and Brun (2008), Kuhn (1981), Oerlemans (2001), Paterson (1994).

The fluxes $S\downarrow$ and LM are not restricted to the surface but may turn over energy within the ice or snow. This is due to solar radiation penetrating into the ice and to melt water percolating and delivering heat to the colder interior, which is included in C. When melt water percolates through the seasonal snow and reaches the impermeable glacier ice, it may refreeze there and form layers of superimposed ice. When the seasonal snow is wasted and the superimposed ice appears at the surface, it displays a different structure and has different optical properties to glacier ice transformed from snow.

The transfer of latent energy of melting by percolation effectively distributes the summer energy surplus from the surface throughout the seasonal snow and gives the snow a mean annual temperature of $0°C$ even at elevations where the mean annual air temperature is below freezing. Temperate ice thus prevails in the Alps up to altitudes of about 3400 m, and cold ice may be found, preferentially on north slopes, above that altitude. Recent warming, however, is upsetting such empirical relations.

High latitude glaciers are more often of the cold type (below freezing temperatures throughout) or polythermal (having both cold and temperate ice). In Arctic and Antarctic ice sheets and glaciers the build-up of superimposed ice by the refreezing of percolating melt water may be an important part of accumulation.

2.5 MASS BALANCE: DEFINITIONS AND KEY PARAMETERS

At any location on a glacier, the resultant of accumulation and ablation is the specific mass balance $b(x,y)$ which is expressed in kg m^{-2} a^{-1}, or after division by the density of water as m of water equivalent (m w.e.) per year. Averaging over the entire glacier area yields the mean specific balance b in the same units. Multiplying b (m w.e. per year) by the total glacier area S gives the volume balance B (m^3 w.e. per year).

The specific mass balance profile $b(h)$ results from averaging specific balance $b(x,y)$ over altitude intervals or elevation bands, (e.g. 2900–3000 m). An example is given in Figure 2.6. The equilibrium line altitude (ELA) is defined as that altitude where $b(h)$ intersects the axis of $b(h) = 0$.

By convention, the accumulation area S_C is defined not as the area above the ELA, but as the area of the glacier surface where $b(x,y) > 0$, and the ablation area is defined correspondingly, the difference between the two approaches being small. The accumulation area ratio

$$AAR = S_C/S \qquad (2.3)$$

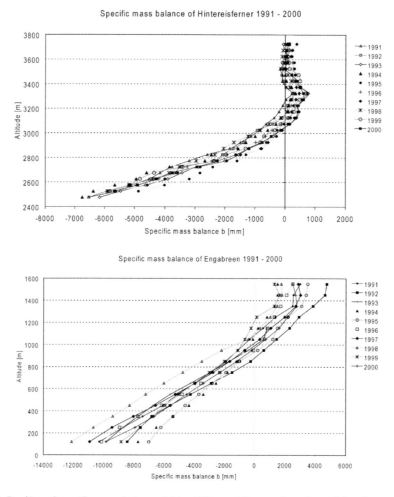

Figure 2.6 Profiles of specific mass balance *b*(*h*) of Hintereisferner, Austria, and Engabreen, Norway, 1990–2000.

is an empirical measure of mean specific balance *b*: in many cases AAR has values between 0.6 and 0.7 on stationary glaciers, and values above or below that for positive or negative mean specific mass balance respectively. (A glacier is stationary if it does not appreciably change its area and shape from one year to another). Annual values of these key parameters are given for Hintereisferner from 1952 to 2006 in Table 2.1.

2.6   ICE FLOW

The motion of ice in a glacier is generally due to deformation in viscous-plastic flow, and to basal sliding (e.g. Hooke 1998, Paterson 1994). The surface velocity can be

Table 2.1 Mass balance parameters of Hintereisferner for 1952/1953 to 2005/2006.

Mass balance year 1.10–30.9	Glacier area km²	Mean specific balance mm water equivalent	Cumulative balance	Equilibrium line altitude m	Accumulation area ratio
1952/1953	10,24	−540	−540	3020	0,53
1953/1954	10,20	−286	−826	2970	0,69
1954/1955	10,15	76	−750	2850	0,75
1955/1956	10,11	−275	−1025	2920	0,69
1956/1957	10,06	−189	−1214	2930	0,65
1957/1958	10,02	−981	−2195	3100	0,35
1958/1959	9,97	−763	−2958	3060	0,34
1959/1960	9,92	−62	−3020	2880	0,72
1960/1961	9,88	−205	−3225	2940	0,63
1961/1962	9,21	−696	−3921	3080	0,39
1962/1963	9,16	−603	−4524	3010	0,53
1963/1964	9,06	−1244	−5768	3180	0,25
1964/1965	9,05	925	−4843	2770	0,81
1965/1966	9,05	344	−4499	2850	0,75
1966/1967	9,03	20	−4479	2920	0,69
1967/1968	9,03	338	−4141	2850	0,73
1968/1969	9,01	−431	−4572	2960	0,56
1969/1970	9,01	−552	−5124	3030	0,49
1970/1971	9,00	−600	−5724	3040	0,49
1971/1972	8,99	−74	−5798	2935	0,66
1972/1973	8,99	−1229	−7027	3250	0,24
1973/1974	8,99	55	−6972	2910	0,68
1974/1975	8,97	65	−6907	2905	0,71
1975/1976	8,96	−314	−7221	2995	0,58
1976/1977	8,88	760	−6461	2840	0,78
1977/1978	8,88	411	−6050	2825	0,77
1978/1979	9,08	−219	−6269	2970	0,59
1979/1980	9,08	−50	−6319	2930	0,67
1980/1981	9,08	−173	−6492	2940	0,64
1981/1982	9,07	−1240	−7732	3260	0,22
1982/1983	9,07	−580	−8312	3075	0,41
1983/1984	9,07	32	−8280	2970	0,63
1984/1985	9,07	−574	−8854	3010	0,50
1985/1986	9,06	−732	−9586	3080	0,40
1986/1987	9,05	−717	−10303	3070	0,45
1987/1988	9,03	−945	−11248	3130	0,29
1988/1989	8,99	−637	−11885	3080	0,38
1989/1990	8,98	−995	−12880	3115	0,32
1990/1991	8,88	−1325	−14205	3260	0,18
1991/1992	8,88	−1120	−15325	3155	0,24
1992/1993	8,75	−573	−15898	3050	0,49
1993/1994	8,74	−1107	−17005	3145	0,31
1994/1995	8,73	−461	−17466	3080	0,53
1995/1996	8,72	−827	−18293	3100	0,41
1996/1997	8,70	−591	−18884	3050	0,48
1997/1998	8,30	−1230	−20114	3160	0,25

(Continued)

Table 2.1 (Continued).

Mass balance year 1.10–30.9	Glacier area km²	Mean specific balance mm water equivalent	Cumulative balance	Equilibrium line altitude m	Accumulation area ratio
1998/1999	8,22	−861	−20975	3105	0,39
1999/2000	8,11	−633	−21608	3050	0,48
2000/2001	7,96	−173	−21781	2955	0,64
2001/2002	7,91	−647	−22428	3050	0,51
2002/2003	7,82	−1814	−24242	>3750	0,03
2003/2004	7,56	−667	−24909	3185	0,32
2004/2005	7,47	−1061	−25970	3225	0,29

Figure 2.7 The accumulation pattern on Hintereisferner, August 21, 1989. To a close approximation, the bright firn areas represent net accumulation, the grey ice indicates areas with net ablation. It is obvious that the firn line separating these areas is not identical with the equilibrium line altitude, conventionally defined as the altitude where the mean specific balance $b(h)$ is zero.

remotely recorded by following the position of natural features like boulders or markers placed on the glacier. In this context, some definitions and clarifications will be given here. The motion of an annual layer or a stone through a stationary glacier can be visualized as by Finsterwalder (1897, Figure 2.8): in a stationary glacier the surface stays in place from one year to the next. The annual surplus of the accumulation is subsumed vertically into the glacier; individual ice volumes or stones then follow a path that is parallel to the surface under the equilibrium line and finally emerge towards the surface to replace the annual deficit of the ablation area. Deviations from this stationary scheme are most strongly manifest near the terminus.

Finsterwalder's view implies that vertical downward velocity in the accumulation area is due to divergent (extending) flow under the condition of mass continuity. For homogeneous and stationary ice density (in a coordinate system with x and u positive in the direction of flow and z and w positive upward) this is

$$\partial u/\partial x + \partial v/\partial y + \partial w/\partial z = 0 \tag{2.4}$$

Emergence, or vertical motion towards the surface, occurs in the convergent (compressive) flow below the equilibrium line.

The rate of change of ice thickness H at a stake moving with the ice is then given by the specific mass balance b and vertically averaged horizontal divergence/convergence

$$(\partial H/\partial t)_L = b - H(\partial u/\partial x + \partial v/\partial y) \tag{2.5}$$

if b and H are given in m of ice equivalent and u and v are vertically averaged.

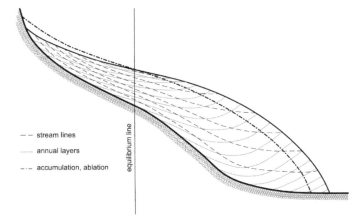

Figure 2.8 Finsterwalder's concept of flow lines and annual layers in a longitudinal section of a glacier (Finsterwalder 1897). The annual net accumulation above the equilibrium line is subsumed into the glacier, travels downward along the stream lines and emerges in the ablation area below the equilibrium line to replace the ice annually lost by net ablation. Note that the accumulation area above the equilibrium line is larger than the ablation area in this figure so that the total mass of accumulation is equal to the total mass of ablation. Annual layers may be made visible by dust deposited each summer.

At a point of fixed position (x, y) ice thickness may change due to advection in basal sliding motion

$$(\partial H/\partial t)_E = b - H(\partial u/\partial x + \partial v/\partial y) - u\partial H/\partial x \tag{2.6}$$

where the indices L and E refer to the Lagrangian and Eulerian reference schemes.

In a stationary glacier the rate of change of surface elevation h at a fixed point $(\partial h/\partial t)_E = 0$ by definition, and the rate of change of surface elevation at a stake moving with the ice is

$$(\partial h/\partial t)_L = u_{sfc}\partial h/\partial x \tag{2.7}$$

In a non-stationary glacier these become

$$(\partial h/\partial t)_E = (\partial H/\partial t)_E \quad \text{and} \quad (\partial h/\partial t)_L = u_{sfc}\partial h/\partial x + (\partial H/\partial t)_L \tag{2.8}$$

The rate of change of elevation of the stake, $\partial s/\partial t$, is given by the change of surface elevation and its velocity with respect to the surface w_r, with $w_r < 0$ in the accumulation zone and emergence $w_r > 0$ in the ablation zone of a glacier, where s, h, and w_r are defined positive upward. Since s is defined following a moving stake, the Lagrangian reference scheme has to be applied.

The stakes that travel down on present alpine glaciers reach speeds of metres or tens of metres per year, noticeably less than in the 1970s and 1980s when glaciers moved at much higher speeds. The velocity field of a glacier is much like that of a river or that of any other viscous fluid displaying maximum values at the surface along the central flow line with nearly parabolic decrease towards the base and the sides.

The glacier may actually move at its base, either by direct sliding over the basal rock or by shearing a water-bearing layer of till between ice and rock (e.g. Paterson 1994, Hooke 1998). The velocity increase from the base u_b to the surface u_{sfc} is due to internal deformation and is usually described as deformation velocity

$$u_{def} = u_{sfc} - u_b = 2A(n+1)^{-1}(\rho g \sin \alpha)^n H^{n+1} \tag{2.9}$$

where A is a parameter of the constitutive relation of ice that is strongly temperature dependent: from $0°C$ to $-48°C$ it decreases by three orders of magnitude so that Antarctic ice is more rigid and flows at a slower rate than temperate alpine ice. ρ is the ice density, g the acceleration of gravity, H is the ice thickness and α is the surface slope. The value of the exponent is generally taken as $n = 3$, which implies that the deformation velocity is proportional to the fourth power of the ice thickness.

The product $(\rho g \sin \alpha H)$ is the basal shear stress τ_b of the glacier. It was empirically found that on most glaciers τ_b ranges between 0.5 and 1.5 bars (1 bar = 10^5 Pa) and 1 bar is thus often taken as the critical yield stress of glacier ice. If that is so, then it follows that the product of ice thickness and slope is constant: steep ice is thin and ice of low surface gradient is thick. Remembering that the deformation velocity of glacier ice increases with the fourth power of ice thickness, we find that flat, thick ice moves much faster than steep, thin ice.

This characteristic velocity difference gives rise to crevasses at the transition from the flat firn basins to the steep ice surrounding them: the so called bergschrund. Examples of bergschrunds are displayed in Figure 2.9. Many impressive examples can be

Figure 2.9 An example of a bergschrund opening between the flat, fast moving firn basin and the steep, nearly stationary ice of the surrounding slopes. Feuerstein, Austria, 3268 m, 46°58′ N, 11°15′ E, September 2003.

found in the beautiful books of Hambrey and Alean (2004), Ohmori and Bonnington (1999) and Post and LaChapelle (2000).

Another site of differential movement and associated crevasses is the transition from slow moving ice margins to the faster flowing central part of the ice body. Here the tensile stress is oriented like the feathers of an arrow flying downglacier and so-called radial crevasses, which always open at right angles to the stress, fan out downglacier.

2.7 METHODS OF MASS BALANCE DETERMINATION

In the present context, four methods of determining the mass balance of a glacier are explained: 1) The geodetic method, where changes of glacier volume are considered from maps or elevation models, 2) the direct, glaciological method, where mass changes are derived from a large number of ground-based spot measurements, 3) the dynamical method where the mass balance of the upper part of a glacier is compared to the ice flow through its lower boundary, and 4) the hydrological method that treats glacier mass balance as the change of the storage in the water balance of a glacierized basin. Of the methods presented here, up to now the direct method has been applied most often. With the advance of remote sensing techniques, it is likely that airborne and satellite applications of the geodetic method will become standard in future mass balance work, with direct density measurements as ground truth.

2.7.1 The geodetic method

First applied more than 100 years ago, the geodetic method used to be carried out by terrestrial photogrammetry, followed by aerial surveys and now often by airborne laser scanning or satellite surveys (e.g. Aberman et al. 2007, Bamber and Kwok 2003,

Favey et al. 1999, Geist and Stötter 2007, Haggrén et al. 2007, Jokinen et al. 2007, Kajuutti and Pitkänen 2007, Pellikka 2007, Sharov and Etzold 2007). In principle, the geodetic method determines the change of altitude in the time interval Δt between two surveys for a closely spaced grid $\Delta h\,(x, y)$, converts these values to mass change with the appropriate density $\rho(x, y)$ and determines the mass balance for Δt by adding over all $[\rho(x, y)\,\Delta h\,(x, y)]$ of the glacier. An example of an orthophoto of Hintereisferner is given in Figure 2.10, while Figure 2.11 shows the hillshade image of the digital elevation model that was constructed from this orthophoto.

The accuracy of determining Δh, however, is limited to a root mean square value of 0.75 m in airborne photogrammetry in the best case, and of 0.25 m in airborne laser scanning (Geist and Stötter 2007). A case study of the agreement of Δh gained by various methods was recently published by Abermann et al. (2007), a comparison for Storbreen Glacier, Norway, was carried out by Andreassen (1999).

More problems arise with the determination of $\rho(x, y)$. While Sorge's Law states that density profiles in the accumulation area of Greenland remain unchanged over years, and while Matzi (2004) has confirmed this for locations of maximum accumulation on Hintereisferner, the recent consumption of old firn in the former accumulation areas of alpine glaciers as the climate warms causes further uncertainties in the application of the geodetic method.

2.7.2   The direct glaciological method

This method is based on direct measurements of specific balance at an optimum number of locations on the glacier surface, optimum meaning a balance between practical

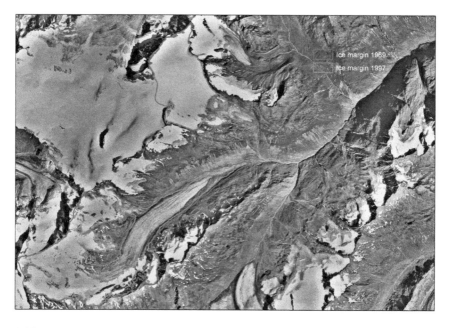

Figure 2.10 Orthophotograph of Hintereisferner, 10°45′ E, 46°48′ N, September 1997. (See colour plate section).

Figure 2.11 Shaded relief image of the digital elevation model of Hintereisferner, corresponding to Figure 2.10. (See colour plate section).

effort of field work and information gained (Fountain and Vecchia 1999). In practical work, this is implemented by determining $b(x, y)$ at a number of strategically located points and mapping isolines of b in between. This interpolation is facilitated by the dominant influence of topography and aspect on accumulation and ablation which produces annual mass balance patterns that are very similar from year to year (Kuhn et al. 1999) and which are evident in Figures 2.7 and 2.10.

From the map produced in this way, values of $b(x, y)$ can be integrated over an entire glacier by various procedures. Comparisons of the direct with the geodetic methods have been carried out on several occasions, for example by Andreassen (1999) and Krimmel (1999). The advantage of the glaciological method is the direct determination of local density and specific mass balance, whereas the geodetic method does not suffer from the uncertainty of interpolation.

2.7.3   The dynamic (balance) velocity method

In order to maintain stationary conditions the volume of ice B added to a glacier each year above a given altitude h needs to be transported downward through the cross section A of the glacier at that altitude:

$$\int_{h}^{top} B \, dh = u \, A \qquad (2.10)$$

where B is given in m³ of ice per year per m of altitude and u is the average velocity over the cross section and is called the balance velocity if h is chosen to be the equilibrium line altitude (Gudmundsson and Bauder 1999, Hooke 1998, Oerlemans 2001, Span and Kuhn 2003).

A first approximation of the mass balance of an otherwise unmeasured glacier can thus be gained from ground penetrating radar measurements of the ice thickness and from a survey of the surface velocity at the equilibrium line altitude. Apart from difficulties in finding the equilibrium line altitude and converting from surface velocities to cross-sectional mean velocity, glaciers are generally far from the steady state so that a mismatch of a factor of two to three was found in certain years at Hintereisferner (Kuhn et al. 1999). Nonetheless, the dynamical method may give the only practical approach to the mass balance of large, remote glaciers.

2.7.4 The hydrological method

The hydrological method (Fountain and Tangborn 1985, Kuhn 2003, Lang 1986) treats the changing storage of ice S in a gauged basin, of which there are rather few, as the residual of the water balance

$$S = P - R - E \qquad (2.11)$$

where P is basin precipitation, R runoff and E evaporation or sublimation. This method basically suffers from a dearth of records and from the basic difficulty of determining basin precipitation in mountain areas to the extent that the annual changes in S are generally smaller than the errors of other terms of the water balance.

2.8 DEBRIS COVER AND MORAINES

Rocks and their debris that fall onto the glacier are incorporated into the ice, transported, and surfaced again along the flow lines or trajectories envisaged by Finsterwalder (1897, Figure 2.8). Once at the surface in the ablation zone, they travel on top of the ice to the glacier margin where they are deposited on bare ground. When the glacier remains in a stationary state for prolonged periods, this deposition will accumulate as moraine ridges and mark the location of the terminus at that time. When a glacier retreats, it still delivers material at its terminus, but that will be spread over the entire area of retreat and will be overridden by the next advance so that, in total, there are distinct moraines marking advanced stages, but no such traces of minimum extents.

The debris cover may reach a density and thickness of a few metres that suffices to protect the underlying ice from melting. This stage is not necessarily the result of the advection of debris: it may also be due to the melting of ice from a glacier tongue interspersed with rocks. In either case, the analysis of remotely sensed images will be impaired: it will be difficult to pinpoint the margin of a retreating, debris-covered glacier, which is a nuisance when making inventories; and it will be difficult to infer climatic conditions from the advance of a glacier tongue well protected by its debris cover.

In periods of glacier retreat, there is no practical limitation to distinguish a rock covered glacier from an ice cored rock glacier. Figure 2.12 displays four stages of rock

Figure 2.12 Four stages in the development of rock glaciers in the Valley of Tux, Austria, 47°07′ N, 11°45′ E.

cover that increase from right to left. The four surfaces may have been glacier ice at the end of the Little Ice Age about 1850 and the debris may have accumulated at the surface in more negative mass balance conditions since that time. In general, rock glaciers and debris covered glaciers are still badly understood and are a very active field of present research (e.g. Mayer et al. 2006, Mihalcea et al. 2008). The longest series of rock glacier measurements exists for Hochebenkar Glacier, Austria, since 1938 (Schneider and Schneider 2001) and shows significant reaction of flow to recent temperature changes.

From a practical view, like that of establishing a glacier inventory by remote sensing methods, we face the problem that a debris-covered glacier tongue retains its morphology even after its ice has gone. In the Alps, this is particularly true for glaciers that remain below an upper altitude of about 3100 m and have rock fall from their surroundings.

2.9   CONCLUSIONS

This chapter has outlined the nature of glacier flow and mass balance. Remote sensing methods have reached a point of advantage over traditional terrestrial surveys, in particular as long as they are supplemented by ground based glaciological work. However, glacier thinning and retreat in the present warming climate leave uncertainties as to the clear identification of their margins that may be mitigated by the application of multiple methods, for example, by repeated airborne laser scanning. As far as the total glacier volume is concerned, these thin, marginal ice areas, however, are of minor concern. The main ice bodies and their volume changes may be sufficiently determined by aerial surveys, by satellite surveys and optimally by airborne laser scanning supported by ground truth measurements.

REFERENCES

Abermann, J., H. Schneider and A. Lambrecht (2007). Analysis of surface elevation changes on Kesselwand Glacier – comparison of different methods. *Zeitschrift für Gletscherkunde und Glazialgeologie* 41, 147–167.

Andreassen, L. (1999). Comparing traditional mass balance measurements with long-term volume change extracted from topographical maps: A case study of Storbreen Glacier in Jotunheimen, Norway, for the period 1940–1997. *Geografiska Annaler* 81A(4), 467–476.

Armstrong, R.L. and E. Brun (2008). *Snow and climate*. Cambridge University Press, Cambridge, 222 p.

Bamber, J. and R. Kwok (2003). Remote-sensing techniques. In: J. Bamber and A. Payne (eds.), *Mass Balance of the Cryosphere: Observations and Modelling of Contemporary and Future Changes*. pp. 59–113, Cambridge University Press, Cambridge, 662 p.

Barry, R.G. (2008). *Mountain Weather and Climate*. 3rd ed., Cambridge University Press, Cambridge, 506 p.

Colbeck, S., E. Akitaya, R. Armstrong, H. Gubler, J. Lafeuille, K. Lied, D. McClung and E. Morris (1992). *The International Classification for Seasonal Snow on the Ground*. International Commission on Snow and Ice (IASH) and International Glaciological Society, 23 p.

Favey, E., A. Geiger, G.H. Gudmundsson and A. Wehr (1999). Evaluating the potential of an airborne laser-scanning system for measuring volume changes of glaciers. *Geografiska Annaler* 81A (4), 555–561.

Finsterwalder, S. (1897). Der Vernagtferner. Seine Geschichte und seine Vermessungen in den Jahren 1888 und 1889. *Wissenschaftliche Ergänzungshefte zur Zeitschrift des Deutschen und Österreichischen Alpenvereins* 1(1), 112 p.

Fountain, A.G. and W. Tangborn (1985). Overview of Contemporary Techniques. In: *Contemporary Techniques for Predicition of Runoff from Glacierzed Areas*. IAHS Publications 149, 27–41.

Fountain, A.G. and A. Vecchia (1999). How many stakes are required to measure the mass balance of a glacier? *Geografiska Annaler* 81A(4), 563–573.

Fuchs, D. (2006). *Temperature analysis of days with precipitation in alpine regions*. M.Sc. thesis, Institute of Meteorology and Geophysics, University of Innsbruck, 140 p.

Geist, T. and J. Stötter (2007). Documentation of glacier surface elevation change with multi-temporal airborne laser scanner data – case study: Hintereisferner and Kesselwandferner, Tyrol, Austria. *Zeitschrift für Gletscherkunde und Glazialgeologie* 41, 77–106.

Gudmundsson, G.H. and A. Bauder (1999). Towards an indirect determination of the mass-balance distribution of glaciers using the kinematic boundary condition. *Geografiska Annaler* 81A(4), 575–583.

Haggrén, H., C. Mayer, M. Nuikka, L. Braun, H. Rentsch and J. Peipe (2007). Processing of old terrestrial photography for verifying the 1907 digital elevation model of Hochjochferner glacier. *Zeitschrift für Gletscherkunde und Glazialgeologie* 41, 29–53.

Hambrey, M. and J. Alean (2004). *Glaciers*. 2nd ed., Cambridge University Press, Cambridge, 376 p.

Hooke, R. LeB. (1998). *Principles of Glacier Mechanics*. Prentice Hall, Upper Saddle River, 248 p.

Jokinen, O., T. Geist, K.-A. Høgda, M. Jackson, K. Kajuutti, T. Pitkänen and V. Roivas (2007). Comparison of digital elevation models of Engabreen glacier. *Zeitschrift für Gletscherkunde und Glazialgeologie* 41, 185–204.

Kajuutti, K. and T. Pitkänen (2007). Close-range photography and DEM production for glacier change detection. *Zeitschrift für Gletscherkunde und Glazialgeologie* 41, 131–145.

Krimmel, R.M. (1999). Analysis of difference between direct and geodetic mass balance measurements at South Cascade Glacier, Washington. *Geografiska Annaler* 81A(4), 653–658.

Kuhn, M. (1981). Climate and glaciers. *IAHS Publications* 131, 3–20.

Kuhn, M. (1987). Micro-meteorological conditions for snow melt. *Journal of Glaciology* 33(113), 24–26.

Kuhn, M., E. Dreiseitl, S. Hofinger, G. Markl, N. Span and G. Kaser (1999). Measurements and models of the mass balance of Hintereisferner. *Geografiska Annaler* 81A(4), 659–670.

Kuhn, M. (2003). Redistribution of snow and glacier mass balance from a hydrometeorological model. *Journal of Hydrology* 282, 95–103.

Kuhn, M. (2007). Using glacier changes as indicators of climatic change. *Zeitschrift für Gletscherkunde und Glazialgeologie* 41, 7–28.

Lang, H. (1986). Forecasting meltwater runoff from snow-covered areas and from glacier basins. In: D.A. Kraijenhoff and J.R.Moll (eds.), *River Flow Modelling and Forecasting*. pp. 99–127, D. Reidel Publishing Company, Dordrecht, 372 p.

Matzi, E. (2004). Zeitreihen der *Dichteentwicklung am Hintereisferner von 1964 bis 2002*. M.Sc. thesis, Institute of Meteorology and Geophysics, University of Innsbruck, 90 p.

Mayer, C., A. Lambrecht, M. Belò, C. Smiraglia and G. Diolaiuti (2006). Glaciological characteristics of the ablation zone of Baltoro Glacier, Karakoram, Pakistan. *Annals of Glaciology*, 43, 123–131.

Mihalcea C., C. Mayer, G. Diolaiuti, C. D'Agata, C. Smiraglia, A. Lambrecht, E. Vuillermoz and G. Tartari (2008). Spatial distribution of debris thickness and melting from remote sensing and meteorological data, at debris-covered Baltoro Glacier, Karakoram, Pakistan. *Annals of Glaciology*, 48, 49–57.

Oerlemans, J. (2001). *Glaciers and Climate*. A.A. Balkema Publishers, Lisse, 148 p.

Ohmori, K. and C. Bonnington (1999). *Himalaya aus der Luft*. Bruckmann, München. 108 p.

Ohmura, A., P. Kasser and M. Funk (1992). Climate at the equilibrium line of glaciers. *Journal of Glaciology* 38(130), 397–411.

Paterson, W.S.B. (1994). *The Physics of Glaciers*. 3rd ed., Butterworth-Heinemann, Oxford, 480 p.

Pellikka, P. (2007). Monitoring glacier changes within the OMEGA project. *Zeitschrift für Gletscherkunde und Glazialgeologie* 41, 3–5.

Post, A. and E.R. LaChapelle (2000). *Glacier Ice*. Revised edition, University of Washington Press, Seattle and International Glaciological Society, Cambridge, 145 p.

Schneider, B. and H. Schneider (2001). Zur 60-jährigen Meßreihe der kurzfristigen Geschwindigkeitsschwankungen am Blockgletscher im Äußeren Hochebenkar, Ötztaler Alpen, Tirol. *Zeitschrift für Gletscherkunde und Glazialgeologie* 37(1), 1–33.

Schwerdtfeger, W. (1976). The atmospheric circulation over Central and South America. In: Schwerdtfeger, W. (ed.), *World Survey of Climatology* 12: Climates of Central and South America. Elsevier Scientific, Amsterdam, 532 p.

Sharov, A.I. and S. Etzold (2007). Stereophotogrammetric mapping and cartometric analysis of glacier changes using IKONOS imagery. *Zeitschrift für Gletscherkunde und Glazialgeologie* 41, 107–130.

Solomon, S., D. Quin, M. Manning, Z. Chen, M. Marquis, K.B. Averyt, M. Tignor and H.L. Miller (eds.) (2007). *Climate Change 2007: the Physical Science Basis. Contribution of Working Group I to the Fourth Assessment Report of the International Panel on Climate Change*. Cambridge University Press, Cambridge and New York, 996 p.

Span, N. and M. Kuhn (2003). Simulating annual glacier flow with a linear flow model. *Journal of Geophysical Research* Vol. 108, No. D10, 4313, doi:10.1029/2002JD002828, pp. ACL 8-1 to 8-9.

Chapter 3

Glacier parameters monitored using remote sensing

Petri Pellikka
Department of Geosciences and Geography, University of Helsinki, Finland

W. Gareth Rees
Scott Polar Research Institute, University of Cambridge, England

3.1 INTRODUCTION

This chapter presents the most typical glacier parameters monitored using remote sensing data as described in this book. It provides a bridge between the glacier characteristics presented in Chapters 2 and 5 and the use of various remote sensing data types presented later in the book. Several glacier parameters, such as albedo, reflectance, surface temperature, melting, glacier zones, glacier area, equilibrium line, mass balance, glacier surface and bed topography, glacier volume, and glacier velocity are indeed possible to detect using terrestrial, airborne or spaceborne remote sensing data. In some cases, the detection is not realistically possible with any means other than remote sensing (e.g. glacier area), but in some cases remote sensing is provided as an alternative to replace labour intensive, though very interesting and sometimes joyful, glaciological field work. In order to use glaciers and their changes as indicators of climate change, or as an early warning signal for sea level rise, remote sensing is the only tool to provide glacier change information from all the continents and from a large number of glaciers and ice sheets, as discussed in Chapter 15. To develop an operational monitoring system was also one of the aims of the OMEGA project. In the past and the present, several initiatives are compiling and distributing glacier data, including the World Glacier Monitoring Service (WGMS) and Global Land Ice Measurements from Space (GLIMS). On the other hand, remote sensing can never totally replace glaciological field work and measurements, which are always needed as ground truth data, but the current and many future instruments may facilitate the work and create new applications and possibilities for discovery.

3.2 GLACIERS IN THE WORLD

Glaciers form from accumulated snow which does not melt during the summer period. When the snow accumulation takes place year by year, then accumulation of snow is transformed into glacier ice, which moves under its own weight as discussed in Chapter 2. Because of the required snowfall and accumulation of snow during consecutive years, glaciers tend to form in wet and cold areas. The main rule is that the winter snowfall should be sufficiently abundant that snow does not melt during the summer period. More snowfall is required in winter if the summer is warm. For example, western Norway receives several metres of precipitation annually, whereas eastern Norway and northern Sweden lie in a precipitation shadow and receive only half a metre (Oerlemans 2001). Glaciers exist in Norway down to mild southern Norway with abundant snowfall. In northern Norway glaciers exist up to Lyngenfjord (e.g. Steindalsbreen) and West Finnmark (e.g. Langfjordjøkelen) at latitude 70° but at the same latitudes 50 km to the east of Lyngenfjord or 100 km south of Langfjord no glaciers form in the Finnish fjells due to smaller precipitation in the more continental climate, even though temperatures are low enough for glacier formation. On the other hand, a small yearly snowfall and accumulation is enough to turn into glacier ice if ablation during the summer is minimal as is the case in cold deserts of Antarctica and Greenland.

The definition of a glacier is that it consists of ice and moves under its own weight, but typically glaciers smaller than one hectare are omitted from glacier inventories. The largest ice masses are the ice sheets of Antarctica and Greenland, representing about 99% of the mass and 97% of the area of land ice (Rees 2006). Ice sheets flow in response to gravitational forces. Faster flowing areas are ice streams within slower moving ice sheet. Similar to ice sheets are ice caps, which are dome-shaped glaciers covering highland areas or high-latitude islands. An ice sheet is classified as an ice cap if it is smaller than 500 000 km^2 (Rees 2006). The major ice caps occur on Iceland, e.g. Vatnajökull, and on the western Canadian archipelago, Svalbard and on Russian Arctic islands. In places, ice flows from the ice sheets or ice caps as outlet glaciers such as Engabreen and several others in Svartisen, Norway (Figure 3.1). Valley glaciers, such as Hintereisferner in Austria or Unteraargletscher in Switzerland, form in mountain valleys, and piedmont glaciers (e.g. Malaspina in Alaska) form when ice spreads out over flatter ground. The main glaciated areas in the world, in addition to Greenland and Antarctica, are in Alaska, Iceland, Svalbard, Norway, the Russian Arctic islands, the Alps, southern Andes, Karakoram and Himalaya mountains.

Glaciers and ice sheets cover 11% of the Earth's land surface, while floating ice shelves cover 50% of the coastline of Antarctica. 11% of the area of the Antarctic ice mass is represented as ice shelves (Rees 2006). Having an albedo which can exceed 90% and covering large areas, glaciers and ice sheets are important components in albedo feedback mechanisms of the climate system. Their low temperature contributes to the global temperature gradient. Ice sheets are the main reservoirs of fresh water and glaciers are important fresh water sources for dry areas in many places, e.g. in Central Asia and India. Glaciers are also stores of hydroelectric energy, exploited particularly in Norway, Iceland, the United States and the Alps. Melting of glaciers would potentially raise the sea level by around 80 metres, but more abrupt hazards are anticipated by disintegration of parts of West Antarctic ice sheet and Greenland ice sheet. Smaller scale glacier hazards, such as *jökulhlaups* (release of ice dammed water) and ice avalanches are locally serious hazards.

There is a need to measure and monitor properties of glaciers and ice sheets, including volume and area, surface zones, glacier dynamics, albedo and mass balance. Part of this monitoring can be carried out by in situ measurements in the field by various glaciological, hydrological or geodetic methods as explained in Chapter 2, but spaceborne and airborne remote sensing provide superior cost-effective and area-effective data and methods for monitoring.

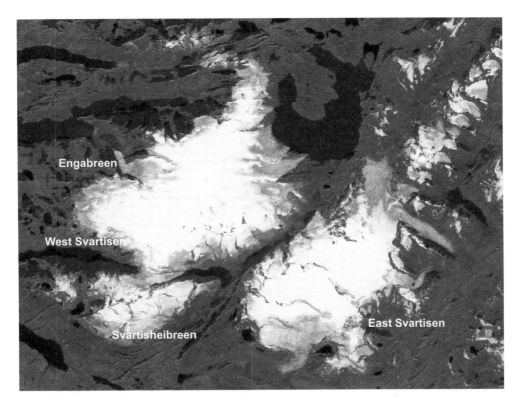

Figure 3.1 Svartisen ice caps in Norway with Engabreen glacier outlet located on the left. Landsat ETM+ satellite image with topographic correction, September 7, 1999 (Heiskanen 2002). The white areas on the glacier are snow surfaces and greyish areas are firn and ice surfaces. (See colour plate section).

3.3   REFLECTANCE AND ALBEDO

The solar radiation reaching the glacier surface may be reflected, absorbed or transmitted from the surface. In optical remote sensing, the reflectance is the main characteristic measured and used for glacier mapping. The various glacier zones – dry snow, wet snow, firn, ice (white and blue) – have their own reflectance characteristics. In addition, ageing of the surface, and various levels of impurities and debris cover on the glacier surface, cause variations in the spectral signatures of various surface types (Hall et al. 1990). Analysis of the spectral reflectance of snow shows a high reflectivity in the visible wavelength regions and a considerable decrease in the near-infrared (Nakamura et al. 2001), and middle and shortwave infrared areas (Chapter 8, Figure 8.2). Freshly fallen dry snow acts like a Lambertian reflector with upward fluxes that are nearly equal in all directions, but ageing contaminated snow is an anisotropic reflector with a significant specular component (Knap and Reijmer 1998, Pellikka and Hendriks 2002, Hendriks and Pellikka 2007a). Glacier surfaces exhibit forward scattering behaviour, with increasing anisotropy from dirty or wet surfaces.

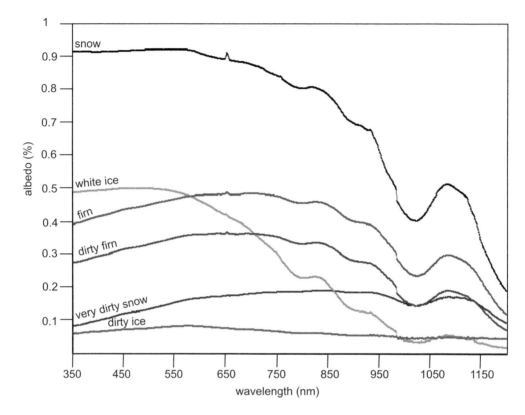

Figure 3.2 Reflectance of snow, firn and ice and contaminated snow measured on September 20, 2002 on Hintereisferner, Austria. The small peak at 650 nm in the curves is caused by the laser pen operating in the red wavelength region. C.f. with Figure 8.2.

Zeng et al. (1984) showed that freshly fallen snow has the highest reflectance in the visible and near-infrared wavelengths. Firn shows a similar pattern, but has a reflectance that is 25–30% less than snow. Glacier ice reflectance remains fairly high in blue (400–500 nm) and green (500–600 nm) wavelength bands, but drops quickly to a near-zero reflectance in the red (600–700 nm) wavelength region. Dirt (dust, volcanic ash, debris) on glacier ice lowers the reflectance considerably and there is no clear variation in reflectance. Figure 3.2 presents reflectance spectra for snow, firn and ice with various contamination levels, measured on Hintereisferner in September 2002 using an ASD FieldSpec® Pro FR spectrometer (Hendriks and Pellikka 2004). Clean snow, firn and ice follow the reflectance patterns described (cf. Figure 8.2), but contamination of snow, firn and ice lowers the reflectance, by 10 to 80% for snow, 10 to 50% for firn and by up to 40% for ice (see Hall and Martinec 1985). Contaminated ice seems to be more reflective than clean ice in the near-infrared. In addition, the variation of reflectance in each wavelength range seems to be much less for contaminated surfaces.

Albedo is the ratio of the radiation reflected from a surface to the radiation incident on that surface, and it varies between 1 and 0. Based on the proportions of electromagnetic radiation from the sun, 41% of the solar radiation is in the visible wavelengths, 9% in the ultraviolet and 50% in the near infrared (Jensen 2000), and albedo is a ratio of all the incident and reflected solar radiation in the whole electromagnetic spectrum, not only in the visible range. A highly reflective surface such as dry snow has a high albedo (0.80–0.97) and a dark surface such as dirty ice has a low albedo (0.15–0.25) (Paterson 1994). Albedo may be a more appropriate parameter to use in various applications (e.g. energy balance modelling) than spectral reflectance in various parts of the electromagnetic spectrum. However, for accurate albedo calculations, one has to take the anisotropic scattering behaviour of the snow surface into account by multiangular measurements, as done by Nolin and Payne (2007) for glacier zone classification in western Greenland.

The smaller the albedo, the higher the transmittance and absorption, but the shares of reflectance, absorption and transmittance depend on the wavelength. The low albedo contributes to the melting of the glacier, especially in the near-infrared wavelengths as more energy is absorbed (Knap 1997). Blue and green light are transmitted more by the glacier surface than red light, which is absorbed, and thus the ice in crevasses and ice caves looks turquoise (Knap 1997). With increasing grain size the albedo decreases, especially in the near infrared wavelengths, as there is much less chance for the photons to be scattered from the snow pack. The impurities decrease the albedo in visible wavelengths, but not as much in near infrared region (Warren and Wiscombe 1980) as seen in Figure 3.2. Interactions of electromagnetic radiation with snow and ice are discussed in more detail in Chapter 5.

The reflectance of the glacier surface types is a main characteristic used in glacier zone mapping using optical remote sensing data. The differences in the reflectance measured as digital numbers or brightness values and converted to reflectance (%) in various wavelength bands assist in delineation of the glacier from its surroundings and also in classification of the glacier into various surface types.

3.4   SURFACE TEMPERATURE AND SURFACE MELTING

The thermally active surface layer of glaciers, where seasonal variations are felt, has a depth of about 10 m, which comes from the penetration of heat into the ice (Chapter 5). The glacier surface is cold. Most of the ablation of the glacier takes place as melting (Figure 3.3), but ablation may also take place as sublimation which is typical for tropical glaciers (Kaser 2001), wind may blow some snow away from the glacier, or glacier avalanche or calving of a glacier may disintegrate parts of it (Chapter 2). The use of radar data for the detection of surface melting in certain elevation zones, for example, is presented in Chapter 9.

Some variation of surface temperature exists on the glacier as liquid water is warmer than snow, firn or ice. The presence of liquid water on the ice surface increases absorption in red and near infrared wavelengths and the wet snow zone can be detected using thermal remote sensing methods (Chapter 5). In addition to liquid water, debris and impurities may cause significant temperature differences on the glacier surface. In the sunlight the temperature of the debris increases, which may speed up the melting

Figure 3.3 Rapid melting of the glacier on a warm day may introduce small river channels on a glacier surface. Hintereisferner, August 12, 2003. Photograph by Petri Pellikka.

process of the glacier if the thickness of the layer does not exceed 2–3 centimetres (Chapter 6, Holmlund and Jansson 2003). A thicker layer will insulate the glacier causing special topographic features, e.g. dirt cones (Figure 3.4), on the glacier.

The fact that the surface temperature of the glacier is lower than that of its surroundings can be used for glacier area mapping by the creation of a glacier mask, which is relatively successful for polar glaciers with little debris and dust on the glacier surface. The temperature can be determined from thermal infrared imaging using the thermal infrared bands of AVHRR or MODIS for coarse scales or band 6 of Landsat TM/ETM+ for finer scales (Figure 1.5). The data must be carefully filtered for clouds and corrected for atmospheric effects.

The creation of a glacier mask typically takes place by assessing a threshold value between the glacier and the surrounding terrain. Heiskanen (2002) tested various methods for creating a glacier mask for the Engabreen glacier outlet of West Svartisen, Norway, using Landsat ETM+ satellite data of 1999: thermal infrared band (TM6), band ratio TM3/TM5, band ratio TM4/TM5 and NDSI index (Chapter 8, Table 8.2, p. 146). The best result was achieved using a thresholding method on the ETM+ thermal band, which showed a clear contrast between the glacier and the surroundings, even in areas of cast shadows (Figure 3.5). The method is more difficult to apply in

Figure 3.4 Dirt cones on Hintereisferner are created when debris insulates the ice, which as a result melts slower that the surroundings. The cone is an ice hummock covered by a few cm of debris. Photograph by Petri Pellikka, 2003.

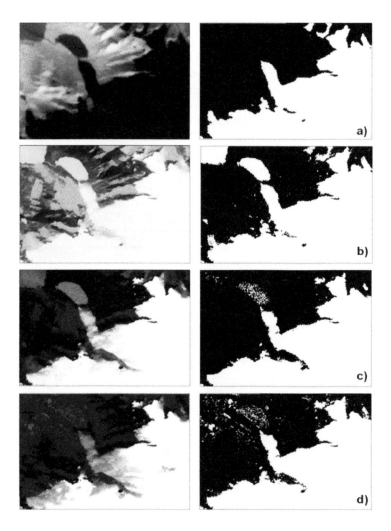

Figure 3.5 Glacier masks over Engabreen, Norway resulting from various methods applied to Landsat TM/ETM+ data (Heiskanen 2002). a) thermal band; b) NDSI; c) band ratio 3/5; d) band ratio 4/5. See location of Engabreen in Figure 3.1.

narrow valley glaciers with debris cover on the glacier or on glaciers with snow cover around the margins. See Chapter 8 for more information on thresholding.

3.5   GLACIER ZONES AND MASS BALANCE

3.5.1   Glacier zones

The surface of a glacier can be roughly divided into accumulation and ablation areas, separated by the equilibrium line where accumulation is exactly equal to ablation. The equilibrium line is generally more or less coincident with the snow line (Chapter 2, Chapter 5). In winter, before any ablation starts, practically the whole glacier is accumulation area and in summer, at the end of the glacier's year, the accumulation area is at its smallest. The end of the glacier's year is the time when the greatest amount of the previous winter's snow has melted (the ablation area is at its maximum) and new snow has not yet fallen. This occurs typically in September or October in the Alps, so that the mass balance year begins on the first of October, by convention, in the so called fixed date system. Further north, in Scandinavia and Svalbard, the moment is approximately a month earlier, while on the glaciers in the southern hemisphere it is in March or April according to the geographical location and altitude (e.g. Anderton and Chinn 1978, Anderson et al. 2006). Tropical glaciers, such as the glaciers of Mount Kilimanjaro, are defined by the global circulation patterns of the Inter Tropical Convergence Zone (ITCZ) and have low thermal seasonality which leads to ablation all year round, while the occurrence of accumulation varies between the humid inner tropics, the outer tropics and the subtropics (Kaser 2001). The investigation of glacier zones is typically carried out at the end of the glacier's year at a time when snow cover on the glacier and in the surroundings is at its minimum.

A glacier can be divided into a number of specific zones (or facies), which are illustrated in Figure 5.1 in Chapter 5. Starting from upper elevations and using Paterson's (1994) nomenclature, the uppermost zone is dry snow zone in which no melting occurs. This zone exists only in the highest elevations of the ice sheets and on the highest (and coldest) mountain glaciers, where the annual average temperature is lower than the threshold value of $-11°C$ (Peel 1992). Below the dry snow zone, the dry snow line separates it from the percolation zone in which some melting takes place during the summer. The wet snow line separates the percolation zone from the wet snow zone in which all the current year's snow melts. Below this, and separated from it by the snow line, is the superimposed ice zone, in which meltwater refreezes onto the colder glacier ice surface. As the superimposed ice is created by melting and refreezing of the current year's snow this zone is still considered to be part of the accumulation area. The equilibrium line (EL) separates the accumulation and ablation areas and is located at the lower boundary of the superimposed ice zone and the bare-ice zone. At the equilibrium line, accumulation is exactly equal to ablation, and the EL is more or less coincident with the snow line (Chapter 2). Below the snow line is the firn zone, which represents last year's snow. Below this, where firn turns into glacier ice at the surface, is the firn line (Figure 3.6). In Figure 3.1 representing Svartisen ice caps in a Landsat ETM+ satellite image, the snow lines and firn lines and respective snow, firn and ice zones can be clearly seen as bright snow, various levels of grey firn and blueish ice.

The lines may sound simple, but as different years create their own snow and firn, there exist several old firn lines on the ablation area of the glacier. In snow-rich years and in cool summers the snow melts only from the lower parts of the glacier, thus hiding the lines of previous years. In such a snow-rich year only the snow line between snow and ice can be detected if the ablation area extends only to very low elevations.

The various zones to be detected by remote sensing were first proposed by Østrem (1975) and they have since been mapped with optical sensors, passive microwave sensors and synthetic aperture radar (SAR). The zones can be detected by various remote sensing data based on the surface wetness, grain size and purity of the glacier

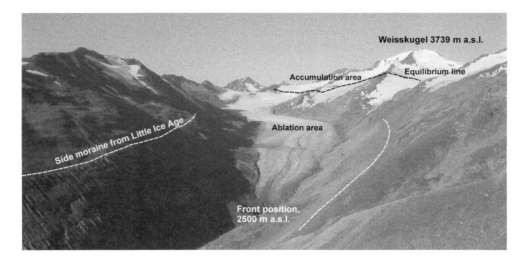

Figure 3.6 The ablation and accumulation areas of Hintereisferner in September 2002. Photograph by
         Norbert Span.

surface. Some attempts at glacier zone mapping have been made using multiangular
satellite imagery (MISR, Multi-angle Imaging Spectroradiometer: Nolin and Payne
2007, Hendriks and Pellikka 2007a). More detail on using optical remote sensing
data is presented later in this chapter and in Chapter 8.

The glacier zones can be mapped using radar data from the ablation period using
the microwave backscattering properties of the surface. In summer, the accumulation area is in general covered with wet snow, surface scattering is controlled by
the water content and surface roughness dominates (Marshall et al. 1995). Typically
the backscattering coefficient is low over a smooth snow surface. Surface scattering
is high in ablation areas controlled by surface roughness, which is typically higher
than in accumulation areas (Brown at el. 1999) and increases downward or throughout the summer as a result of melting. Crevassed areas (Figure 3.7, Chapter 5) give
high backscatter values depending on the orientation of the crevasses relative to the
look-direction of the radar. This phenomenon, called the cardinal effect and known
particularly in urban areas (Jensen 2000), complicates the study of snow surfaces in
crevassed areas. The use of summer SAR data for glacier zone mapping is based on
distinguishing areas of contrasting backscatter on glaciers. Rau et al. (2000) and Braun
et al. (2000) have described so called radar glacier zones: the dry snow zone, frozen-
percolation radar zone, wet snow radar zone, phase 2 melt radar zone, superimposed
radar zone and the bare glacier ice radar zone. However, these zones do not necessarily
coincide with the classical glaciological zones described earlier (Paterson 1994), but
have certain unique applications (Chapter 9). Winter radar images may be applied
to glacier zone mapping as the radar radiation is able to penetrate through dry snow
(Langley et al. 2008). The principles of detecting snow and ice using radar data are
presented in Chapter 5 and further reading about glacier zone detection is given in
Chapter 9.

Figure 3.7 Crevassed area in the upper parts of Hintereisferner, 3100 m a.s.l. Photograph by Pertti Parviainen, 2003.

A fairly new and promising approach to assessing surface zones is to use laser scanner data. Modern laser scanners also retrieve an intensity signal from the surface, as well as the travel time which is used for elevation determination, and this can be used for mapping surface characteristics. Some preliminary classifications using the intensity signal have been made by Lutz et al. (2003), who studied Svartisheibreen in Norway (Figure 3.1) and found some artefacts in the data hampering the classification (Chapter 10), and by Arnold et al. (2006) in Svalbard.

3.5.2 Glacier mass balance

The mean specific mass balance b (expressed in metres water equivalent per year) results from averaging the accumulation and ablation of the various glacier locations. Usually b is averaged over 100 m elevation bands as $b(z)$. Multiplying $b(z)$ by the area of the respective elevations bands $s(z)$ and adding over all elevations gives the volume balance $B(m^3$ w.e. per year). The equilibrium line altitude (ELA) is assessed from glaciological field studies as that altitude where $b(z)$ is zero. The equilibrium line is close to the snow line at the end of glacier's year. However, in this case one has to use term averaged snow line, since in Figure 2.7 and 3.8 it can be seen that the snow line separating snow and firn does not follow the contour lines, so that it does not have a unique altitude.

Figure 3.8 The Hintereisferner in August 2003 was entirely ablation area to the top of Weisskugel at 3739 m a.s.l. Photograph by Petri Pellikka.

As seen in Table 2.1 the mean specific mass balance for Hintereisferner has varied since 1952/1953 between 925 mm w.e in 1964/1965 and −1814 mm w.e in 2002/2003. The ELA defined was 2770 m for 1964/1965 and more than 3750 m for 2002/2003. The average values for b and ELA for the same period were −490 mm w.e. and 3023 m respectively.

The accumulation area S_c is defined not as the area above the EL, but as the area of the glacier surface where $b > 0$, and the ablation area is defined correspondingly. The accumulation area ratio (AAR, Equation 3.1) is characteristic of the glacier's status.

$$AAR = S_c/S \qquad\qquad (3.1)$$

(S_c is the accumulation area and S is whole glacier area.)

AAR has values between 0.6 and 0.7 on stationary glaciers. Values above or below are for positive or negative mean specific mass balance values, respectively (Chapter 2). Based on Table 2.1, an AAR of 0.81 was calculated for 1964/1965 and 0.03 for 2002/2003. The average AAR for the observation period for Hintereisferner was 0.5. In effect, in 2003 the whole of Hintereisferner was ablation area as the highest part of the glacier is at 3739 m, which is the peak of Weisskugel (Figure 3.8). The summer of 2003 was the warmest in Tyrol during the observation period. Using the data from Table 2.1, the AAR correlates almost linearly in Hintereisferner with the mean specific mass balance ($R^2 = 0.91$) and with ELA ($R^2 = 0.85$).

Using remote sensing methods, the AAR can be determined at the end of the glacier year by mapping the glacier zones and dividing the snow covered area of the glacier by the whole glacier area (Equation 3.1). Imprecision in this method may be caused by the non-optimal date of the remote sensing image (taken before the timing of the maximum ablation area) or inaccuracies in the classification itself (Pellikka 2007). The EL may also be determined by detecting averaged snow line, a border between snow and firn, simply by manual delineation. Based on the high correlation between AAR and mean specific mass balance, classification of glacier surface characteristics into snow, firn and ice and calculation of AAR from the results gives an approximation of the mean specific mass balance for the glacier in certain year. Some uncertainties exist since same AAR can still give values of specific mass balance (b) spreading over several hundred mm of water equivalent as presented for Hintereisferner in Kuhn et al. (1999).

The snow line and AAR can be monitored by various optical remote sensing data based on spectral signature differences between snow and firn. Pesu (2007) applied object oriented classification to the classification of snow, firn and ice areas of Landsat TM or ETM+ satellite images at the end of glaciological year (August–September) in 1990, 1991, 1997, 1999 and 2002 over Hintereisferner, Kesselwandferner and Vernagtferner in Austria. The resulting AAR and snow line altitude were compared with glaciologically determined AAR and ELA. The results were satisfactory as the correlation between the remote sensing and glaciological methods for AAR was 0.64 and for snow line and ELA was 0.56. (Figure 3.9). Heiskanen et al. (2003) produced AAR by pixel-based classification and studied the location of the snow line using Landsat TM/ETM+ imagery in Engabreen, Norway. The classification produced a measure of AAR that considered spatial variations of ablation better than the one calculated from field measurements. On Engabreen the boundary between firn and snow corresponded well to the equilibrium line derived from mass balance measurements. Also AAR showed a good fit in the mass balance years studied. Another study on Engabreen was performed by Braun et al. (2005), in which unsupervised classification of Landsat TM/ETM+ images was used for mass balance modelling.

Assessing mass balance by AAR is useful for remote glaciers for which mass balance, ELA and AAR are difficult to assess using conventional methods. Aniya et al. (1996) used a Landsat TM mosaic of the Southern Patagonian Icefield, South America, and supervised classification methods for TM bands 1, 4, and 5, for dividing various glacier drainage basins into accumulation and ablation areas, thereby determining the position of the snow line. It was found that the snow line could be taken as the equilibrium line and by comparing topographic maps the ELA was determined. Kulkarni et al. (2004) developed a regression between AAR and specific mass balance using field data over Baspa basin, India, which suggested a correlation of 0.80. Applying LISS III and WiFS satellite images of IRS satellite, AAR was calculated for 19 glaciers in the basin and converted to mass balance using the determined regression equation.

Mass balance is most typically determined by the glaciological method, but also using hydrological and geodetic methods as explained in Chapter 2. A comparison of these methods is presented in a case study of Tuyuksu glacier region in Central Asia by Hagg et al. (2004), in which the glaciological and hydrological methods were inspected. The use of the geodetic method, applying elevation models with greater accuracy than previously available, is gaining popularity especially with the success of

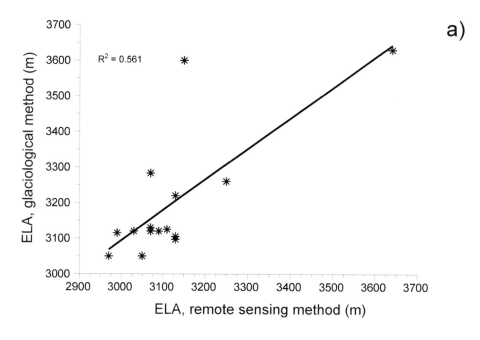

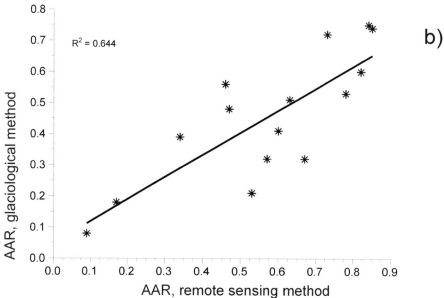

Figure 3.9 Regression lines between ELA (a) and AAR (b) assessed from the glaciological data and using object oriented classification of satellite images of 1990, 1991, 1997, 1999, 2002 over Hintereisferner, Vernagtferner and Kesselwandferner, Austria. The glaciological data provided by Institute of Meteorology and Geophysics, University of Innsbruck and Commission for Glaciology, Bavarian Academy of Sciences and Humanities.

airborne laser scanning as presented in Chapter 10. The use of remote sensing data for the geodetic method or using the determination or ELA and AAR by classification procedures would increase the number of glaciers for which the mass balance is estimated, even though the results may be less accurate than from the glaciological method. Mass balance studies using different remote sensing mechanisms have been thoroughly reviewed by Bamber and Rivera (2007).

3.6   GLACIER AREA

The area of the glacier may be the simplest parameter to determine, since the bright glacier surface has a higher albedo compared to the vegetation, bedrock or moraines surrounding it. After orthorectification of the optical remote sensing data the glacier area can be determined by several methods: manual delineation of the glacier perimeter, using band ratioing techniques, unsupervised or supervised pixel-based classification algorithms or object-oriented classification as described in detail in Chapter 8. However, debris cover on the glacier surface and its surroundings, snow cover on the glacier margins and its surroundings, and also cloud cover introduce uncertainties to glacier area mapping. Several methods have been developed to overcome these uncertainties. In automated and semiautomated glacier delineation procedures various masks can be created using various indices and threshold values calculated from the satellite data. Hendriks and Pellikka (2007b), for instance applied Landsat ETM+ data to create masks for water, cloud and finally for the Hintereisferner itself. Thick debris on the glacier surface is difficult if not impossible to take into account on the glacier surface (Figure 3.10). Another problem is caused by so called debris covered dead ice, which is very slowly if at all moving part of the glacier covered so heavily by insulating debris that it melts slowly. In order to reveal the debris-covered areas and dead ice zones from valley slopes and moving glacier, elevation models can be used. This technique is based on studying the change of the slope angle or change in the microtopography between the slope and the debris-covered glacier.

SAR data have not been used as extensively for glacier area mapping as optical remote sensing data, but several features can be detected by different bands and techniques. C-band imagery, e.g. from Radarsat and ERS-1/2, can be used to map wet snow and ice-free surfaces, but provides poor discrimination of snow, ice and bare rock. Using texture parameters, the discrimination of rocks may be improved (Sohn and Jezek 1999). L-band imagery is useful for distinguishing snow from other surfaces especially after correction for incidence angle variations (Shi and Dozier 1993). The usability of SAR data may be improved by using optical satellite imagery (Sohn and Jezek 1999). See Chapters 1 and 11 for details of wavelength bands of SAR data.

Glacier area can be also mapped using very high resolution digital elevation models (DEM), which can be produced by airborne laser scanner data (e.g. Spikes et al. 2003). Knoll and Kerschner (2009) applied an automated glacier delineation algorithm to a DEM created from laser scanner data for mapping glaciers in South Tyrol, Italy. The method worked well except for heavily debris-covered glaciers. Abermann et al. (2009a, 2009b) applied DEMs and shaded reliefs created from laser scanner data to visually delineate several glacier margins in Tyrol, including the snout of Hintereisferner. For debris-free glaciers the delineation was made based on roughness

Figure 3.10 The snout of Hintereisferner has two tongues: on the left is the dead ice in which the jökulport (glacier portal) is located, and on the right is the moving Hintereisferner with less debris cover. The debris on the dead ice originated from the valley slopes as a result of erosion, but also from the material brought by the glacier (ref. Chapter 2). Photograph by Petri Pellikka, 2003.

differences between the glacier and its surroundings. For the detection of debris-covered areas and dead ice areas, the elevation changes on the glacier caused by the ice movement or melting were measured from multitemporal datasets. Another work in this application field is by Rees and Arnold (2007).

Changes in glacier area are considered to be an indicator of (regional) climate change, and change monitoring is a natural progression from mapping of glacier area and volume for a particular glacier or glaciated area, such as Ötztal, Austria. Furthermore the land surface covered by glaciers plays an important role in climate models, and glacier volume is an important resource in hydro-power production and irrigation in some areas of the world. Several methods for glacier change detection and visualization of the changes are presented in Chapter 12.

3.7 GLACIER TOPOGRAPHY

Glacier volume is the product of glacier area and its mean thickness, and therefore glacier change studies should address both parameters. Glacier thickness can be determined if both glacier topography and bed topography are measured, but the

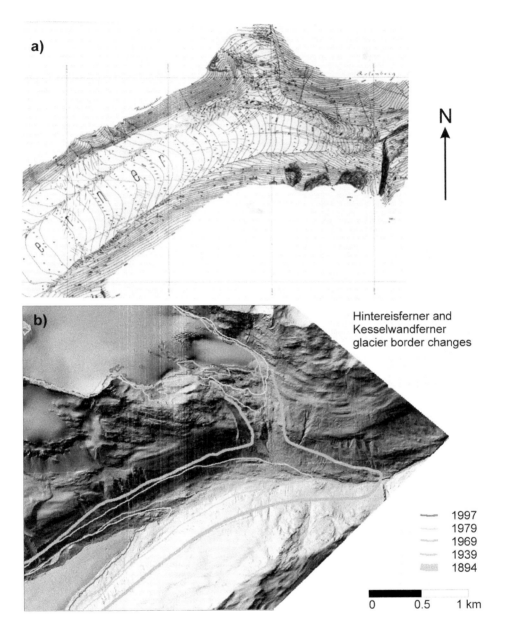

Figure 3.11 a) the map of 1894 by Blümcke and Hess (1899) shows that Hintereisferner reached an elevation of 2275 m in 1894 and thinned 200 m in a century in the location of current snout; b) the position of the snout of Hintereisferner and Kesselwandferner since 1894 digitized from maps various glaciological maps (1894, 1939, 1969, 1979, 1997) and overlaid on a shaded relief model derived from airborne laser scanner data from 2002. The connection between the two glaciers was rapidly lost between 1894 and 1939. Figure material provided kindly by Institute of Meteorology and Geophysics, University of Innsbruck. (See colour plate section).

latter is known for only a few glaciers in the world, at least with fine spatial resolution. For glaciers for which bed topography is not known, but surface topography is monitored, relative volume changes (thickness loss or gain) can be determined.

A demonstration of the importance of the determination of bed topography is given by considering the glacier changes of Hintereisferner since the Little Ice Age from the mid of 19th century. As seen in Figure 3.11, Hintereisferner retreated approximately two kilometres between 1894 and 2004. Simultaneously, a glacier thinning of at most 200 metres took place, which can be measured from the valley from which the glacier

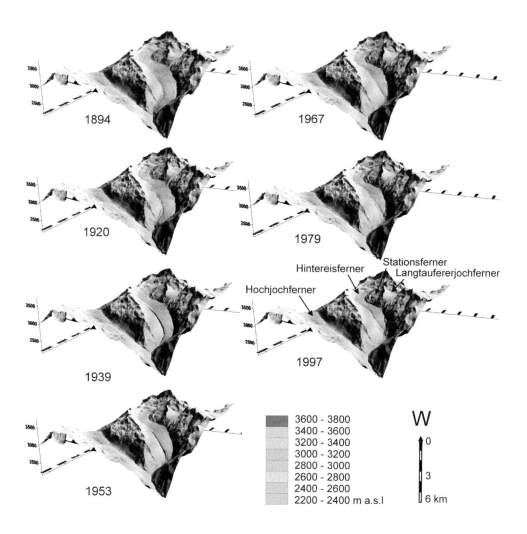

Figure 3.12 The changes of Hintereisferner length and topography between 1894 and 1997 digitized from maps by Norbert Span. The status for Hochjochferner is from 1997 map. Langtaufer-erjochferner lost its connection to Hintereisferner in 2000. The figure material provided kindly by Institute of Meteorology and Geophysics, University of Innsbruck. (See colour plate section).

melted away using topographic maps from 1894 and modern DEMs or maps. Even older glacier margins can be found by detecting side moraines and other evidence in photographs or in shaded relief images generated by laser scanner data. For example, Knoll and Kerschner (2009) studied the Little Ice Age glacier margins of South Tyrol, Italy, by detecting the side moraines. In Hintereisferner, the Little Ice Age moraines from around 1850 (based on the erosion rate, lichen on the rocks and vegetation pattern) are visible on the left hand slopes on the southern side of the valley in Figure 3.11 and in Figure 3.6.

Glacier topography is a very important parameter to monitor as it is a key to glacier volume and glacier mass balance studies using the geodetic method. As expressed throughout this book, topography can be mapped using various data types; terrestrial photography, aerial photography, digital camera data, airborne radar data, satellite radar data (SAR), very high resolution satellite imagery and laser scanner data. In recent years airborne laser scanner data has become superior to other types in terms of accuracy (Chapters 10, 13 and 14). During the OMEGA project, Hintereisferner was monitored ten times (Geist and Stötter 2007) and Engabreen three times using airborne laser scanner (Geist et al. 2005). Historical glacier topography can be reconstructed using old glaciological or other topographic maps. Magnificent examples of elevation models and glacier change visualizations since 1894 over Hintereisferner made by Norbert Span of the University of Innsbruck are presented in Figure 3.12.

The change in glacier elevation (Δz) is studied simply by subtracting the elevation value at one time (t_1) from an elevation value at another time (t_2) as in Equation 3.2.

$$\Delta z = t_2 - t_1 \tag{3.2}$$

Figure 10.3 presents surface elevation changes on Hintereisferner for the glaciological year 2001/2002, derived from multi-temporal airborne laser scanner data. For the respective period, the overall elevation change of Hintereisferner was -1.3 m, but the thinning occurred very unevenly on the glacier being more than 5 metres in the snout, but less than one metre in the accumulation area (Chapter 10).

3.8 BED TOPOGRAPHY AND GLACIER VOLUME

The glacier bedrock is typically mapped using ground penetrating radar, discussed in detail in Chapter 11. On small and narrow valley glaciers, such as Hintereisferner, it is possible to map the bedrock with dense spatial resolution producing detailed bed topography (Figure 11.7), but for larger glacier and ice sheets the measured data need to be interpolated. The results of the data analysis with existing DEMs may be used for production of a landscape visualisation without the glacier, as in the case of Hintereisferner (Figure 3.13). Compare the figure with Figure 3.6. For glaciers for which both the bedrock topography and surface topography are mapped, the glacier thickness and volume can be determined. The results of the bed topography studies give a volume of 0.5 km^3 for Hintereisferner (Span and Kuhn 2003).

The signals used in ground penetrating radar provide other information than the echo from the bedrock. Several boundaries, e.g. between firn and ice, can be detected and used for assessing the internal layers of the glacier or ice sheet. In addition, density,

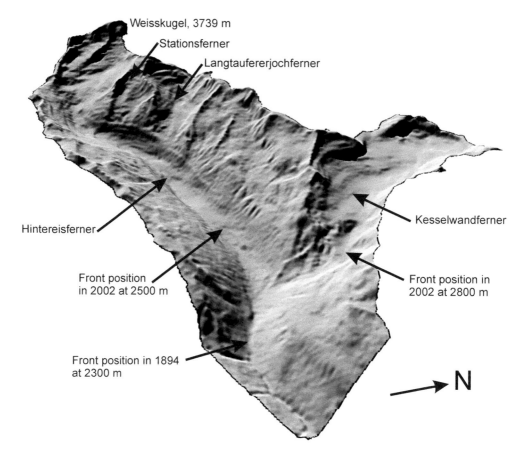

Figure 3.13 The topography of Hintereisferner valley without a glacier. Figure provided kindly by Institute of Meteorology and Geophysics, University of Innsbruck (Span et al. 2005).

water content, other hydrological aspects, and crevasses and englacial meltwater channels in the glacier can be derived, as described in Chapter 11.

3.9   GLACIER VELOCITY

A glacier is like a river, flowing faster in the middle and closer to the surface compared to the sides and bottom of the glacier. The glacier motion and varying flow speed within the glacier between the mid-stream and the sides cause radial crevasses on the glacier surface. Glacier velocity is an important parameter to monitor, since it reflects the mass gain in the accumulation area. The greater the mass gain in the accumulation area is, the faster the glacier flows. In Figure 3.14 presenting the glacier velocity changes of Hintereisferner from 1930s up to 2000s it can be seen that maximum velocities of

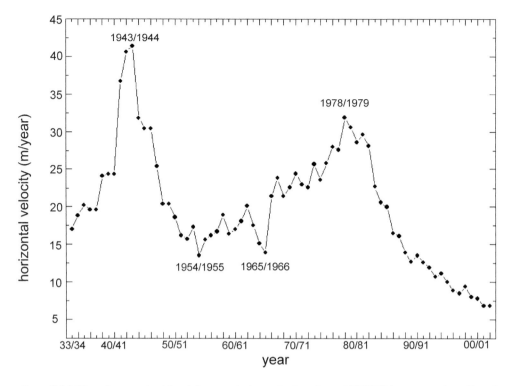

Figure 3.14 The changes in Hinteisferner glacier velocity since 1933/1934 measured at line 6 approximately at an elevation of 2650 metres (the elevation is lowering due to glacier thinning).

40 metres per year were achieved in the early 1940s, while after mid 1990s during the 2000s the speed has been less than 10 metres per year, which indicates a retreating glacier (Span et al. 1997). The velocity of a glacier determines also how rapidly it responds to a change in regional climate.

Glacier motion is conventionally surveyed by putting stakes on a glacier and measuring their x, y and z coordinates over one or several periods. The use of accurate differential global positioning systems (GPS) has made coordinate measurements for glacier motion assessment simpler than with the use of traditional surveying methods (Hinze and Seeber 1988, Tremel et al. 1993, King 2004). In crevassed areas, the use of GPS devices is still somewhat difficult (Figure 3.7).

The use of remote sensing in glacier motion studies is based on feature tracking from terrestrial, airborne or satellite remote sensing data or by measuring the surface change by interferometric SAR (synthetic aperture radar). The use of terrestrial photographs is presented in Chapter 6 and aerial photographs in Chapter 7. Common to both methods is that a stereopair is needed for the photogrammetric evaluation from two or more timesteps from which the movement of the feature, e.g. flowlines, shear margins, bed features, crevasses or dirt cones (Figure 3.3) is measured. Crevasses are mentioned to be one of the feature tracking targets. They are created at the same location on the glacier as a result of glacier speed differences caused by variation in the bedrock height below the glacier or a narrow passage between valley slopes, for example. Crevasses open and close after the topographic feature causing them, so using them as an indicator of motion should be done with care.

The timestep for feature tracking may be only a few days for fast moving alpine glaciers (Rasmussen 1988) or years for ice streams from continental ice sheets (Whillans and van der Veen 1993). If the targets on the glacier are large, velocity mapping is possible even from optical satellite imagery, but it necessitates co-registered satellite imagery (Li et al. 1999). See Figures 7.11 and 7.12 for glacier velocity results using aerial photography. The very detailed multitemporal shaded relief images resulted from high resolution laser scanner data may reveal interesting topographic features, such as meltwater channels, which can be used for velocity studies (Rees and Arnold 2007).

The interferometric SAR technique uses two or more synthetic aperture radar (SAR) images acquired at different times and from slightly different orbit configurations to look at surface deformation, and hence velocity. SAR interferometry was first applied to velocity studies by Goldstein et al. (1993), who studied the velocity of the Rutford Ice Steam in Antarctica. Since then, it has been applied also to smaller glaciers (e.g. Strozzi et al. 2003). The period separating the images is critical: too long a period causes decorrelation due to too fast velocities and surface melting, and no useful information can be obtained. Figure 9.5 shows the velocity of the fast-flowing Kronebreen in Svalbard determined using ERS-1/ERS-2 three-day Tandem mission data obtained on March 8 and 11, 1994. The fastest part of the glacier flows 1.6 m per day, while for comparison Hintereisferner flows approximately 7 metres per year. See Chapter 9 for more detailed methods for ice velocity measurements using interferometry.

3.10 SUMMARY

The list of glacier parameters that can be studied by remote sensing methods is encouragingly long. The most straightforward cases are those in which the link between the parameter and the measured property of the EMR is reasonably direct, such as surface reflectance and surface temperature. Surface reflectance, or albedo, can be determined from calibrated optical data provided that atmospheric effects are corrected for. Surface temperature is similarly directly obtained from thermal infrared data, again after correction for the atmosphere. Both of these parameters are important in themselves, for example in energy balance modelling, but can also form the basis of less direct inferences about the glacier. An obvious example of this is the way in which the contrast in reflectivity between a glacier surface and its surroundings can be used to delineate the area of the glacier. In general, the situation with regard to the delineation of a glacier's two-dimensional geometry is satisfactory: optical or radar imagery, either airborne or spaceborne, can be used in most cases to map the area and to track features for surface velocity measurements. The task becomes somewhat more difficult when it is necessary to map the glacier surface at a particular point in the balance year, particularltly for the delineation of glacier surface zones or facies. Spaceborne optical imagery may not be able to provide sufficient temporal resolution to achieve this aim, while it is still not completely clear how radar surface zones correspond to the glaciological facies. However, the main reason for mapping these facies is as a means to estimating mass balance, and this is a task for which a very promising technology has emerged over the last few years. The precision and data volume offered by airborne laser profiling have both increased dramatically over the last few years, and a number of recent studies, including those reported in this book, have drawn attention to the scope of such datasets for measuring the surface topography, surface velocity and mass balance of glaciers. The ability to measure the subglacial bedrock topography through ground-penetrating radar, combined with the ability to measure the surface topography by a number of possible methods, means that even glacier volume can now be determined.

It continues to be true that neither spaceborne nor airborne remote sensing methods are adequate on their own, and that a combination of the two is required as a result of their complementary characteristics. It is also still true that remote sensing methods usually have to be applied in conjunction with fieldwork, for tuning algorithms and verifying results if for no other reason.

REFERENCES

Abermann, J., A. Fischer, A. Lambrecht and T. Geist (2009a). Multi-temporal airborne LIDAR-DEMs for glacier and permafrost mapping and monitoring. *The Cryosphere Discussions (TCD)*.

Abermann, J., A. Lambrecht, A. Fischer and M. Kuhn (2009b). Quantifying changes and trends in glacier area and volume in the Austrian Ötztal Alps (1969–1997–2006). *The Cryosphere Discussions* 3, 415–441.

Anderson, B., W. Lawson, I. Owens and B. Goodsell (2006). Past and future mass balance of 'Ka Roimata o Hine Hukatere', Franz Josef Glacier, New Zealand. *Journal of Glaciology* 52(179), 597–607.

Anderton, P.M. and T.J. Chinn (1978). Ivory Glacier, New Zealand, an I.H.D. representative basin study. *Journal of Glaciology* 20(82), 67–84.

Aniya, M., H. Sato, R. Naruse, P. Skvarca and G. Casassa (1996). The use of satellite and airborne imagery to inventory outlet glaciers of the southern Patagonia icefield, South America. *Photogrammetric Engineering and Remote Sensing* 62(12), 1361–1369.

Arnold, N.S., W.G. Rees, B.J. Devereux and G.S. Amable (2006). Evaluating the potential of high-resolution airborne LiDAR data in glaciology. *International Journal of Remote Sensing* 27, 1233–1251.

Bamber, J.L. and A. Rivera (2007). A review of remote sensing methods for glacier mass balance determination. *Global Planetary Change* 59(1–4), 138–148.

Blümcke, A. and H. Hess (1899). Untersuchungen am Hintereisferner. *Wissenschaftliche Ergänzungshefte zur Zeitschrift des Deutschen und Österreichischen Alpenvereins* 1(2), 87 p.

Braun, M., F. Rau, H. Saurer and H. Goßmann (2000). Development of radar glacier zones on the King George ice cap, Antarctica, during austral summer 1996/1997 as observed in ERS-2 SAR data. *Annals of Glaciology* 31, 357–363.

Braun, M., T.V. Schuler, R. Hock, I. Brown and M. Jackson (2005). Comparison of remote sensing derived glacier facies maps with distributed mass balance modelling at Engabreen, northern Norway. *IAHS Publications* 318, 126–134.

Brown, I.A., M.P. Kirkbride and R.A. Vaughan (1999). Find the firn line! The suitability of ERS-1 and ERS-2 SAR data for the analysis of glacier facies on Icelandic icecaps. *International Journal of Remote Sensing* 20(15), 3217–3230.

Geist, T., H. Elvehøy, M. Jackson and J. Stötter (2005). Investigations on intra-annual elevation changes using multi-temporal airborne laser scanning data – case study Engabreen, Norway. *Annals of Glaciology* 42, 195–201.

Geist, T. and J. Stötter (2007). Documentation of glacier surface elevation change with multi-temporal airborne laser scanner data – case study: Hintereisferner and Kesselwandferner, Tyrol, Austria. *Zeitschrift für Gletscherkunde und Glazialgeologie* 41, 77–106.

Goldstein, R.M., H. Engelhardt, B. Kamb and R.M. Frolich (1993). Satellite radar interferometry for monitoring ice sheet motion: application to an Antarctic ice stream. *Science* 262, 1525–1530.

Hagg, W., L. Braun, V.N. Uvarov and K.G. Makarevich (2004). A comparison of three methods of mass balance determination in the Tuyuksu Glacier region, Tien Shan. *Journal of Glaciology* 50(171), 505–510.

Hall, D.K. and J. Martinec (1985). *Remote Sensing of Ice and Snow*. Chapman and Hall, London, 189 p.

Hall, D.K., R.A. Bindschadler, J.L. Foster, A.T.C. Chang and H. Siddalingaiah (1990). Comparison of *in situ* and satellite-derived reflectances of Forbindels Glacier, Greenland. *International Journal of Remote Sensing* 11(3), 493–504.

Heiskanen, J. (2002). *Assessment of glacier changes and massbalance using Landsat satellite data in Svartisen, Northern Norway*. M.Sc. thesis, Department of Geography, University of Turku, Finland. In Finnish, English abstract, 148 p.

Heiskanen, J., K. Kajuutti, M. Jackson, H. Elvehøy and P. Pellikka (2003). Assessment of glaciological parameters using Landsat satellite data in Svartisen, northern Norway. *EARSeL eProceedings* 2, 34–42. Proceedings of the EARSeL-LISSIG-Workshop Observing our Cryosphere from Space, Berne, Switzerland, March 11–13, 2002.

Hendriks, J. and P. Pellikka (2004). Estimation of reflectance from a glacier surface by comparing spectrometer measurements with satellite-derived reflectances. *Zeitschrift für Gletscherkunde und Glazialgeologie* 38(2), 139–154.

Hendriks, J. and P. Pellikka (2007a). Using multiangular satellite and airborne remote sensing data to study glacier surface characteristics in Hintereisferner, Austria. *Proceedings of the First International Circumpolar Conference on Geospatial Sciences and Applications*, Yellowknife, N.W.T., Canada, August 20–24, 2007, 8 p.

Hendriks, J.P.M. and P.K.E. Pellikka (2007b). Semiautomatic glacier delineation from Landsat imagery over Hintereisferner glacier in the Austrian Alps. *Zeitschrift für Gletscherkunde und Glazialgeologie* 41, 55–75.

Hinze, H. and G. Seeber (1988). Ice-motion determination by means of satellite positioning systems. *Annals of Glaciology* 11, 36–41.

Holmlund, P. and P. Jansson (2003). *Glaciologi*. Vetenskapsrådet & Stockholms universitet, Stockholm, 176 p.

Jensen, J.R. (2000). *Remote Sensing of the Environment – an Earth resource perspective*. Prentice Hall, Upper Saddle River, 544 p.

Kaser, G. (2001). Glacier-climate interaction at low latitudes. *Journal of Glaciology* 47, 195–204.

King, M. (2004). Rigorous GPS data processing strategies for glaciological applications. *Journal of Glaciology* 50(171), 601–607.

Knap, W.H. (1997). *Satellite derived and ground-based measurements of the surface albedo of glaciers*. Ph.D. dissertation, University of Utrecht, The Netherlands, 175 p.

Knap, W.H. and C.H. Reijmer (1998). Anisotropy of the reflected radiation field over melting glacier ice: measurements in Landsat TM bands 2 and 4. *Remote Sensing of Environment* 65, 93–104.

Kuhn, M., E. Dreisetl, S. Hofinger, G. Markl, N. Span and G. Kaser (1999). Measurements and models of the mass balance of Hintereisferner. *Geografiska Annaler* 81(A)(4), 659–670.

Kulkarni, A.V., B.P. Rathore and S. Alex (2004). Monitoring of glacial mass balance in the Baspa basin using accumulation area ratio method. *Current Science* 86(1), 185–190.

Knoll, C. and H. Kerschner (2009). A glacier inventory for South Tyrol, Italy, based on airborne laser scanner data. *Annals of Glaciology* 53(50), 46–52.

Langley, K., S.-E. Hamran, K.A. Høgda, R. Storvold, O. Brandt, J. Kohler and J.O. Hagen (2008). From glacier facies to SAR backscatter zones via GPR. *IEEE Transactions on Geoscience and Remote Sensing* 46(9), 2506–2516, DOI: 10.1109/TGRS.2008.918648.

Li, Z., W. Sun and Q. Zeng (1999). Deriving glacier change information on the Xizang (Tibetan) plateau by integrating RS and GIS techniques. *Acta Geographica Sinica* 54, 263–68.

Lutz, E., T. Geist and J. Stötter (2003). Investigations of airborne laser scanning signal intensity on glacial surfaces – utilising comprehensive laser geometry modelling and orthophoto surface modelling. *International Archives of Photogrammetry, Remote Sensing and Spatial Information Sciences* 35(3/W13), 143–148.

Marshall, G.J., W.G. Rees and J.A. Dowdeswell (1995). The discrimination of glacier facies in ERS-1 data. In: J. Askne (ed.), *Sensors and Environmental Applications in Remote Sensing*. A.A. Balkema Publishers, Rotterdam, 500 p.

Nakamura, T., O. Abe, T. Hasegawa, R. Tamura and T. Ohta (2001). Spectral reflectance of snow with a known grain-size distribution in successive metamorphism. *Cold Regions Science and Technology* 32, 13–16.

Nolin, A.W. and M.C. Payne (2007). Classification of glacier zones in western Greenland using albedo and surface roughness from the Multi-angle Imaging SpectroRadiometer (MISR). *Remote Sensing of Environment* 107, 264–275.

Oerlemans, J. (2001). *Glaciers and Climate Change*. A.A. Balkema Publishers, Lisse, 148 p.

Østrem, G. (1975). ERTS data in glaciology, an effort to monitor glacier mass balance from satellite imagery. *Journal of Glaciology* 15, 403-415.

Paterson, W.S.B. (1994). *The Physics of Glaciers*. 3rd ed., Butterworth-Heinemann, Oxford. 480 p.

Peel, D.A. (1992). Spatial temperature and accumulation rate variations in the Antarctic Peninsula. In: E.M. Morris (ed.), *The Contribution of Antarctic Peninsula Ice to Sea Level Rise*. pp. 11–15, British Antarctic Survey, Cambridge.

Pellikka, P. and J. Hendriks (2002). Application of multiangular remote sensing data for detecting glacier zones. *Proceedings of the 3rd International Workshop on Multiangular Measurements and Modelling*, Steamboat Springs, Colorado, U.S.A., June 10–12, 2002.

Pellikka, P. (2007). Monitoring glacier changes within the OMEGA project. *Zeitschrift für Gletscherkunde und Glazialgeologie* 41, 3–5.

Pesu, M. (2007). *Thematical mapping of glacier environment and defining the mass balance of a valley glacier from a satellite image with pixel oriented- and object oriented classification methods*. M.Sc. thesis, Department of Geography, University of Helsinki, In Finnish, English abstract, 70 p.

Rasmussen, L.A. (1988). Bed topography and mass-balance distribution of Columbia Glacier, Alaska, U.S.A., determined from sequential aerial photography. *Journal of Glaciology* 34(117), 208–216.

Rau, F., M. Braun, M. Friedrich, F. Weber and H. Goßmann (2000). Radar glacier zones and their boundaries as indicators of glacier mass balance and climatic variability. *EARSeL eProceedings* 1, 317–327. Proceedings of the EARSeL-LISSIG-Workshop Land Ice and Snow, June 16–17, 2000 Dresden.

Rees, W.G. (2006). *Remote Sensing of Snow and Ice*. CRC Press, Taylor & Francis Group, Boca Raton, 285 p.

Rees, W.G. and N.S. Arnold (2007). Mass balance and dynamics of a valley glacier measured by high-resolution LiDAR. *Polar Record* 43(227), 311–319.

Shi, J. and J. Dozier (1993). Measurement of snow- and glacier-covered areas with single polarization SAR. *Annals of Glaciology* 17, 72–76.

Sohn, H.-G. and K.C. Jezek (1999). Mapping ice sheet margins from ERS-1 SAR and SPOT imagery. *International Journal of Remote Sensing* 20(15), 3201–3216.

Span, N., M. Kuhn and H. Schneider (1997). 100 years of ice dynamics of Hintereisferner, Central Alps, Austria, 1894–1993. *Annals of Glaciology* 24, 297–302.

Span, N. and M. Kuhn (2003). Simulating annual glacier flow with a linear reservoir model. *Journal of Geophysical Research* 108(D10), 4313.

Span, N., A. Fischer, M. Kuhn, M. Massimo and M. Butschek (2005). Radarmessungen der Eisdicke Österreichischer Gletscher. *Österreichische Beiträge zu Meteorologie und Geophysik* 33, 145 p.

Spikes, V.B., B. Csatho, G. Hamilton and I. Whillans (2003). Thickness changes on Whillans ice stream and ice stream C, West Antarctica, derived from laser altimeter measurements. *Journal of Glaciology* 49(165), 223–230.

Strozzi T., G.H. Gudmundsson and U. Wegmüller (2003). Estimation of the surface displacement of Swiss alpine glaciers using satellite radar interferometry. *EARSeL eProceedings* 2(1), 3–7.

Tremel, H., H. Hornik and H. Rentsch (1993). Bestimmung der geodätischen Position von Gletschersignalen mit Hilfe des Global Positioning System (GPS). *Zeitschrift für Gletscherkunde und Glazialgeologie* 29(2), 173–178.

Warren, S.G. and W.J. Wiscombe (1980). A model for the spectral albedo of snow. II: Snow containing atmospheric aerosols. *Journal of Atmospheric Sciences* 37, 2734–2745.

Whillans, I. and C. van der Veen (1993). New and improved determinations of velocity of Ice Streams B and C, West Antarctica. *Journal of Glaciology* 39(133), 438–490.

Zeng, Q., C.M. Cao, X. Feng, F. Liang, X. Chen and W. Sheng (1984). Study on spectral reflection characteristics of snow, ice and water of northwest China. *IAHS Publications* 145, 451–462.

Chapter 4

The early history of remote sensing of glaciers

Christoph Mayer
Commission for Glaciology, Bavarian Academy of Sciences and Humanities,
Munich, Germany

4.1 INTRODUCTION

The history of monitoring and investigating glaciers by remote sensing techniques is basically divided into two distinct periods: before and after the International Geophysical Year (IGY) 1957/1958. For more than 100 years before the IGY, remote sensing was used almost exclusively for mapping purposes and for the determination of glacier surface velocities. It was only with the massive increase in glacier investigations during the IGY that the development of new sensors during the 1950s and 1960s and the advent of space technology opened up the broad field of remote sensing techniques covering almost the entire range of the electromagnetic spectrum. Now not only geometry changes were detectable from a distance; other glaciological parameters also became the focus of remote sensing applications. These new techniques in particular are covered in more detail in other chapters of this book. The historical part focuses on the earlier history of glacier remote sensing.

4.2 EARLY GLACIER OBSERVATIONS

In the mountain areas of the world, people have been in contact with glaciers for thousands of years. Traders and herders in particular had to find ways to cross the ice-covered passes in order to reach neighbouring valleys. The scientific interest in glaciers, however, really began only about 230 years ago, when Horace Bénédict de Saussure became fascinated by the large glaciers of the Mont Blanc massif in France. But even then another four decades passed before Ignatz Venetz, in the 1820s, realised that glaciers are highly dynamic entities, which once filled even the largest valleys in the Alps and also extensive areas of the lowlands. Jean de Charpentier and later Karl Friedrich Schimper were the only ones at that time who pursued the ideas of Venetz. They realised the existence of past widespread glaciation, a concept which Schimper promoted in the scientific community as *Eiszeit* (ice age), although the expression was already used earlier by some people (e.g. J.W. Goethe). This new perception, and the postulation of the existence of ice ages by Louis Agassiz at a conference in Neuchatel, Switzerland in 1837, mark the beginning of modern glaciology.

The question of who, in fact, was the first scientist to develop the theory of ice ages is not easily answered. The tireless efforts of Agassiz with his fundamental publication

Études sur les glaciers (Agassiz 1840), however, finally inspired other scientists as well to concentrate on the investigation of glaciers (e.g. Edward Forbes, John Tyndall). This increased interest in glaciers naturally led to the development of suitable observations techniques, which, even in the early stages, already involved the acquisition of remotely sensed data.

At that time glaciers were even more difficult to reach and especially to travel on than today. Therefore it is not surprising that observations from a distance were a common method for glacier investigations without taking high risks. In general, glaciers and ice sheets and remote sensing are an ideal combination, even more so today where satellite observations are readily available at high temporal and spatial resolution. In addition, many glaciers, and especially ice caps and ice sheets, are very large. This makes it difficult, if not impossible, to obtain data at a reasonable spatial resolution by *in situ* investigations. For these reasons, remote sensing techniques were used in glacier research almost from the beginning.

In fact, even the naturalistic paintings of glaciers from the 17th and 18th century (e.g. Figure 4.1), E. Handmann in 1748 and many more thereafter are the remotely sensed documentation of glacier conditions at that time. Some of these images have been used to reconstruct the glacier limits and their variations during this period (Zumbühl and Holzhauser 1988). In some cases individual glaciers were the subjects

Figure 4.1 One of the earliest artistic representations of a glacier; Unterer Grindelwaldgletscher, Switzerland, by J. Plepp, before 1642.

of artists more or less regularly over longer time spans. For example, the tongue of the Rhone glacier, Switzerland, was documented by artists for at least 11 different years between 1770 and 1848. If the paintings are made from similar locations and enough realistic details are detectable, the delineation of the glacier boundaries is possible with acceptable errors. Nussbaumer et al. (2005/2006) were able to reconstruct the length variations of Mer de Glace in France during the 18th and 19th century based on artistic representations. Because glaciers usually leave only clear signs of their advances (terminal moraines), these early documents are important sources for interpreting glacier retreat and prevailing climatic conditions (Oerlemans 2005). The artistic representation of glaciers and their scientific use was not restricted to the Alps and comparable naturalistic drawings and paintings are known e.g. from Nigardsbreen in Norway (1822 by J. Flintoe, 1851 by J.D. Forbes and others) and many other regions. Even

Figure 4.2 Map of the tongue of Hintereisferner, Austria in 1817 in a scale of 1:28 800 by F. von Hauslab. This is a section of the original map of Obergurgl.

in New Zealand sketches of Franz Josef Glacier date back to 1865 (McKinzey et al. 2004).

Maps produced around the end of the 18th century usually gave only a rough overview of glacier extents, but no clear representation of the glacier and its features (Figure 4.2), because the focus of these maps was general topography and landscape, not ice covered regions. Often the scale was about 1:100 000 which is not ideal for delineating glacier boundaries (e.g. the Dufour Atlas in Switzerland).

4.3   THE SCIENTIFIC APPROACH

The first glaciological application of remote sensing was the determination of ice velocity on glaciers. Franz Josef Hugi observed the displacement of his hut on Unteraargletscher, Switzerland, in the years from 1827 to 1830. In the following years prominent stones, or arranged rows of marked stones, were observed over defined time intervals in order to estimate the surface velocity. But only in 1841 were systematic velocity measurements initiated by Agassiz, whereas Johannes Wild was using a theodolite for his precise survey of wooden poles drilled into the ice. This type of measurement was subsequently also conducted by James Forbes (1842) and John Tyndall (1860) on Mer de Glace. Early scientific glacier observations are also known from Norway, where Nigardsbreen was already described in 1750 (Foss 1803). In other parts of the world such activities mostly started in the second half of the 19th century. In North America glaciers were described by Russell in 1885 and Shaler and Davis in 1881.

For a long time, the main focus of glacier research was the documentation of glacier geometry at the time of observation. Ground surveys required the erection of signals on the glacier or the identification of stable features, such as supraglacial boulders. Smooth and uniform ice surfaces which were difficult to access, as well as dangerous areas, therefore, were automatically excluded from the investigations. Also, surveying enough points to generate a reliable glacier geometry was a very time-consuming task

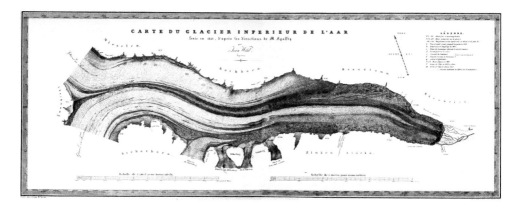

Figure 4.3 The first scientific representation of a glacier as a map, by J. Wild in 1842: Unteraargletscher, Switzerland.

in the field. The efforts of Agassiz and Wild, however, led to the first scientific map of a glacier in 1842: the *Carte du glacier inferieur de l'Aar*, at a scale of 1:10000 (Figure 4.3).

4.3.1 Glacier mapping from point observations

During the following decades a number of glaciers were mapped mainly by the use of tachymetry or plane-table cartography. As an example, Hermann and Adolf Schlagintweit produced a map of Pasterze glacier in Austria at a scale of 1:14400 from observations made in 1846 and 1848. They also published a map of the southern Ötztal Alps, Austria, at the same time, showing the glacier extent in that region (Schlagintweit and Schlagintweit 1850).

The brothers Hermann and Adolf Schlagintweit were among the first researchers to systematically investigate glaciers, especially by cartographic means in the Eastern Alps (Figure 4.4). At the Pasterze glacier they also measured glacier velocities. Their results were published in the book *Untersuchungen über die physikalische Geographie der Alpen* in 1850. Together with their brother Robert, they also applied their experience and knowledge to glacier investigations during their extensive travels through the high mountains of Asia. Most of the Asian glaciers in India, Nepal and China had their maximum extent during the period of travels of the Schlagintweit brothers, as in the European Alps. Observations of the glacier situation and the relevant sketches,

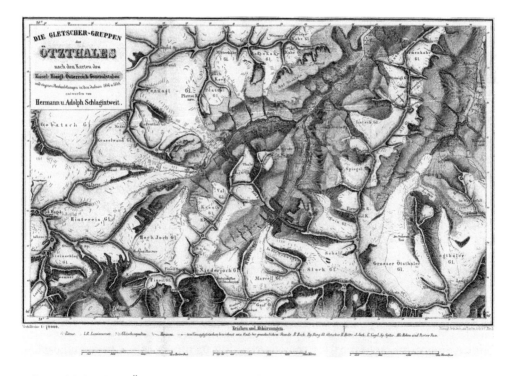

Figure 4.4 Southern Ötztal Alps, Austria, in 1847 by Hermann and Adolf Schlagintweit (1850).

very often taken from remote viewpoints, are of a remarkable accuracy (Kick 1967). In 1860, Karl von Sonklar published maps of several glaciers in the Ötztal valley, at a scale of 1:28800, based on the Austrian National Survey (Sonklar 1860). Eduard Richter used a newer Austrian National Survey (1870–1873) at a scale of 1:25000 as the basis for the construction of several glacier maps by plane-table cartography (Richter 1888).

In this period systematic measurements of glacier terminus variations started in many mountain regions from the European Alps, the Scandinavian Mountains, the Rocky Mountains and the New Zealand Alps. The foundation of the International Glacier Commission in 1894 soon coordinated these efforts (www.geo.unizhch/wgms). Many of these length records are today continued by the use of modern remote sensing methods. At roughly the same time an initiative was started in Switzerland which, for the first time, aimed at monitoring the geometric and kinematic changes of an alpine glacier in detail. This also was the beginning of plane-table tachymetry, surveying the entire area of the Rhone glacier in the Bernese Alps from the tongue up to the accumulation zone. This glacier was chosen due to its rather simple geometry and comparatively easy access. The survey of the lower part started in 1874 and velocity measurements were repeated thereafter every year by Philippe Gosset. Finally the survey was also extended to the upper part of the glacier in 1882 (Mercanton 1916). As a result of these surveys, detailed maps of the glacier and continuous velocity measurements over a period of thirty years were published in 1916 by Mercanton (Figure 4.5). These early mapping activities were not restricted to the Alps and simple glacier maps produced during the last decades of the 19th century are available from a number of countries. Douglas and Harper compiled the first map of Franz Josef glacier in New Zealand in 1893, the glaciers of Mt. Rainier in the United States were surveyed in 1896 (Russell 1898), and Sexe had produced a map of Folgefonna in Norway some 30 years earlier (Sexe 1864).

4.4   THE DAWN OF PHOTOGRAMMETRY

Glaciers continued to be surveyed from remote observation points, but it was still necessary to identify and locate local targets on the glacier. This method was rather time-consuming in the field and required a large number of local control points. Ideally, it should be possible to preserve the view from a distance of the glacier with a known perspective, so that the quantitative interpretation can be transferred back to the office. In principle this can even be achieved by making a very precise drawing, if a constant projection is maintained. A better solution would be an automated technique to preserve the scene in a repeatable way. For this purpose photography, invented not very long before, seemed a very promising technique. The mathematical formulation for reconstructing the geometry from central perspective images was described much earlier than the invention of photography. As early as 1759, the mathematician J.H. Lambert presented geometrical solutions of this problem (Finsterwalder and Hofmann 1968). After the introduction of photography in the 1830s by J.N. Nièpce and J.L.M. Daguerre it took another two decades before surveyors used the geometric information of photographs for the construction of topographic maps and architectural plans. Oblique photographs from that time still serve as a kind of baseline for comparisons with modern pictures, demonstrating in an impressive way the temporal

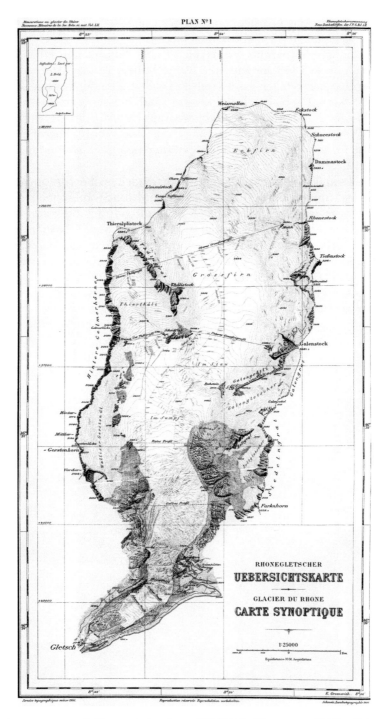

Figure 4.5 Accurate map of Rhone glacier from 1874/1882 (Mercanton 1916). (See colour plate section).

changes of ice masses during the last century (Figure 4.6). The quest to obtain a picture with the least distortions possible, and the best, unobstructed view soon resulted in photographs being taken from above the object.

The first aerial photograph was taken by G.F. Tournachon (alias Nadar) from a tethered balloon in 1858 showing the Bievre Valley, France. The oldest preserved aerial photography, also taken from a tethered balloon, is a photograph depicting part of the city of Boston, US, by Black and King in 1860. In later years kites were also used (e.g. by Arthur Batut in France). Aerial photography was further developed during the First World War by the military, now using fixed wing aircraft. Sherman M. Fairchild developed the first camera suitable for aerial mapping (high shutter speeds, accurate geometry) towards the end of the war.

In the 1870s and 1880s, the technique was extended further for topographic survey work. The ideal applicability to the generation of topographic maps in mountainous terrain was recognised during that time, and considerable parts of the Italian Alps were surveyed using early photogrammetric methods. Sebastian Finsterwalder finally extended the application to glaciers by producing a map of Vernagtferner in the Ötztal Alps, Austria, derived from plane-table photogrammetric surveys in 1889 (Finsterwalder 1897). This method allowed the identification of a large number of survey points without the need for direct access. This was the first map of an entire glacier, produced for the purpose of documenting the state of a glacier at a certain time and observing temporal changes on subsequent visits. This method was also successfully used for observations at the Schneeferner in the Wetterstein (Germany), Suldenferner in the Ortler Alps (Italy) by Finsterwalder and the Hochjochferner by Otto Gruber in 1907, as well as at Hintereisferner (Austria) by Blümcke and Hess in 1894 (Figure 4.7, Blümcke and Hess 1899).

Figure 4.6 A daguerreotype photograph by D. Dollfuss-Ausset in 1849 of the Rhone glacier is probably the first photograph of a glacier.

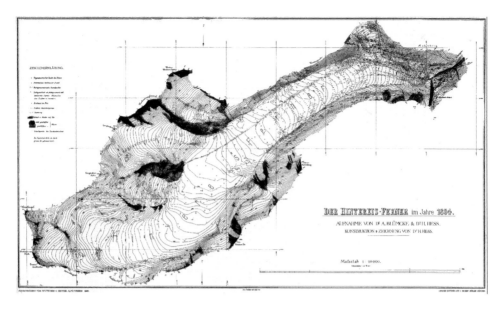

Figure 4.7 An early map of Hintereisferner in Austria in 1894 (Blümcke and Hess 1899).

4.5 THE GOLDEN AGE OF TERRESTRIAL GLACIER MAPPING

S. Finsterwalder also improved and further developed the mathematical basis of terrestrial photogrammetry (Finsterwalder 1899). Until then only the information acquisition was different from classic geodetic surveys. The position angles have been determined in the images; the construction of the projected geometry, however, was done in the classical style. Now tools were available through stereo-photogrammetric methods (by simulating three-dimensional models from planar stereographic images) which allowed the direct determination of point positions by stereographic analysis. An example of this application is the map of Gepatschferner in the Ötztal Alps with a scale of 1:10000 and 10 m contour lines, made by Finsterwalder in 1922. Also the technology had to be developed according to the improved theoretical methods. In 1901, Carl Pulfrich presented his *Stereokomparator* which allowed much faster point measurements in the image pairs. The next step was the introduction of the *Stereo-autograph* by Eduard von Orel in 1911, which allowed a much faster analysis of line measurements. This technique of terrestrial photogrammetry combined with the analysis using the *Stereoautograph* remained the state of the art for documenting glacier geometries during the first three decades of the 20th century.

After the detailed documentation of several of the important glaciers in the Alps, this method was also used in remote glacierized regions of the world. Surveys were accomplished e.g. in Norway (Hoel and Werenskiold 1962), Svalbard (G. Isachsen in 1906–1910), the Caucasus (by L. Distel and H. Burmester in 1911) and Pamir (by R. Finsterwalder and H. Biersack in 1928). Also the first measurements of ice velocities by photogrammetric methods were achieved, for example the velocity measurements by Richard Finsterwalder, son of S. Finsterwalder, in the Himalaya in 1934 (Finsterwalder 1935). The lightweight phototheodolite, developed by S. Finterwalder and Max Ott, proved to be especially useful for glacier surveys in remote and difficult terrain (Finsterwalder 1896). This instrument was later improved by Zeiss in 1926,

Figure 4.8 The lightweight phototheodolite TAF developed by S. Finsterwalder.

Figure 4.9 Velocity analysis from terrestrial photogrammetry of Rakhiot glacier, Pakistan (Pillewitzer 1938).

which also changed the image format from 12×16 cm to the more standard format of 13×18 cm (Figure 4.8). Similar instruments were also developed in Switzerland and other countries. The terrestrial photographs were used for a number of applications, for example for velocity measurements (Figure 4.9).

4.6   THE AERIAL PERSPECTIVE AND THE STEP INTO A NEW AGE

An important role in the development of aerial photogrammetry was played by S. Finsterwalder. Using photographs taken from balloons he very early described the theory of their geometry and the problems of aerial photogrammetric analysis (Finsterwalder 1900). The solution for image pairs with unknown location and orientation was a particularly big step forward for aerial photogrammetry. Rapid development in aviation also led to considerable progress on technical issues of aerial photogrammetry after the First World War. In most cases this was now the method of choice for large-scale surveys, including those over glacierized areas. Nevertheless, terrestrial photogrammetry still remained in use for special applications on smaller glaciers and especially in remote mountain regions, where the use of aerial surveys was not possible for several reasons. Even at that time, remote sensing of glaciers was still restricted to the observation of ice velocities (now also from photogrammetric surveys) and the production of high resolution topographic glacier maps.

During the Second World War remote sensing techniques were further developed especially for intelligence and navigational purposes. Analysis methods for aerial photogrammetry were refined and also infrared sensitive film was tested for improved image quality during haze conditions. In relation to glaciology, the main activities were directed towards improving reconnaissance of sea-ice covered ship routes in the Arctic Ocean. In this context, brightness variations of sea ice were assumed to provide information about ice thickness and navigability (Moss and Glen 1947). Another development which in later decades proved to revolutionise remote sensing, especially of polar regions, was the successful construction of rockets, able to reach above the troposphere. But still during the 1950s aerial photographs remained the major tool of remote sensing in glaciology, restricting the analysis to visual interpretation, delineation of glacier boundaries and the construction of digital elevation models. In particular the remote and wide spread glaciers of North America could now be mapped in an efficient way (e.g. Ommanney 1980 and Figure. 4.10). But also the glaciers in the former Soviet Union and China were now systematically mapped by the use of aerial photogrammetry (Dolgushin 1961, Kotlyakov 1980).

One peculiar observation during that time, which led to the development of another very successful tool for remote sensing of ice masses, was the potential malfunction of aircraft radar altimeters while passing the Greenland Ice Sheet. Unrealistic high flight elevations above ground were regularly observed which indicated to penetration of radar waves through the ice cover (see Chapter 11, and Waite and Schmidt 1962). Specially designed radar instruments subsequently allowed large scale surveys of ice thicknesses over Greenland, Antarctica and many other ice covered regions.

The development of multispectral film and sensors, the concept of radar interferometry and Synthetic Aperture Radar (SAR) and the first rocket launch into space during the end of the 1950s finally started a new era of remote sensing. Most of these observation techniques are discussed in the different chapters of this book and thus makes it unnecessary to reiterate here. In glaciology, due to the often difficult access to the subject of investigation, most remote sensing techniques were readily

Figure 4.10 Early aerial photogrammetric image of Athabasca glacier, Canada (http://airphotos. nrcan.gc.ca/).

included in the investigation methods. To date, five symposia were organised by the International Glaciological Society (IGS) alone, dedicated especially to mapping and remote sensing of the cryosphere (Ottawa 1965, Reykjavik 1985, Cambridge 1986, Boulder 1993, College Park 2001), with increasing interest and impact on glaciology.

REFERENCES

Agassiz, L. (1840). Ètudes sur les glaciers. Jent & Gassmann, Neuchâtel, 346 p. & Atlas part.

Blümcke, A. and H. Hess (1899). Untersuchungen am Hintereisferner, *Wissenschaftliche Ergänzungshefte zur Zeitschrift des Deutschen und Österreichischen Alpenvereins* 1(2), 87 p.

Dolgushin, L.D. (1961). Main particularities of glaciation of central Asia according to the latest data. *IAHS Publications* 54, 348–358.

Finsterwalder, R. (1935). Alpenvereinskartographie auf Forschungsreisen. In: R. Finsterwalder (ed.), *Alpenvereinskartographie und die ihr dienenden Methoden*. Sammlung Wichmann 3, 77–81.

Finsterwalder, R. and W. Hofmann (1968). *Photogrammetrie*. Walter de Gruyter, Berlin, 455 p.

Finsterwalder, S. (1896). Zur photogrammetrischen Praxis. *Zeitschrift für Vermessungswesen* 25(8), 225–240.

Finsterwalder, S. (1897). Der Vernagt Ferner im Jahre 1889. Map 1:10000, supplement to *Wissenschaftliche Ergänzungshefte zur Zeitschrift des Deutschen und Österreichischen Alpenvereins* 1(1), 1–112.

Finsterwalder, S. (1899). Die geometrischen Grundlagen der Photogrammetrie. *Jahresbericht der deutschen Mathematiker-Vereinigung* 6.

Finsterwalder, S. (1900). Die Konstruktion von Höhenkarten aus Ballonaufnahmen. *Bayerische Akademie der Wissenschaften, Mathematisch-Physikalische Klasse, Sitzungsberichte 1900 9,* 149–164.

Forbes, J.D. (1842). First letter on glaciers, Courmayeur, Piedmont, 4 July 1842. *Edinburgh New Philosophical Journal* 33(66), 338–341.

Foss, M. (1803). Justedalens kortelige beskrivelse. *Magazin for Danmarks og Norges Topografiske, Oeconomiske og Statistiske Beskrivelse* 2, 1–42.

Hoel, A. and W. Werenskiold (1962). Glaciers and snowfields in Norway. *Norsk Polarinstitutt, Skrifter* 114, 291 p.

Kick, W. (1967). Schlagintweits Vermessungsarbeiten am Nanga Parbat 1856. *Deutsche Geodätische Kommission bei der Bayerischen Akademie der Wissenschaften Reihe* C 97, 1–146.

Kotlyakov, V.M. (1980). Problems and results of studies of mountain glaciers in the Soviet Union. *IAHS Publications* 126, 129–136.

McKinzey, K.M., W. Lawson, D. Kelly and A. Hubbard (2004). A revised Little Ice Age chronology of the Franz Josef Glacier, Westland, New Zealand. *Journal of the Royal Society of New Zealand* 34(4), 381–394.

Mercanton, P.-L. (1916). Vermessungen am Rhonegletscher 1874–1915. *Neue Denkschriften der Schweizerischen Naturforschenden Gesellschaft* 52. Basel, Genf, Lyon, 190 p.

Moss, R. and A.R. Glen (1947). Investigations on ice during the war. *Journal of Glaciology* 1(1), 8–9.

Nussbaumer, S.U., H.J. Zumbühl and D. Steiner (2005/2006). Fluctuations of the *Mer de Glace* (Mont Blanc area, France) AD 1500–2050: an interdisciplinary approach using new historical data and neural network simulations. *Zeitschrift für Gletscherkunde und Glazialgeologie* 40, 3–183.

Oerlemans, J. (2005). Extracing a climate signal from 169 glacier records. *Science* 308, 675–677.

Ommanney, C.S.L. (1980). The inventory of Canadian glaciers: procedures, techniques, progress and applications. *IAHS Publications* 126, 35–44.

Pillewitzer, W. (1938). Photogrammetrische Gletscherforschung. *Bildmessung und Luftbildwesen* 13(2), 66–73.

Richter, E. (1888). Gletscher der Ostalpen. In: *Handbücher der Deutschen Landes- und Volkskunde* 3, Stuttgart, 306 p.

Russell, I.C. (1885). *Existing Glaciers of the United States.* 5th Annual Report of the U.S. Geologic Survey, 303–355.

Russell, I.C. (1898). Glaciers of Mount Rainier. *U.S. Geological Survey Annual Report* 18(2), 349–415.

Schlagintweit, H. and A. Schlagintweit (1850). *Untersuchungen über die Physikalische Geographie der Alpen in ihren Beziehungen zu den Phänomenen der Gletscher, zur Geologie, Meterologie und Pflanzengeographie.* Verlag V.J.A. Barth, Leipzig, 600 p.

Sexe, S.A. (1864). *Om Sneebræen Folgefond.* Universitetsprogram for annet halvår 1864. Christiania, 36 p.

Shaler, N.S. and W.M. Davis (1881). *Illustrations of the Earth's surface: Glaciers.* J.R. Osgood and Co., Boston, 198 p.

Sonklar, K. (1860). *Die Oetzthaler Gebirgsgruppe mit besonderer Rücksicht auf Orographie und Gletscherkunde.* Justus Perthes, Gotha, 292 p.

Tyndall, J. (1860). *The glaciers of the Alps, being a narrative of excursions and ascents, an account of the origin and phenomena of glaciers, and an exposition of the physical principles to which they are related.* Royal Institution, London, 550 p.

Waite, A.H. and S.J. Schmidt (1962). Gross errors in height indication from pulsed radar altimeters operating over thick ice or snow. *Proceedings of the Institute of Radio Engineers* 50, 1515–1520.

Zumbühl, H.J. and H. Holzhauser (1988). Alpengletscher in der Kleinen Eiszeit. *Die Alpen* 64(3), 129–322.

Chapter 5

Physics of glacier remote sensing

Matti Lepparanta
Department of Physics, University of Helsinki, Finland

Hardy B. Granberg
Département de géomatique appliqué, Université de Sherbrooke, Canada

5.1 INTRODUCTION

Our interest in glaciers stems mostly from their role as stores of fresh water and from the observation that glaciers vary in extent, retreating and advancing in response to climate variations, to the potential peril of alpine and coastal habitations. Also the climate history records of glaciers have become extremely valuable in understanding the Earth system. The massive ice sheets of Antarctica and Greenland play a major role in the global climate by regulating the sea level and modifying the atmospheric general circulation. Other glaciers account altogether for 1% of the total volume of ice on land, the largest of these ice massifs being Southern Patagonia ice field between Argentina and Chile and Vatnajökull in Iceland. Their influence is limited to the local climatic and hydrological conditions, and also they serve as good regional climate indicators.

Glaciers are part of the hydrologic cycle. They accumulate from snowfall and evolve further through transformation of seasonal snow into firn and ice and through slow flow and deformation. Ablation takes place as evaporation, runoff and calving of icebergs. By consequence the winter glacier is mostly snow-covered, while summer surface is either snow or ice, depending on the position respectively above or below the equilibrium line, which separates the accumulation zone from the ablation zone (Figure 5.1). Glaciers also contain impurities. They enclose atmospheric gases and particles and meteorites at the upper surface, and rock particles at the bottom and side boundaries. Although the ablation zone is mostly located at lower, warmer elevations, this is not exclusively so. Wind erosion and evaporation can create spots of net ablation within the accumulation zone. In such 'blue-ice areas' we find bare, often deep-blue (hence the name), ice at the surface throughout the year.

Some blue ice spots of the Antarctic ice sheet have contributed greatly to our understanding of the solar system, since through isostatic adjustment to continuous local mass loss meteorites are brought to the surface. In the Antarctic summer, shallow ice-covered supra-glacial lakes may form through internal melting in blue ice areas (Winther et al. 1996). Such lakes may empty into crevasses and produce meltwater runoff even though they are situated well above the dry snow line, which is the limit up to where snow melt takes place in the surface layer in summer. At the base of the

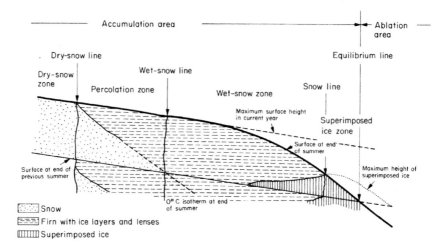

Figure 5.1 Cross-section of the structure of a glacier (Paterson 1994). Printed with permission from Elsevier/Butterworth-Heinemann.

Antarctic ice sheet many large sub-glacial lakes have been found (e.g. Siegert et al. 2001), further indicating the complexity of glacier thermodynamic processes.

In this chapter we shall first discuss the physical properties of glaciers relevant to the signal generation, and thereafter we shall deal with the interaction of glaciers with these signals. The chapter concludes with a discussion of the present and potential future use of remote sensing technology to examine and monitor ice sheets and glaciers.

5.2   GLACIER ICE AND SNOW

5.2.1   Formation of glaciers

High latitudes and altitudes favour glaciation, mostly due to low ablation rates. In mountain areas glaciers are often the result of high accumulation, often orographically induced, but the rate of ablation also decreases with altitude. In the dry glacier zone snow metamorphosis is due to compression only, and the snow becomes fairly homogeneous apart from the density increase with depth. A glacier forms wherever ablation is insufficient to remove the mass brought by snow accumulation, and the size and shape of the glacier is then a result of accumulation, ablation and flow. In a glacier in equilibrium there is slow and continuous flow from the accumulation zone towards the ablation zone to satisfy the overall mass balance.

Glaciers are built from snow, one flake after the other, which gradually turns into firn and eventually clear ice under the ever-increasing weight of snowflakes added on top. The transformation is driven by microscale variations in surface tension, by gravity, and by temperature variations related to variations in surface energy fluxes. The result is a laminated deposit, which grades from snow at the surface to ice, which with increasing depth becomes quite transparent along with the shrinkage of the gas bubbles. The reason for the increasing transparency is a reduction in the amount of reflecting air/ice interfaces per unit volume.

Although much air is initially expelled as the snow densifies, some air becomes trapped in pockets of the snow and firn layers and compressed into bubbles. According to Schytt (1958) the densification in the dry snow zone results in an exponential density profile $\rho(z)$,

$$\frac{\rho_i - \rho(z)}{\rho_i - \rho(0)} = \exp(-Cz) \tag{5.1}$$

where ρ_i is the density of pure ice (equal to $917 \, \text{kg m}^{-3}$ at $0°\text{C}$) and C is a constant, $C \sim 0.025 \, \text{m}^{-1}$. The transformation from snow to ice, i.e. when the aggregate ceases to be permeable, is considered to occur near the snow density of $830 \, \text{kg m}^{-3}$ (Paterson 1994), and the relative air volume trapped in the ice is then 9.5%. The depth of this density level is 50–100 m in the dry snow zone and the age of the snow there varies widely depending on location. The deep glacier ice reaches densities of about $900 \, \text{kg m}^{-3}$. In the basal layers of a 4-km thick glacier an air bubble is only about 1/360th of its initial volume when it closed up at a few tens of meters below the surface. Such bubbles are mostly invisible to the unaided eye.

In addition to ice and air, glaciers contain small amounts of other substances, which were brought by precipitation, by erosion of the substrate due to glacier flow, and by long-distance wind transport of terrestrial materials (Figure 5.2). These particles are buried within the accumulation zone and later brought to the surface in

Figure 5.2 Release of particles in the ablation zone (Vatnajökull, Iceland). Photograph by Matti Leppäranta.

the ablation zone by ice flow. Micrometeorites, and occasionally larger ones, are also incorporated into glaciers, as are materials originating from the substrate. Some particles have secondary effects. Dust from volcanic eruptions can enhance solar absorption in the surface layers, enhancing temperature gradient metamorphism and evaporation crusts, or augment melting. Depending on the type of the contaminants and the mode of ablation, they may be washed away by melt water or accumulate on the glacier surface, sometimes covering the glacier snout.

Seasonal variations influence the mass and energy balance at the glacier surface and a laminated structure develops where growth increments can be distinguished, much like the case is with tree rings. In deposits of wind-blown snow, every gust of wind shows up as a variation in grain size and density, since wind causes fragmentation of drifting snow. Evaporation and solar absorption produce characteristic crusts and changes in the snow structure. Similarly, melt produces distinctive changes in the snow as a result of wetting and melt water infiltration. As the factors which produce these different features vary seasonally, annual accumulation layers are usually readily distinguishable from variations in snow structure alone. Additionally, there are seasonal variations in atmospheric fallout and in the temperature at which the precipitation forms. This enables chemical and isotopic distinction of annual layers in the snow accumulation zone. The structural variations are gradually destroyed as the snow is compressed into ice. However, because most materials diffuse very weakly in ice, the chemical and isotopic layering decreases slowly and gradually with increasing depth.

The thermally active surface layer of glaciers, where seasonal variations are felt, has a thickness of about 10 m. This scale comes from the penetration of temperature signals into the ice according to the classical heat conduction law. The penetration depth of solar radiation is less than the penetration depth of seasonal surface temperature variations, and consequently thermal conduction determines the thickness of the thermally active layer. The damping is exponential with the e-folding depth of $z_1 = [\kappa(\pi\omega)^{-1}]^{1/2}$ here κ is thermal diffusivity and ω is frequency; for the annual cycle $z_1 \approx 3$ m. Where we refer to this 10-m layer ($\approx 3z_1$) as *the glacier surface layer*. In the accumulation zone the surface layer consists of snow (seasonal snow and firn). The key material properties of snow are density, hardness, grain size and shape, and liquid water content. The relative air volume of snow is simply $1-\rho_s/\rho_o$, where ρ_s is snow density. In remote sensing the surface roughness is an additional key property of the snow layer.

5.2.2   Glacier surface layer: snow

The glacier surface layer is usually snow and firn. The snow properties needed for interpretation of remote sensing data and the need of snow information in glaciological research are:

Remote sensing	Information needed for glaciers
Grain size	Extent
Density	Snow stratigraphy
Temperature	Surface temperature
Liquid water content	Melting stage
Surface roughness	Surface geometry
Impurities	Elevation
Elevation	

The extent of snow cover can be relatively well mapped because of the contrast between snow and other surfaces. When snow is present the surface temperature and melting stage are critical to identify the different zones in the glacier (see Figure 5.1). In ground truthing the liquid water content is the most difficult property of snow cover to obtain (see Denoth et al. 1984 for measurement techniques).

In calm conditions, upon landing, the snow crystal creates sharply concave surfaces of very high surface tension at the points of contact with the surface snow. These are sites of strong net mass gain by sublimation. Accordingly the new snow crystals become attached to the snow surface by rapidly growing ice bonds, forming a skeletal, rather than granular, deposit. The density of this initial deposit may be as low as 40 kg m^{-3} for so-called 'wild snow' formed out of spatial, i.e. three-dimensional, dendrites. A typical initial snow density for calm conditions is $\rho_s \approx 80\text{--}100$ kg m^{-3}. However, on glaciers the snowfall is usually accompanied by wind. The atmospheric snow crystals therefore hit the surface with an additional velocity, as compared to calm conditions, and often continue to slide and bounce along the surface for a long distance before they become attached to the snow skeleton. In the process they are broken into smaller fragments, smaller the higher the wind speed. At higher wind speeds fragments are brought into suspension by air turbulence and carried along for longer distances as drifting and blowing snow. Thus, wind transport mechanically destroys the original delicate structure of the crystals. The degree of fragmentation depends on the wind speed and the transport distance, and this controls the initial density, which becomes just over 400 kg m^{-3} for well-fragmented snow (Kotlyakov 1961, Kärkäs et al. 2002). Figure 5.3 shows an example of the surface layer density profile from the Dronning Maud Land, Antarctica.

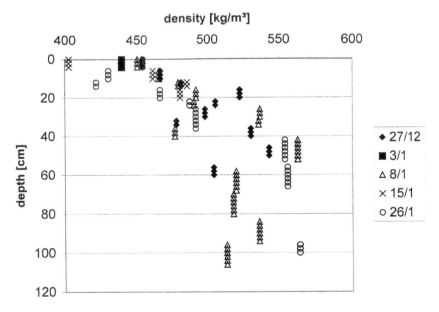

Figure 5.3 Stratification of the snow surface layer in a glacier, Dronning Maud Land in austral summer 1999/2000 (Rasmus et al. 2003). The symbols shown on the right refer to date in December 1999–January 2000.

As elaborated in greater detail by Granberg (1998), the degree of fragmentation also controls the hardness and the rate of hardening of the snow deposit. The hardness is essentially a function of the total number of bonds per unit volume, for it is these bonds which generate the resistance to deformation of the snow skeleton. In wind-fragmented snow the temperature gradients are much steeper than in snow consisting of unbroken crystals and the heat transfer rate and thereby bond growth rate are therefore much greater than in snow that has accumulated in calm conditions. The heat transfer and snow hardening rates both increase as the size of the snow fragments decreases. This is important to the generation of snow surface forms (Granberg 1998).

Although snowfall tends to exhibit only gradual spatial variations, snow drifting rearranges the new snow spatially. On open, semi-infinite snow fields the new snow is amassed into dunes, or snow-barchans, which travel along the surface by erosion on their windward sides and snow trapping on their lee sides. The spacing of the dunes depends on snow availability; often light snowfalls can be amassed into dunes covering perhaps only 20–30 percent of the total surface area. By their modes of snow trapping, dunes produce a complex snow stratification consisting of inclined layers generated by the lee accumulations. These layers exhibit considerable density variations on the micro-scale as a result of variations in the degree of fragmentation of the deposited snow.

In general, as the storm progresses, the grain size decreases and therefore the rate at which inter-granular bonding occurs is accelerated. This leads to increased hardness of the windward side particularly in the oldest part of the dune. Slight variations in hardness lead to variations in erosion rates and etching of this side. The etching leads to the development of channels eroded into the hardened drift forms creating surface forms, which are known as sastrugi (Figure 5.4). They consist of steep-sided semi-parallel ridges and valleys with a vertical extent of typically about 0.2 m, but in particular settings their vertical extent may be greater. Large, hardened sastrugi are formidable obstacles to glacier travel. Sastrugi formed in one storm provide the micro-topographic setting for the next snow accumulation event. Snowfalls with winds of a different direction fill the sastrugi fields with weakly fragmented snow and, if sufficiently abundant, cap this deposit of low-density snow with a new high-density layer of dunes and sastrugi (Granberg 1998).

The deposited snow undergoes continuous changes, which are initially driven by the large micro-scale variations in surface tension in the new snow. Under the influence of wind pumping these variations are rapidly reduced in a process which has been called 'destructive metamorphism', for it results in the rapid disappearance of the delicate shapes of the atmospheric crystals. The general rule of destructive metamorphism is to minimize the specific surface area of the ice skeleton, for this minimizes the micro-scale spatial variations in surface tension.

In a non-melting snow cover there are nearly always also temperature and vapour pressure gradients of a larger scale. Near the surface, these gradients can vary rapidly and even change sign on a seasonal, daily, or even shorter time scale causing convergence and divergence of moisture, and this influences the development of the ice skeleton. While a seasonal snow cover generally exhibits an upward flux of moisture, the opposite is often true for the summer snow cover on glaciers. In winter the flux is directed upwards, but because of the overall colder temperatures these fluxes are weak.

Figure 5.4 Sastrugi, etched by evaporation, showing the inclined layer structure and the rough surface. The hardness is indicated by the ski tracks made by a person weighing 120 kg with the equipment. (Granberg 1998).

Densification of the snow on glaciers is therefore the result of both depth compaction and downwards diffusion and sublimation of water vapour.

The snow grains are initially small (sub-millimetre size) and rounded from wind transport, but they rapidly bond together into an ice skeleton, which coarsens over time, i.e. with depth. In the dry snow zone the grain size is less than 1 mm at the surface and increases with depth by of the order of 1 mm over 100 m (Gow 1971).

Below the dry snow line, the densification of snow is also the result of infiltration of melt water generated in the near surface layer or in radiation-melt wedges created on the side of sastrugi that are exposed to the afternoon sun. The depth to which the melt water penetrates is function of its amount, the temperature gradient, and the porosity of the snow. Infiltrating water tends to follow a path of least resistance. Hence, water spreads laterally on top of layers where the permeability is lower than that of the layer above. At the lower end of the percolation zone (see Figure 5.1), the underlying snow from previous years has become sealed by re-freezing of infiltrating water and this causes the snow cover to become saturated with melt water. This part of the glacier, sometimes called the saturation zone, produces runoff.

The volume of liquid water in the near-surface layer has a large influence on many remote sensing signals. As the solar radiation increases, the so-called cold skin phenomenon develops first. Due to long-wave radiation losses at the surface, the surface temperature remains below zero but melting of snow commences a few centimetres beneath the surface (of the order of the e-folding depth of light attenuation). The melt layer then grows both upward and downward and may reach the surface. Thus in the dry snow zone melt–freeze metamorphosis may take place beneath the surface.

5.2.3   Glacier surface layer: ice

The surface of glaciers is ice below the snow line or in the blue ice regions (Figure 5.5). This surface differs in quality from snow surface. The key material properties of ice are the crystal size and shape, gas bubble content, and other impurities. In blue ice areas the flow of ice is upward, and consequently the surface ice comes from deeper inside the glacier where it was originally formed. In the ablation zone below the snow line the surface ice may be deep glacier ice as in blue ice regions, but in places also superimposed ice occurs. Superimposed ice is the ice layer, which grows upward on glacier surface from meltwater and slush (see Chapter 2). This ice forms from slush, or, in the autumn, as the surface temperature descends below the freezing point, water drainage from the saturation zone can continue for a while, creating superimposed ice which freezes onto the glacier below (e.g. Palosuo 1985).

Figure 5.5 A photograph of Lake Suvivesi taken from the top of Basen nunatak in Dronning Maud Land, Antarctica. A low moraine ridge is seen in the middle of the lake, and Plogen nunatak shows up on the left at the distance of 20 km. The dark area in the foreground is bare ground. Photograph by Matti Leppäranta. (See colour plate section).

In polar summer supraglacial lakes form in blue ice regions (Winther et al. 1996, Kanto et al. 2008). Due to the cold skin phenomenon, they have first a thin (10–20 cm) ice cover and a water body of 0.5–2 m thickness, and later in summer they may in places reach open surface conditions (Figure 5.5).

The surface is therefore ice in blue ice areas and in the ablation zone. Blue ice is smooth, but the ablation zone may be very rough with cracks and crevasses. Also in the ablation zone debris is surfacing because of the surface melt. The surface can be therefore very dirty near the snout of glaciers (see Figure 5.2).

5.2.4 Glacier flow

Due to the weight of the snow and ice, pressure increases with depth in a glacier. This pressure always causes compression of the ice, and because of the pressure-dependence of the rheological properties of ice, the lower layers of a glacier are capable of deforming and flowing. The internal stress often puts the upper few hundred metres under strong tension, as seen from crevassing that develops, not only on land-based glaciers but also on floating ice shelves and tabular icebergs floating at sea. Thus, one important characteristic to the evolution of glaciers is their flow and deformation. The flow velocity is of the order of 10–1000 m per year, with the maximum at the surface and in the centre of valley glaciers, see Chapter 2.

Glacier deformation is due to differential internal flow, while at the base sliding motion may normally be observed if the bed is at the pressure melting point. Up to a certain critical strain rate, ice exhibits viscous-plastic deformation in response to a sustained stress. Above the critical rate it exhibits brittle behaviour, i.e. it fractures. Ice streams tend to form where the glacier is thickened by valleys in the underlying bedrock. The flow also creates shear zones along the sides of the ice streams, seen as areas of heavy crevassing. Similarly, strain rate variations associated with increasing slope and/or decreasing friction against the substrate also produce crevasse zones through tension as the glacier thins and accelerates. The depth of the crevasses is limited by the increase in the critical strain rate with depth. Crevasses are efficient snow traps. However, rather than becoming snow-filled, they tend to become bridged by accumulations of drifting and blowing snow. Such snow-bridged crevasses constitute an important travel hazard. While open, crevasses also influence snow accumulation and thereby snow cover development locally and for long distances downwind. In sea ice remote sensing, Goldstein et al. (2000) showed how ordered structures of fractures can be found from remote sensing imagery and interpreted for the mechanical processes which created them.

A glacier is normally taken as a highly viscous incompressible medium. The viscous rheology employed is described by a sublinear power law (see Paterson 1994)

$$\sigma = A\dot{\varepsilon}^n \tag{5.2}$$

where σ is the internal deviatoric stress, $\dot{\varepsilon}$ is strain rate, and A and n are the flow parameters. The exponent is normally taken as $n = 1/3$. The plastic law is obtained as the limiting case as $n \to 0$, and then the yield criterion needs to be additionally established. In one-dimensional flow Eq. (5.2) gives the relation between shear strain rate and shear stress, and in two-and three-dimensional cases Eq. (5.2) gives the relation between the tensor invariants equal to the sum of squares of the components and assuming then that the deviatoric stresses and strain rates are parallel. Due to the sub-linear rheology of glacier flow the boundary layers are thin. Internal deformation is intensive at side and bottom boundaries as well as in internal boundaries due to geometrical constraints but the bulk of the ice mass deforms weakly.

At the grounding line the bottom of ice shelves is at the oceanic shoreline level and at this line, the friction against the substrate becomes essentially zero. The glacier is therefore free to expand horizontally. Gravity gives the glacier its weight, which is perfectly counter-balanced by the buoyancy force. Because of its lower density, the glacier floats in the water with a keel that displaces its equivalent weight in ocean water. At the bottom of the glacier, the pressure in the ice is equal to that of the water there. However, with decreasing depth there is an increasing difference between the two. This difference reaches its maximum value at the water surface and decreases with height above it where the internal pressure of the ice continues to decrease while that of the surrounding air only decreases by a very small amount over the same height interval and can be ignored.

The pressure difference strains the ice and rapid thinning occurs as the glacier expands into a floating ice shelf. The tensile strength of ice is low. Therefore crevassing, perpendicular to the main direction of expansion, occurs in the near-surface, low-pressure part of the glacier. In the high-pressure lower strata the deformation is plastic. Therefore ice shelves maintain their integrity until their thickness has diminished to about 500 m. In thinner ice shelves, the lowest strata are no longer sufficiently plastic to deform in response to the strains imposed by tidal action leading to fracture, giving birth to tabular icebergs. These icebergs become increasingly prone to breakup by wave action as their underside melts and the plastic layer thins.

5.3   INTERACTION OF ELECTROMAGNETIC RADIATION WITH ICE AND SNOW

5.3.1   General

Electromagnetic radiation is, as its name implies and as described by Maxwell's equations of the electromagnetic field, of both electric and magnetic nature. Electromagnetic waves are transverse, i.e. they do not involve motion across the direction of propagation but consist of variations in an electrical and associated magnetic field, one orthogonal to the other and both perpendicular to the direction of propagation of the wave. Thus, wavelength, λ, the length from peak to peak of one wave (m), is inversely related to the frequency, ν, cycles per second (Hz), by

$$c = \nu\lambda \qquad (5.3)$$

where c is the velocity of signal propagation. While frequency remains constant, the wavelength and the velocity both vary with the square root of the dielectric constant of the medium through which the wave propagates. In vacuum the speed of propagation is constant at $c_0 \approx 3.00 \cdot 10^8$ m s^{-1}. In common use, the term 'wavelength' refers to the length of the wave in vacuum. This relation may also be expressed in terms of the relative permeability μ_r and the dielectric constant ε_r as

$$c = \frac{c_0}{\sqrt{\mu_r \varepsilon_r}} \tag{5.4}$$

The ratio of the speed in vacuum to that in a particular substance is known as the refractive index of that substance. This ratio is a property of the material and varies with both wavelength and temperature. In ice, which is anisotropic, it also varies in relation to the principal crystal axes.

The wavelength bands employed in remote sensing are classified as

Ultraviolet	0.1–0.4 μm
Optical	0.4–0.7 μm
Near infrared	0.7 μm–3 μm
Thermal infrared	3–30 μm
Far infrared	30 μm–1,000 μm
Microwaves	1 mm–1 m

The exact boundaries of these bands show small variations in the literature. The main discrepancy is in the infrared band. The above divison is common in physics and astronomy, while the International Commission of Illumination (CIE) classifies infrared radiation as

Near infrared (IR-A)	0.7 μm–1.4 μm
Short-wavelength infrared (IR-B)	1.4 μm–3 μm
Mid-wavelength infrared (IR-C)	3 μm–8 μm
Long wavelength infrared (IR-C)	8 μm–15 μm
Far infrared	15 μm–1,000 μm

In remote sensing the penetration depth varies with the wavelength and the target medium, and this indicates from how deep within the medium information can be obtained directly. Optical wavelengths and microwaves have significant penetration depth into snow and ice, but other wavelengths are informative only about the medium very close to the surface.

5.3.2 Optical and near infrared signals

In the light transfer through snow and ice, radiance $L = L(z; \lambda)$ is attenuated by absorption and scattering. For a more complete treatment of the radiation transfer see, e.g. Arst (2003). Absorption and scattering are inherent optical properties, which depend only on the medium in question. The scattering function is usually integrated for forward scattering and backscattering $b_f(\lambda)$ and $b_b(\lambda)$, respectively, and the attenuation of a light beam due to scattering is then $b = b_f + b_b$. Linear attenuation laws are employed for the absorption and scattering, and additively for the total light attenuation. Then we have

$$\frac{dL}{dz} = -cL \tag{5.5}$$

where $c = a + b$ is the (total) attenuation coefficient, $a = a(\lambda)$ is the absorption coefficient and z is the direction of light propagation.

At each z-level, the radiance reaching an infinitesimal area element from different directions is given in local spherical direction coordinates (θ, φ), where $\theta(0 \leq \theta \leq \pi)$ is the zenith angle and $\varphi(0 \leq \varphi \leq 2\pi)$ is the azimuth (the angle in the xy-plane measured anticlockwise from the x-axis), i.e. the functional form of radiance is then $L = L(z, \theta, \varphi; \lambda)$. These radiances are integrated into downwelling and upwelling planar irradiances E_d and E_u defined by (e.g. Arst 2003).

$$E_d(z, \lambda) = \int_0^{2\pi} d\varphi \int_0^{\pi/2} L(z, \theta, \varphi, \lambda) \cos\theta \, \sin\theta \, d\theta \tag{5.6a}$$

$$E_u(z, \lambda) = -\int_0^{2\pi} d\varphi \int_{\pi/2}^{\pi} L(z, \theta, \varphi, \lambda) \cos\theta \, \sin\theta \, d\theta \tag{5.6b}$$

The planar irradiances specify the radiance arriving at a horizontal plane from the upper (downwelling irradiance) or lower (upwelling irradiance) hemisphere. Irradiances are apparent optical properties since they depend on the illumination conditions.

The amount of backscattered radiation is given by the surface reflectance, defined as

$$r(\lambda) = \frac{E_u(0; \lambda)}{E_d(0; \lambda)} \tag{5.7}$$

The penetration depth of optical signals at 400–700 nm wavelength is 10–20 cm in snow and 1–5 m in ice, depending on the number of gas pockets. At shorter or longer wavelengths the penetration depth is much smaller. Radiation is lost by absorption, whereby it is changed to heat and chemical energy, or by scattering back to the atmosphere. When the irradiances are integrated over all wavelengths, their ratio gives the albedo of the surface (albedo comes from latin and means 'whitenesss').

Backscattering depends on the surface roughness, size and shape of gas inclusions, size and shape of ice crystals, volume of liquid water, and the quality and quantity of impurities other than gases (Warren 1982, Warren and Clarke 1990). Since the gas pockets are good scatterers they increase the level of reflectance, and since they are large compared with optical and near infrared wavelengths they flatten the reflectance spectrum. This is particularly evident from the bright and white signature of snow in the optical band. Towards and in the near infrared band, absorption by water molecules becomes more significant, and the reflectance of ice and snow decreases. Surface roughness influences the surface reflection of the optical and near infrared bands. It is described in terms of the rms (root-mean-square) roughness ϑ and correlation length ℓ:

$$\vartheta = \sqrt{(h - \bar{h})^2}, \quad \ell = \int_0^{\infty} c(l) \, dl \tag{5.8}$$

where h is surface elevation, overbar stands for spatial averaging, and c is the autocorrelation coefficient of h. The rms roughness may also be defined over a given spectral band.

The reflectance contains contributions from the surface reflection and volume backscatter. The latter contains information about the properties of the medium. The snow surface is often assumed to be Lambertian, i.e. a perfectly diffuse reflectance. However, in the remote sensing of snow this assumption is not always justified, and instead bi-directional reflectance distribution function (BRDF) models are employed (e.g. Voss et al. 2000, Hendriks and Pellikka 2007, Chapter 8).

5.3.3 Thermal infrared signals

Thermal infra-red (TIR) is a powerful remote sensing tool for revealing surface objects with different temperatures or emissivities. In fact, since the surface temperature depends on the thermal conductivity, buried objects can be found by this technique (Del Grande et al. 1991). Terrestrial thermal radiation peaks at about 10 μm over cold surfaces. The TIR technique possesses potential for glacier mapping since the surface temperature is a function of the heat conducted from below to the surface and the heat exchange of the glacier with the atmosphere. TIR mapping produces so-called radiative surface temperature or brightness temperature T_R, which can be transformed into real surface temperature T by

$$L(\lambda, T_R) = \varepsilon L(\lambda, T) \tag{5.9}$$

where L is the Planck's black body radiation, λ is wavelength, and $\varepsilon = \varepsilon(\lambda)$ is the emissivity of the surface. The emissivity at 10 μm is 0.998 for snow, 0.997 for ice, and 0.995 for liquid water (see Steffen 1985). The radiative temperatures are therefore lower than the real temperatures typically by 0.10 K, 0.15 K or 0.25 K for snow, ice or water, respectively.

When the surface temperature is 0°C, liquid water may be present and phase changes take place at the surface. The presence of liquid water has a major impact on backscattering as it absorbs radiation well in certain spectral bands; this is the case for red light and near infrared bands. Identification of the wet snow zone is possible by thermal remote sensing methods. Over cold surfaces ($T < 0°C$), the surface boundary condition is written as

$$k \left. \frac{\partial T}{\partial z} \right|_{z=0} = Q_0 \tag{5.10}$$

where T is temperature, k is thermal conductivity, z is the vertical coordinate (defined positive downward), and Q_0 is the net upward surface heat flux. Thus the surface temperature feels the structure of the underlying material through the thermal conductivity. The thermal conductivity is sensitive to the density of snow, functionally of the form $k \propto \rho^2$ (see Male 1980). This principle was used for sea ice mapping by Leppäranta and Lewis (2007). Because the temperature gradient in the ice and snow changes easily, care needs to be taken when thermal mapping is performed.

The thermal radiation loss at the surface can be very large, which sometimes results in the cold skin phenomenon. The depth scale of the subsurface "hot point" equals the e-folding depth of the light attenuation. Surface temperature information is of key importance in energy balance investigations, and it also allows monitoring of the melting front in summer.

5.3.4 Microwave signals

Microwave methods are passive (radiometer) or active (radar). For mapping the Earth's surface the microwave window has the valuable property of containing bands for which the atmosphere is transparent in all weather conditions. Microwave radiometers detect radiation from the top centimetre or less, but with short-wavelength satellite radars (Ku to L band) the penetration depth may be several metres in dry glaciers and much less in wet or moist glaciers. Consequently, the satellite radars provide information mostly about the glacier surface layer. Airborne and surface-operated radars of longer wavelengths, so-called ground penetrating radars (see Chapter 11), can penetrate the whole glacier and are consequently useful for collecting information about the glacier's internal stratigraphy and substrate. Surface roughness influences the microwave signals via reflection and the surface layer by the volume backscatter.

Glaciers consist mostly of water, in all its phases, and air. At low frequency microwave wavelengths the sharp change in dielectric constant from about 80 to a mere 3.2 from the liquid to the solid states, to essentially 1.0 for air, and the additional dependence of the dielectric constant on the volumetric concentration of the three, play important roles in the detectability of glacier properties by microwave remote sensing. The dielectric constant is also temperature-dependent. Additional materials, such as rocks and dust falling onto the glacier may also influence the signals generated by the glacier.

Radar remote sensing is now capable of sub-metre spatial resolution using synthetic aperture techniques (SAR) and samples the reflection of coherent illuminating radiation at centimetre to metre wavelengths. Airborne and satellite radars furnish information on the amplitude, polarization, and phase of the returned radiation. Through interferometric processing the phase information also enables highly precise change detection. Ground-penetrating radars (sometimes called georadars) may be airborne but are usually deployed on the surface, either vehicle- or sled-mounted. They are capable of sampling reflections from great depths below the surface and are typically used for studying snow and ice stratigraphy, following particular reflecting layers (isochrones) for estimating net accumulation, determining glacier thickness, bottom topography, and detecting the presence of subglacial lakes.

Passive microwave remote sensing operates across the whole microwave window and samples thermal (non-coherent) radiation emitted by the surface. This enables sensing of the temperature of the near-surface region of the glacier. Satellite sensors have a spatial resolution of typically 20–30 km. Airborne sensors can operate at resolutions of a few hundred metres.

At microwave wavelengths the emissivity of liquid water is much different from that of ice, and this enables detection of the dry or melting state of the glacier surface. However, caution should be exercised in the interpretation of such signals, because dry ice is quite transparent in the microwave region as compared to liquid water and this renders difficult the distinction between surface and sub-surface melt. Their respective microwave signals are indistinguishable but their climatological significance is very different. The subsurface melt is the result of metamorphic changes in the properties of the surface snow which occur over periods without significant snowfall or snow drifting and which enable sub-surface melting to altitudes well above the climatological dry snow line. It therefore cannot serve as indicator of surface temperature. Indeed, it may indicate cooling rather than the warming that is associated with surface melt. Recent headlines about melting at high altitudes on the Antarctic ice sheet are the likely result of such confusion. Passive microwave also provides all-weather information about the presence/absence of liquid water, temperature, and surface layer properties. The low spatial resolution of passive microwave sensing systems limits their usefulness in glacier remote sensing.

The surface topography and thereby accumulation and ablation rates can be obtained using satellite-borne radars either as altimeters or interferometers. Because of the large difference between the dielectric constants of the liquid and solid phases of water, radar is also highly sensitive to the wetness of the glacier surface. Interferometric SARs with night-and-day orbits are therefore capable of monitoring the spatial extent of the melt zone through measurement of interferometric coherence. Melting also produces ice inclusions, which provide strong dielectric contrast against snow, and the dry snow limit is therefore readily mappable by satellite radars at any time of the year. Aside from enabling ready mapping of zones where melt has occurred (with the caveat expressed above), the ice inclusions enable snow accumulation to be estimated from a time-series of radar images, as the backscatter is attenuated by subsequent snow accumulation. Crevasses generate strong and characteristic backscatter in radar imagery. The backscatter depends strongly on crevasse alignment in relation to the radar beam, however, and this limits the usefulness of satellite radars in crevasse mapping. Glacier flow rates are measureable by differential interferometric SAR and speckle correlation.

Because of the large difference between the dielectric constants of the liquid (ca.80) and solid (ca.3.2) phases of water, radar is highly sensitive to the wetness of the glacier surface. Radar interferometric coherence is also highly sensitive to snow wetness and satellite SARs with night-and-day orbits are therefore capable of monitoring the spatial extent of the melt zone.

5.4   POTENTIAL USES FOR REMOTE SENSING OF GLACIERS

Remote sensing is a far-reaching, extremely powerful information-gathering tool. From the office the scientist can gather data from the other side of the planet in near real-time. A wide variety of aspects of glaciers and their environment can be gathered without the scientist ever visiting the object of their research. Remote sensing scientists are therefore sometimes jokingly, but with some truth, called 'remote scientists' and their results 'remotely sensible'.

Remote sensing has evolved to become a highly versatile set of tools to map a wide variety of glacier features, such as their surface and bottom topography, structure, flow, and mass and energy balance. Optical and thermal infrared signals originate in the very surface layer, microwave radiometers see effects from a few decimetres' depth, but at radar wavelengths the penetration depth may be several metres in dry glaciers. Consequently, the information satellites provide concerns mostly the glacier surface layer. Ground penetrating radar, on the other hand, has the capacity to penetrate through the whole glacier. Although such radars are often employed with the antennas posed directly on the surface, they also function well in airborne mode and have been used for mapping the internal structure and bottom topography of glaciers.

The surface topography and thereby accumulation and ablation rates can be obtained using satellite-borne radars either as altimeters or interferometers or through high resolution radar optical stereograms. Surface elevations can additionally be determined by airborne laser scanning or satellite-borne laser profiling. The presence of contaminants can be sensed by hyperspectral sensors, and in some cases by laser fluorometry. Snow optical grain size is also sensed by hyperspectral sensors and its wetness is, at least theoretically, also determinable by radar. Sequences of aerial photographs or high resolution satellite images are helpful in determining variations in the surficial extent of glaciers.

Although we have mostly considered electromagnetic radiation here, potentially useful information can also be obtained from acoustic, seismic and positioning networks. New data logger systems, 'smart-dust' technology, satellite navigation systems and direct satellite communication render such systems potentially useful in the study of crack-propagation, glacier motion, shelf-forming and iceberg calving and break-up processes.

Gravity-sensing offers interesting possibilities with regard to mass-balance studies. The results so far obtained by GRACE (Gravity Recovery and Climate Experiment) are encouraging. However, its poor spatial resolution challenges accurate mass balance estimation. Surface elevations can additionally, with the aid of increasingly precise navigation techniques, be determined by airborne or satellite-borne laser profiling or scanning. Sequences of aerial photographs or high resolution satellite images are helpful in determining changes in the superficial extent of glaciers. However, there are often difficulties at the lower end of glacial streams where the snout is often buried beneath an accumulation of moraine.

Surface temperatures are detectable by thermal infrared remote sensing which pick up skin temperatures, and by passive microwave remote sensing which, if multi-band, can cover a wider depth range and yield much information about near-surface snow properties and, in particular, the presence or absence of wet snow in the near-surface layers. Multi-band radars can also potentially provide such information, particularly if operated in interferometric mode.

The topography of the underlying bedrock, the internal stratigraphy of the glacier itself, spatial variations in accumulation rates, and past altitudinal variations of the dry snow line can all be measured using ground penetrating radar. Active microwave signals do not distinguish well between ice inclusions resulting from respectively surface and sub-surface melt. High resolution radars, such as Radarsat-2, will probably be able to do so from their respective spatial characteristics and possibly also from their differing influences on the polarization of the return signal. The quality and strength of the return signal depend mostly on the surface roughness and dielectric properties of the medium, and in particular the dielectric contrast they generate. Radar interferometric coherence has shown promises for remote sensing of snow topography (e.g. Gazkohani 2008).

Perhaps the most important challenge in satellite remote sensing is still that which remote sensing is intended to eliminate – field observations. The precise quantitative description of the properties of the ice skeleton that is the snow cover and eventually becomes the glacier, and the understanding of how its properties relate to present and past variations of environmental factors, is still a major challenge in glaciology.

REFERENCES

Arst, H. (2003). *Optical properties and remote sensing of multicomponental water bodies.* Springer, Praxis-Publishing, Chichester, 231 p.

Del Grande, N.K., G.A. Clark, P.F. Durbin, D.J. Fields, J.E. Hernandez and R.J. Sherwood (1991). Buried object remote detection technology for law enforcement. *SPIE Surveillance Technologies* 1479, 335–351.

Denoth, A., A. Folgar, P. Weiland, C. Mätzler, H. Aebischer, M. Tiuri and A. Sihvola (1984). A comparative study of instruments for measuring the liquid water content of snow. *Journal of Applied Physics* 56, 2154–2160.

Gazkohani, A.E. (2008). *Exploring snow information content of interferometric SAR data.* Ph.D. thesis, University of Sherbrooke, Canada.

Goldstein, R., N. Osipenko and M. Leppäranta (2000). Classification of large-scale sea-ice structures based on remote sensing imagery. *Geophysica* 36(1–2), 95–109.

Gow, A.J. (1971). Depth-time-temperature relationships of ice crystal growth in polar glaciers. *CRREL Research Report* 300, 1–19.

Granberg, H.B. (1998). The snow cover on sea ice. In M. Leppäranta (ed.), *Physics of Ice-covered Seas.* Helsinki University Printing House, Vol. 2, 605–649.

Hendriks, J. and P. Pellikka (2007). Using multiangular satellite and airborne remote sensing data to study glacier surface characteristics in Hintereisferner, Austria. *Proceedings of First International Circumpolar Conference on Geospatial Sciences and Applications*, Yellowknife, N.W.T., Canada, August 20–24, 2007, 8 p.

Kanto, E., M. Leppäranta and O.-P. Mattila (2008). Seasonal snow in Antarctica II. Data report. *Report Series in Geophysics* 55. Department of Physics, University of Helsinki.

Kärkäs, E., H.B. Granberg, C. Lavoie, K. Kanto, K. Rasmus and M. Leppäranta (2002). Snow conditions in the Queen Maud Land, Antarctica. *Annals of Glaciology* 34, 89–94.

Kotlyakov, V.M. (1961). Snezhnyy pokrov Antarktidy, i yego rob'v sovremmon oledemevie materika [Snow cover of Antarctica and its role in the present glaciation of the continent]. *Rezultaty Issledovaniy po Programme Mezbidunarodnogo Geofizicheskogo Goda Glyatsiologiya. IX razdel programmy MGG* [Results of studies in the IGY Programme. Glaciology. Section IX of the IGY Programme], No. 7.

Leppäranta, M. and J.E. Lewis (2007). Observations of ice surface temperature in the Baltic Sea. *International Journal of Remote Sensing* 28(17), 3963–3977.

Male, D.H. (1980). The seasonal snow cover. In: S. Colbeck (ed.), *Dynamics of Snow and Ice Masses.* pp. 305–395, Academic Press, New York, 468 p.

Palosuo, E. (1985). Ice layers and superposition of ice on the summit and slope of Vestfonna, Svalbard. *Geografiska Annaler* 69(A), 289–296.

Paterson, W.S.B. (1994). *The Physics of Glaciers*, 3rd ed. Butterworth-Heinemann, Oxford. 480 p.

Schytt, V. (1958). The inner structure of the ice shelf at Maudheim as shown by core drilling. *Norwegian-British-Swedish Antarctic Expedition, 1949–52*, Scientific Results 4, *Glaciology* 2, 115–151. Norsk Polarinstitutt, Oslo.

Siegert, M.J., J.C. Ellis-Evans, M. Tranter, C. Mayer, J.R. Petit, A. Salamatin and J.C. Priscu (2001). Physical, chemical and biological processes in Lake Vostok and other Antarctic subglacial lakes. *Nature* 414(6864), 603–609.

Steffen, K. (1985). Surface temperature and sea ice of an Arctic polynya: North Water in winter. *Zürcher Geographische Schriften* 19. Geographisches Institut, ETH.

Voss, K.J., A. Chapin, M. Monti and H. Zhang (2000). Instrument to measure the bidirectional reflectance distribution function of surfaces. *Applied Optics* 39(33), 6197–6206.

Warren, S.G. (1982). Optical properties of snow. *Reviews of Geophysics and Space Physics* 20(1) 67–89.

Warren, S.G. and A.D. Clarke (1990). Soot in the atmosphere and snow surface of Antarctica. *Journal of Geophysical Research* 95, 1811–1816.

Winther, J.-G., H. Elvehøy, C.E. Bøggild, K. Sand and G. Liston (1996). Melting, runoff and the formation of frozen lakes in a mixed snow and blue-ice field in Dronning Maud Land, Antarctica. *Journal of Glaciology* 42(141), 271–278.

Chapter 6

Terrestrial photogrammetry in glacier studies

Kari Kajuutti
Department of Geography, University of Turku, Finland

Tuija Pitkänen & Henrik Haggrén
Department of Surveying, Helsinki University of Technology, Finland

Petri Pellikka
Department of Geosciences and Geography, University of Helsinki, Finland

The possibilities of terrestrial photogrammetry for glacier mapping were already recognized in the late 19th century, a few decades after the invention of photography. Those early photographs can be considered as the first attempts to use remote sensing in glaciology. The term terrestrial photogrammetry is used here to mean surveying or mapping from photographs taken by cameras placed on the ground.

In terrestrial photogrammetry, there are two alternative concepts of photography which are applied. Stereo images are generally used for topographic mapping, whereas multi-station photography is used for accurate point measurement (see, e.g. Luhmann et al. 2006, Fryer et al. 2007). On each station, the photography can be recorded as single frames or panoramic image sequences. The glacier surface is interpreted from photographs by means of stereoscopy, and upon that, the topography is reconstructed as an elevation model. A digital elevation model (DEM) is defined and measured either as contour lines or as a three-dimensional network of distinct points.

Terrestrial photogrammetry can be based on two or more overlapping photographs, from which the elevation models are derived. Figure 6.1 presents the principle of stereophotography in terrestrial applications on Hintereisferner glacier in Austria. The direction of photography is parallel from both camera positions. The baseline, or the distance between the camera positions, is 50 cm and the distance from the camera positions to the sediment-covered ice ridge, which is the main target of the photography, is 10 m. The dotted rectangle shows the area which is common to both photographs allowing 3D stereo viewing. If the photographs are to be used for DEM production, reliable georeferencing is essential.

Terrestrial photogrammetry today is practised using cameras equipped with digital sensors, but modern photogrammetric methods may use old analogue photographs as well, by digitizing them and using appropriate image processing techniques. In this way the information recorded on old photographs can be used as primary data for glacier studies with a time span of more than a century. Terrestrial photogrammetry can thus usefully complement other data sources, and be used to evaluate their reliability, especially in connection with analyses of climate change.

Present-day terrestrial photogrammetry is a modern tool for glacier studies in particular situations, as presented later in this chapter. Using digital cameras that have been accurately geometrically calibrated, glacier surface images can be recorded

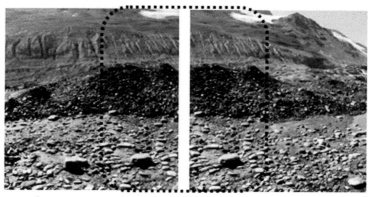

The image from camera station A   The image from camera station B

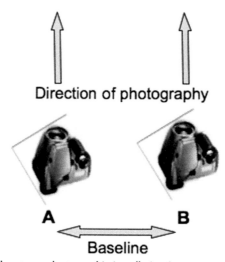

Figure 6.1 The stereophotographic installation in a terrestrial application.

to produce accurate digital elevation models. Terrestrial photogrammetry has some advantages compared to other data sources. The primary advantage is that the surface topography can be determined in much finer detail from close-range than from airborne or spaceborne sensors. With repeated photography and close-range photogrammetry during a season or between years even very small changes can be detected and analyzed, often even more detailed than with other remote sensing techniques over small areas. In the case of glacier dynamics, permanent camera stations can be arranged on site and used for accurate determination of local deformation and movement of the glacier surface. The equipment is also light-weight and transferrable to places with difficult access.

In this chapter the use of terrestrial photogrammetry in glacier studies is presented, with a practical case study from Hintereisferner glacier in Austria. In Hintereisferner, the aim of the terrestrial photogrammetry was to study the glacier surface by building

DEMs and furthermore to evaluate the accuracy of DEMs produced from other remote sensing data between 2002 and 2003. The comparisons between the DEMs constructed using various data types are discussed in more detail in Chapter 14.

6.1   THE EARLY DAYS OF TERRESTRIAL PHOTOGRAMMETRY

As interest in glaciers and glacier mapping increased throughout the 19th century, there was an obvious need for a mapping device and method that would be easy to carry out in the field. The newly developed photogrammetric mapping was applied to glaciology in the late 19th century as one of the earliest applications (see Chapter 4). Sebastian Finsterwalder's innovation of lightweight phototheodolite, also called a photogrammeter, was a significant step in the application (Finsterwalder 1896, Weiss 1913). Finsterwalder studied the geometry of photogrammetry and introduced the basis of absolute and relative orientation for photographs. The phototheodolite was like a metric camera standing on a tripod and usually levelled. It worked without a theodolite, and the necessary vertical and horizontal angles were gained from the ocular and a circle mounted to the camera. Canadian Edouard Deville used a panoramic camera and a theodolite mounted on the same tripod (Burtch 2008). In addition to photography part of the glacier was surveyed using tacheometry in order to define a network of fixed points for more precise interpolation of contour lines. In practice, tacheometric stations were established on the snow-free ablation area with relatively gentle topography for easy access. The orientation required several fixed control points on the bedrock nearby. After measuring possibly hundreds of points a relatively good accuracy was achieved for the surveyed part of the glacier (Collier 2002, Haggrén et al. 2007, Burtch 2008). In addition, Deville created a projective grid (Canadian Grid Method) to be able to compile the map based on the photographs. His panoramic camera approach was adopted by the U.S. Geological Survey for mapping purposes in Alaska. In the early 20th century much attention was paid to the development of stereoscopic photogrammetry. Among the many inventions was the plotter from Ritter von Orel, which was the first stereoautograph to be particularly practical in mountainous areas, allowing direct delineation of elevation contours. However, stereoscopic interpretation did not really begin to be developed until the 1920s. Many others, such as Otto von Gruber, Reinhard Hugershoff and Heinrich Wild, also introduced successful innovations (Collier 2002, Burtch 2008).

In the 1950s aerial photogrammetry of glaciers entered a new era, and terrestrial photogrammetry declined in relative importance (Collier 2002, Burtch 2008). One of the latest studies to use traditional terrestrial photogrammetry was the mapping of the Neh-Nar glacier in Himalaya in 1979 (Agarwal 1989). In the 1950s and 1960s the basis for digital photogrammetry was established by the pioneering work of Gilbert Hobrough and Uuno Helava (Bethel 1990). Recently digital cameras and digital image processing have opened up new possibilities. The development of close-range photogrammetry has been closely bound to the development of computer technology, digital cameras and image processing. This is also the case with close-range photogrammetry, which was started in Germany by Albrecht Meydenbauer as early as the 1880s (Albertz and Wiedemann 1996). However, it was not until the beginning of the 1990s that digital close-range photogrammetry began to be used in many different applications (e.g. Bernard et al. 1999, Rönnholm et al. 2004), ultimately including glaciology (cf. Kajuutti and Pitkänen 2007).

6.2 THE NEW ERA OR DIGITAL TERRESTRIAL PHOTOGRAMMETRY

The use of terrestrial photogrammetry was limited for several decades since first airborne and later on spaceborne data sets offered many possibilities to satisfy the needs of glacier monitoring. However, there have been a number of applications in which terrestrial photogrammetry has represented an adequate choice (Theakstone 1997). On the other hand, attempts to use terrestrial photogrammetry for research have met with some difficulties. Evidently, considerable development work is still required to define work procedures and routines. Here a few terrestrial photogrammetry projects are presented in order to give an overview of different applications as well as the potential problems.

Existing old photographs over glaciers compared with present day photographs offer valuable information about glacier changes and also about local climate change. The scientific importance of the old and new photography is high when the differences in position and elevation in the photographs are reliably measurable in a digital format.

Haggrén et al. (2007) tried to resurvey old photographs and field measurements using present-day digital image processing techniques for terrestrial photographs taken over Hochjochferner, Austria, in 1907. In total 89 glass plates were scanned and analyzed (Figure 6.2). Eleven tie points were measured, even though the variation from image to image was quite large, ranging from a few tie points to more than 20 per image. From the tie points 23 were common with the original geodetic measurements. The processing work faced numerous challenges starting from the many uncertainties concerning the original photography as the interior orientation was not exactly known. Furthermore, the aim was to produce stereo models, which was difficult since the original photographs were convergent and even though they formed panoramic sequences their overlap was only about 5–10%. The control points for relative and absolute orientations were employed from the tie points of the block adjustment and stereo models were successfully produced. Two profiles from the models were compared with the digitized and georeferenced topographic map of 1907 indicating the accuracy of the map contours to be about 10 m. At the terminus of Hochjochferner more precise interpolation of the contour lines was possible with the tacheometric measurement, which improved the accuracy to 1–2 m.

Kaufmann and Ladstädter (2004) wanted to quantify the retreat of the Goessnitzkees glacier in Austria using terrestrial photographs from 1988, 1997 and 2003. The photographs of 1988 were taken with a Zeiss TAL phototheodolite, which was also used for measuring seven control points, which were measured again for better accuracy during the 2003 campaign. In 2003, they applied a semi-metric Rolleiflex 6006 reseau camera, which was also used in 1997, and tested the usefulness of a calibrated Nikon D100 digital camera. The distance from the camera to the glacier snout was several hundred metres and the DEMs were produced from stereo pairs using a digital photogrammetric workstation. All the images were digitised and film unflatnesses, film distortions and chromatic aberrations were compensated.

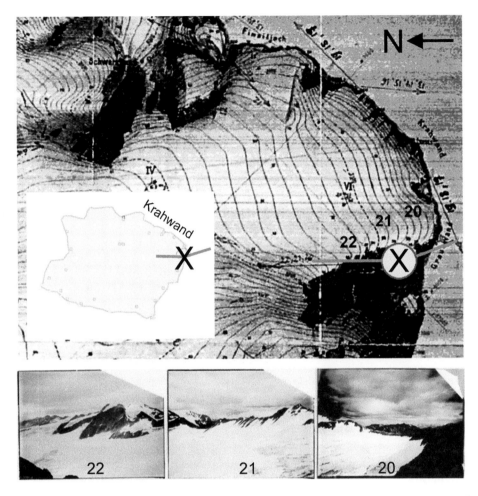

Figure 6.2 Panoramic sequence of the images 20, 21 and 22 towards Krahwand (known today as Gra Wand, 3251 m, located at the southern edge of Hochjochferner). The continuous view covers 160 degrees. The station point is marked both on the index map and the topographic map as X, as well as the total angle of view as lines.

Photogrammetric orientation was carried out using Z/I Imaging software and absolute orientation with geodetically measured control points. A total of 55 tie points was measured around the glacier and the DEMs were produced manually with a 5 m sampling distance. However, there were some difficulties in stereoscopic viewing especially with the older stereo pairs. The outcome of the study was that the accuracy of the changes in ice thickness was around 20 cm as evidenced using a DEM constructed with the Nikon D100 images. Another attempt to use terrestrial photogrammetry for glacier monitoring was made by Kaufmann et al. (2006) at the Doesen glacier, Austria. Rolleiflex photography from 1997 and Rolleiflex and Nikon D100 photography were to be used, but the authors encountered difficulties in photogrammetric evaluation and chose to use airborne data instead of terrestrial photography.

Gruber and Slupetzky (2002) wanted to compare changes in topography of the Cathedral Massif Glacier in Alaska from surveys of 1976, 1985 and 1999. From the years 1976 and 1985 there were traditional paper maps available without any digital information, and for 1999 a total of 78 terrestrial photographs from 13 camera stations were taken with a calibrated medium-format camera (Mamiya RB 67). In addition, several photographs were recorded from a helicopter and a field survey with GPS was carried out. More than 3000 3D-coordinates were determined, old paper maps were scanned and manually digitised, and a DEM was produced for the comparison with the DEM made from the terrestrial photographs. Gruber and Slupetzky (2002) concluded that even though the process is possible, difficulties may occur in digitising analogue materials as well as combining materials from different sources.

A challenging test was carried out on Triglav Glacier in Slovenia (Triglav et al. 2000, Triglav-Čekada 2006). This tiny glacier has been photographed several times per year since 1976, providing plenty of photographic material. However, the photographs have been taken with a rather complicated non-metric Horizon panorama camera without proper field surveying. In 1999, a metric Rolleiflex 6006 was applied and with common control points the photographs from both cameras were linked. The Horizon camera was also calibrated and GPS surveys performed, but still the production of accurate DEMs turned out to be difficult. After the 1999 campaign photographs have also been taken from a helicopter and subsequently glacier changes have been detected from airborne models.

6.3 GLACIER DEMs FROM TERRESTRIAL CLOSE-RANGE PHOTOGRAPHS: CASE STUDY OF HINTEREISFERNER

In 2002 and 2003 limited areas from the ablation area of Hintereisferner in Austria were photographed in order to produce detailed DEMs of the glacier surface as a part of the OMEGA project (Pellikka and Kajuutti 2005, Pellikka 2007). In this study a calibrated digital camera was used for close-range photographs as well as a tacheo-meter for accurate location measurements. The objectives of the study were to develop a procedure from photographing in the field to producing DEMs from different years in order to be able to detect and analyze changes on the glacier surface, and to produce DEMs of high accuracy for comparison with DEMs derived from other remote sensing data. This was expected to be especially interesting with airborne laser scanner data, which was at that time a relatively new technique in glacier studies.

6.3.1 The glacier surface as an object for terrestrial photography

Hintereisferner is a gently sloping valley glacier in Tyrolia, Austria (Figure 6.3). The uppermost part of the glacier is rather steep, but the snout at about 2500 m elevation as well as the whole ablation area under 2900 m is flat or gently sloping providing an easy

Figure 6.3 The snout of Hintereisferner, shown on a digital camera image mosaic constructed by ALTM images (Parviainen 2006), presents the locations for the acquisition of the terrestrial photographs (triangles) and also the fixed geodetic point on the glacier-free foreground (circle) to which the measurements were tied. An inset photograph shows Hintereisferner in 2003. (See colour plate section).

access, which makes Hintereisferner a suitable area for terrestrial photogrammetry tests (Greuell 1992, Kuhn et al. 1999).

The glacier flow together with changing weather conditions in the mountainous areas formulates the shape and structure of a glacier surface (see Chapter 2). The glacier surface is full of features, ranging from large crevasses down to tiny holes and mounds. During the summer, melting and refreezing of ice are common phenomena strongly affecting the build up and formation of new surface features as well as the reforming the old ones. Some of these features can persist for several years, but many of them will last only days or even hours (Benn and Evans 1998, Betterton 2001, Benn et al. 2003). The amount of supraglacial debris also has a major influence on the surface topography, because the debris absorbs more sunlight than clean ice. As a consequence, a thin debris layer or a small particle will enhance melting beneath. On the other hand, a debris layer more than 1–3 cm or particles more than 20 cm in diameter will insulate the ice and reduce melting correspondingly (Østrem 1959, Kajuutti 1989, Kirkbride 1995, Nakawo and Rana 1999, Purdie and Fitzharris 1999). With repeated close-up photography it is possible to follow the changes in the amount and location of the debris and ice features when the photography is well-timed.

Erosion processes have resulted in the deposition of thick debris layers on both sides of Hintereisferner. Close to the snout the amount of supraglacial debris increases as both melting and ice flow deposit more sediment. Some parts have been disconnected from the ice flow and form a very slowly melting section called *dead ice*. Common phenomena on the sides of Hintereisferner are so-called glacier tables, which are large stones that have protected the ice beneath from melting, thus ending up perched on ice well above the glacier surface. There are also some ridges, hillocks and dirty cones consisting mostly of ice but being covered with debris (cf. Drewry 1972, Ferguson 1992, Kirkbride 1995, Benn and Evans 1998, Benn et al. 2003).

The terrestrial photography on Hintereisferner extended over an ice ridge and its foreground area, covered almost entirely by supraglacial debris. The particle size varied from clay up to stones with a diameter of around 20 cm. During the cool melting period in 2002 the effect of supraglacial debris was insignificant in relation to melting, but enhanced the melting during the warm melting period in 2003. The mean temperature in August 2003 was 13.1°C in the nearby village of Vent making it the warmest August since the start of the measurements in 1961. In any case the biggest stones were close to the critical particle size where the insulation effect starts causing the formation of glacier tables, which in turn would considerably change the surface topography (cf. Kajuutti 1989).

With close-range photogrammetry it is possible to produce detailed DEMs over limited glacial areas. However, the glacier surface is an object for an ongoing process of changes caused by different climatic elements and ice flow. As the changes can happen in a minute, it makes the photography especially challenging depending on the scale in focus in the research. The more detailed surface information is needed, the more important is the correct timing of the photography. Furthermore the differences in reflectance of the light from the clean and sediment-covered ice surfaces can easily cause strong contrasts in the images making both the photography itself and the image processing demanding.

6.3.2   The setting and equipment

The locations presented in Figure 6.3 were selected for the acquisition of terrestrial photography following the criteria that the camera location should be stable, the photography could be done uphill, the area was not strongly crevassed, and the distances from the fixed control points were not too long for tacheometer measurements. Due to the lack of visible bedrock and the existence of unstable and loose gravel by the glacier edges it was not possible to locate the camera on firm ground, nor even to carry out tacheometer measurements from the edges of the glacier. Therefore the camera stations and control points were located on the glacier ice (Figure 6.4) even though it meant that their location would move with the glacier. However, the glacier's velocity at the location from which the photography were acquired, close to the glacier edge, was not considered fast enough to cause errors during the few hours of the photography session. The tacheometer measurements were begun at the nearest fixed control point located on the bedrock close to the glacier terminus and ended at the reference points (marked with red paint) at the photography site (Figure 6.3). The first photography site was located close to the terminus where the slope angle is rather steep (app. 20°), while the second site higher up consisted of a debris-covered hillock on a gentle glacier slope (<10°). During the DEM processing the geometry of the upper site DEM appeared to be much better than the lower one and therefore the DEM of the upper site is mostly employed here.

Figure 6.4 The camera station set up on debris-covered ice at Hintereisferner, viewing an ice surface covered by debris. Photograph by Kari Kajuutti, 2002.

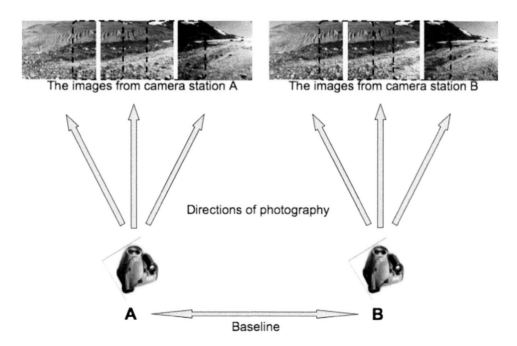

Figure 6.5 The photogrammetric setting applied. The dotted lines indicate the areas common to consecutive photographs.

Figure 6.6 A panoramic image from the Hintereisferner terrestrial photography site combined from a concentric sequence of three images.

The photogrammetric setting applying two series of photographs (A and B) used by Kajuutti and Pitkänen (2007) is presented in Figure 6.5. The photographs were taken with a calibrated Olympus Camedia C-1400L digital camera mounted on a tripod. The camera produces an image format of 1280 × 1024 pixels and the zoom lens was set to its minimum focal length, corresponding to 1400 pixels. The photographs were taken from each station as panoramic sequences, consisting of three images with a sequential overlap of approximately 30% corresponding to a horizontal angle of view of about 120°. Because the camera was rotated when taking the three photographs, a cross-slide controlling the concentricity of the images and preserving the central projection during the camera rotation was used between the camera and the tripod. The three photographs were combined to form one panoramic image and furthermore the panoramas from the different camera stations could then be used as stereo pairs (Figure 6.6). The baseline between the camera locations for the stereo pairs was less than a metre. The lens distortions of the images produced were corrected after which the images were combined using an algorithm of two dimensional projective transformation (Pöntinen 2002).

6.3.3 Orientations and DEM production

The absolute orientation of the terrestrial DEM was solved using existing ground control points and the orientations for the panoramic images were calculated by applying standard Intergraph Z/I digital workstation software. The coordinates of the points were measured using zone M28 of the local reference system of the Military Geographic Institute of Austria, then reprojected to a UTM projection using the WGS 84 datum (Kajuutti and Pitkänen 2007, Kajuutti et al. 2007).

The control points on the ice surface were distributed in relation to each other and the direction of photography in order to avoid problems in image orientation experienced on Engabreen glacier in Norway (Kajuutti and Pitkänen 2007). At the upper study site the distribution of the control points appeared to be good for model production.

The DEMs were digitized manually from the absolutely oriented stereo models using Intergraph Z/I software. The average point separation of the 2002 model is 25 cm, which is more than in DEMs produced from other remote sensing sources (cf. Gao and Liu 2001). The 2003 model had an average point spacing of 88 cm, which was considered dense enough for the comparison of the geometry between the data sets.

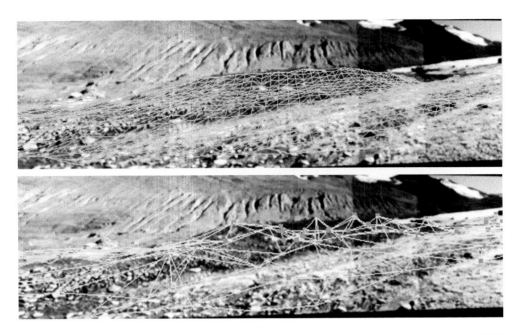

Figure 6.7 The terrestrial photography DEM with 25 cm grid size (above) and a laser scanner DEM (below) over terrestrial photography acquired on September 18, 2002.

The points of models with low point density can be projected to a terrestrial stereo model and used either for visualizing the differences or for estimating the accuracy of the other DEM (Kajuutti et al. 2007, Kajuutti and Pitkänen 2007). Two digital elevation models from the upper site at Hintereisferner were produced based on close-range terrestrial photography performed in 2002 and 2003. The DEM resulting from 2002 survey is presented in Figure 6.7.

6.3.4   Digital elevation models generated from terrestrial photogrammetry

The 2002 DEM showed the glacier surface in such detail that objects with a diameter of 3–4 centimetres, like gravel and ice hollows, could be detected and their changes would be possible to record from DEMs produced with an adequate time interval. The terrestrial photography DEMs were also used for the verification of DEMs derived from airborne laser scanning data (Figure 6.7). The DEMs constructed with terrestrial photogrammetry on September 18–19, 2002 and August 12, 2003 were compared with the DEMs constructed with airborne laser scanner data from the same dates. Due to problems in georeferencing, the elevation difference appeared to be about 1–2 metres, but when using a surface matching algorithm presented by Jokinen (2000) the maximum noise level was less than 20 cm. It is evident that even though laser scanning data does not achieve the accuracy of terrestrial photography, its accuracy can be considered excellent for airborne data (Jokinen et al. 2007, Kajuutti et al. 2007).

Terrestrial photogrammetry also has some limitations. Close-range DEMs are the most suitable for modelling relatively small areas, while large areas would be too labour-intensive and time consuming. Rough and unstable terrain gives a challenge for terrestrial photogrammetry. The camera station (projection centre) should be selected carefully in order to avoid distortions in the DEMs, since the accuracy of relative orientation depends highly on the geometry of panoramic images. The geometry of the central image can still be corrected, but at the cost of distorted adjoining images. The accuracy of the relative orientation as well as the accuracy of the ground control points mostly determines the accuracy of the absolute orientation, and in general the quality of images, accuracy of tacheometer measurements as well as adequacy of a ground control network all have a major effect on the accuracy of the DEM.

The importance of reliable ground control is emphasized in the case of glacier mapping, due to glacier movement and melting. On fast-moving glaciers and during warm weather conditions the possible gap between the photography and tacheometer measurements is a potentially significant source of error. On Hintereisferner the velocity at the edge of the glacier is very slow and in 2002 the weather was cool during the photographic campaign. In 2003, the weather was much warmer and surface melting of about 14 centimetres per day was measured on clean ice. On the sediment-covered photography site the melting was slower in general but due to the uneven distribution of the sediments the melting varied from place to place. Therefore, in 2003, the possibility to perform the photography and tacheometer measurements simultaneously was an evident advantage. For comparisons and change detection of the glacier topography between successive years the DEMs have to be matched.

Detailed DEMs offer a chance to follow changes of even small objects on the glacier surface, but detailed analysis of the processes forming and reforming the ice surface requires analysis together with meteorological data and in many cases *in situ* meteorological measurements are required.

6.4   PROSPECTS FOR TERRESTRIAL PHOTOGRAMMETRY

Within the OMEGA project different kinds of remote sensing data and methods were tested and their suitability for operative monitoring was evaluated (e.g. Pellikka and Kajuutti 2005, Geist and Stötter 2007, Jokinen et al. 2007, Sharov and Etzold 2007). It is very clear that terrestrial photogrammetry cannot be considered as an operative system for large glacial areas. The main application for terrestrial photogrammetry in glaciological studies is considered to be analysis of changes in small areas and in small time windows, which are not possible with airborne or spaceborne remote sensing data for logistical and economical reasons.

However, the different cases presented here show the advantages of terrestrial photography in particular situations. With old photographs the challenge was to look for the possibilities of current technology in photogrammetry to renew the old surveys (Haggrén et al. 2007). A hundred years ago photogrammetric map production did not mean continuous interpretation and plotting of contour lines, but measurement of elevation points. The highest accuracy was achieved by convergent photography with long baselines, which is a rather different method from that used later for stereoscopy. Modern close-range terrestrial photography was used for detailed DEM production in small areas for detecting glacier changes and validation of the accuracy of DEMs produced from other remote sensing data. Different applications are possible, but there are some critical points. The movement of the projection centre during the photography should be avoided and the need for a stable camera setting sets some limitations on the study areas. The amount of work required in DEM production is still high and needs specially developed software for its reduction. New developments of terrestrial photography in glaciological research are obvious and will enhance the number of choices when an appropriate methodology is chosen.

REFERENCES

Agarwal, N.K. (1989). Terrestrial photogrammetric mapping of the Neh-Nar glacier in Himalaya, India. *ISPRS Journal of Photogrametry and Remote Sensing* 44, 245–252.

Albertz, J. and A. Wiedemann (1996). From analogue to digital close-range photogrammetry. In: O. Altan and L. Gründig (eds.), *Proceedings of the First Turkish-German Joint Geodetic Days*, Istanbul, Turkey, September 27–29, 1995, pp. 245–253.

Benn, D.I. and D.J.A. Evans (1998). *Glaciers and Glaciation*. Edward Arnold, Oxford, 734 p.

Benn, D.I., M.P. Kirkbride, L.A. Owen and V. Brazier (2003). Glaciated valley landsystems. In: D.J.A. Evans (ed.), *Glacial Landsystems*. pp. 372–406, Hodder Arnold, London, 544 p.

Bernard, E.S., R. Coleman and R.Q. Bridge (1999). Measurement and assessment of geometric imperfections in thin-walled panels. *Thin-Walled Structures* 33, 103–126.

Bethel, D.J. (1990). Digital image processing in photogrammetry. *The Photogrammetric Record* 13(76), 493–504.

Betterton, M.D. (2001). Theory of structure formation in snowfields motivated by penitentes, suncups and dirt cones. *The American Physical Society, Physical review* E 63, 1–12.

Burtch, R. (2008). *History of Photogrammetry*. An unpublished teaching document: The Center for Photogrammetric Training, Surveying Engineering Department, Ferris State University, U.S.A., 25 p. http://www.ferris.edu/htmls/academics/course.offerings/burtchr/sure340/notes/history.pdf

Collier, P. (2002). Impact on topographic mapping of developments in land and air survey: 1900–1939. *Cartography and Geographic Information Science* 29(3), 155–174.

Drewry, D.J. (1972). A quantitative assessment of dirt-cone dynamics. *Journal of Glaciology* 11(63), 431–446.

Ferguson, S.A. (1992). *Glaciers of North America. A field guide*. Fulcrum publishing, Golden, U.S.A., 176 p.

Finsterwalder, S. (1896). Zur photogrammetrischen Praxis. *Zeitschrift für Vermessungswesen* 25(8), 225–240.

Fryer, J., H. Mitchell and J. Chandler (eds.) (2007). *Application of 3D Measurement from Images*. Whittles Publishing, Dunbeath, 384 p.

Gao, J. and Y. Liu (2001). Applications of remote sensing, GIS and GPS in glaciology: a review. *Progress in Physical Geography* 25(4), 520–540.

Geist, T. and J. Stötter (2007). Documentation of glacier surface elevation change with multi-temporal airborne laser scanner data – case study: Hintereisferner and Kesselwandferner, Tyrol, Austria. *Zeitschrift für Gletscherkunde und Glazialgeologie* 41, 77–106.

Greuell, W. (1992). Hintereisferner, Austria: mass-balance reconstruction and numerical modelling of the historical length variations. *Journal of Glaciology* 38(129), 233–244.

Gruber, W. and H. Slupetzky (2002). Remapping of the Cathedral Massif Glacier (B.C., Canada) – from traditional mapping to digital techniques. *Proceedings of the 2nd ICA Mountain Cartography Workshop*, Mt. Hood, Oregon, U.S.A., May 15–19, 2002.

Haggrén, H., C. Mayer, M. Nuikka, L. Braun, H. Rentsch and J. Peipe (2007). Processing of old terrestrial photography for verifying the 1907 digital elevation model of Hochjochferner glacier. *Zeitschrift für Gletscherkunde und Glazialgeologie* 41, 29–53.

Jokinen, O. (2000). Matching and modeling of multiple 3-D disparity and profile maps. Doctoral thesis, *Acta Polytechnica Scandinavica, Mathematics and Computing Series* 104, Espoo, 117 p.

Jokinen, O., T. Geist, K.-A. Høgda, M. Jackson, K. Kajuutti, T. Pitkänen and V. Roivas (2007). Comparison of digital elevation models of Engabreen Glacier. *Zeitschrift für Gletscherkunde und Glazialgeologie* 41, 185–204.

Kajuutti, K. (1989). The effects of supraglacial debris on the melting of glacier ice. *Terra* 101(3), 261–264, In Finnish, English abstract.

Kajuutti, K., O. Jokinen, T. Geist and T. Pitkänen (2007). Terrestrial photography for verification of airborne laser scanner data on Hintereisferner in Austria. *Nordic Journal of Surveying and Real Estate Research* 4(2), 24–39.

Kajuutti, K and T. Pitkänen (2007). Close-range photography and DEM production for glacier change detection. *Zeischrift für Gletscherkunde und Glazialgeologie* 41, 131–145.

Kaufmann, V. and R. Ladstädter (2004). Documentation of the retreat of a small debris-covered cirque glacier (Goessnitzkees, Austrian Alps) by means of terrestrial photogrammetry. *Proceedings of the 4th ICA Mountain Cartography Workshop*, Vall de Nuria, Spain, September 30–October 2, 2004, 12 p.

Kaufmann, V., R. Ladstädter and G. Kienast (2006). 10 years of monitoring of the Doesen rock glacier (Ankogel Group, Austria) – a review of the research activities for the time period 1995–2005. *Proceedings of the 5th ICA Mountain Cartography Workshop*, Bohinj, Slovenia, March 30–April 1, 2006, 17 p.

Kirkbride, M.P. (1995). Processes of transportation. In: J. Menzies (ed.), *Modern glacial environments – Processes, Dynamics and Sediments* 1, 261–292, Butterworth-Heinemann, Oxford, 392 p.

Kuhn, M., E. Dreiseitl, S. Hofinger, G. Markl, N. Span and G. Kaser (1999). Measurements and models of the mass balance of Hintereisferner. *Geografiska Annaler* 81A(4), 659–670.

Luhmann, T., S. Robson, S. Kyle and I. Harley (2006). *Close Range Photogrammetry: Principles, Methods and Applications*. Whittles Publishing, Dunbeath, 528 p.

Nakawo, M and B. Rana (1999). Estimate of ablation rate of glacier ice under a supraglacial debris layer. *Geografiska Annaler* 81A(4), 695–701.

Østrem, G. (1959). Ice melting under a thin layer of moraine and the existence of ice cores in moraine ridges. *Geografiska Annaler* 41(4), 228–230.

Parviainen, P. (2006). *Detection of glacier facies and parameters using aerial false colour digital camera data – case study Hintereisferner, Austria*. M.Sc. thesis, Department of Geography, University of Helsinki, 74 p. + appendices.

Pellikka, P. and K. Kajuutti (2005). Remote sensing of glacier area, topography and changes on Svartisen and Hintereisferner – Scientific results of the OMEGA project. *Turku University Department of Geography Publications* B6. Digipaino, Turku, 83 p.

Pellikka, P. (2007). Monitoring glacier changes within the OMEGA project. *Zeitschrift für Gletscherkunde und Glazialgeologie* 41, 3–5.

Pöntinen, P. (2002). Camera calibration by rotation. *International Archives of Photogrammetry, Remote Sensing and Spatial Information Sciences* 34(5), 585–589.

Purdie, J. and B. Fitzharris (1999). Processes and rates of ice loss at the terminus of Tasman Glacier, New Zealand. *Global and Planetary Change* 22, 79–91.

Rönnholm, P., J. Hyyppä, H. Hyyppä, H. Haggrén, X. Yu and H. Kaartinen (2004). Calibration of laser-derived tree height estimates by means of photogrammetric techniques. *Scandinavian Journal of Forest Research* 19(6), 524–528.

Sharov, A.I. and S. Etzold (2007). Stereophotogrammetric mapping and cartometric analysis of glacier changes using IKONOS imagery. *Zeitschrift für Gletscherkunde und Glazialgeologie* 41, 107–130.

Theakstone, W.H. (1997). Mapping changing glaciers. *Aarhus Geoscience* 7, 101–111.

Triglav, M., M.K. Fras and T. Gvozdfanovic (2000). Monitoring of glacier surfaces with photogrammetry, a case study of the Triglav Glacier. *Acta Geographica* 40(1), 7–30.

Triglav-Čedaka, M. (2006). Photogrammetrical monitoring of the Triglav Glacier in Slovenia. *Proceedings of the 5th ICA Mountain Cartography Workshop*, Bohinj, Slovenia, March 30–April 1, 2006, 8 p.

Weiss, M. (1913). Die geschichtliche Entwicklung der Photogrammetrie und die Begründung ihrer Verwendbarkeit für Meß- und Konstruktionszwecke. Strecker & Schröder, Stuttgart, 94 p.

Chapter 7

Aerial photogrammetry in glacier studies

Andreas Kääb
Department of Geosciences, University of Oslo, Norway

7.1 INTRODUCTION

Aerial photogrammetry retrieves geometric information about objects and processes from airborne analogue or digital image data. Such data stem from airborne sensors such as analogue frame (i.e. metric) cameras, analogue amateur (i.e. non-metric) cameras, digital frame cameras, digital linear array sensors (pushbroom principle), digital scanners, and digital amateur cameras. Typically, these sensors are mounted on platforms such as (surveying) aeroplanes, helicopters, and unmanned airborne vehicles (e.g. model helicopters, airplanes, or balloons).

The capability of photogrammetry to obtain quantitative, spatially distributed information over large areas without requiring direct ground access, and the large potential of image data for information storage and retrieval in general, led to the early application of photogrammetry in glaciology (for an overview see Pillewizer 1938 and Chapter 2). Photogrammetry, like other airborne and spaceborne remote sensing techniques, transfers large parts of the work to the office, which is particularly important for glaciological research in mountain and polar areas where the fieldwork season is usually short and often interrupted by adverse weather conditions.

Aerial photogrammetry is usually based on single original image data, overlapping image data, or transformed image data. Single image data with original geometry (so-called mono-techniques) are used where individual photos are analysed. This group of techniques allows usually only for qualitative analyses, but can be very valuable for various inspection tasks, and for preparation of further quantitative analyses. Computer-aided measurements on single images are possible when the image orientation is resolved and the image is combined with a digital elevation model (DEM) (digital monoplotting).

Overlapping image data (stereo-techniques) allow for three-dimensional vision and measurement. This group of techniques is the most widely used in glaciological photogrammetry. It allows for reconstruction of the surface of glaciers, three-dimensional measurement of point, linear, and spatial surface features, and spatio-temporal changes of these objects.

Transformed image data are for instance orthoprojected images (also called orthorectified images, orthoimages, or orthophotographs). Orthoprojection is the process of transforming images from the original acquisition geometry, which is commonly a central projection through a camera lens system, into an orthogonal projection (i.e. map geometry). Orthoprojection removes, among others, topographic distortions in the original image data. Orthophotographs represent a basic prerequisite of many digital glaciological analyses. They can easily be integrated in a glaciological geographic information system (GIS) and be combined with any other georeferenced information (Baltsavias 1996).

Aerial photogrammetry of glaciers is for the most part used for interpretation and mapping of surface features of glaciological interest, generation of DEMs, often repeat DEMs, and measurement of surface displacements.

7.2 INTERPRETATION AND MAPPING

Planimetric measurements (e.g. within a GIS and based on orthoimages) or stereo-compilations (using specialized hard- and software) are applied to any surface feature of glaciological interest, such as glacier boundaries, terminus positions, crevasses, ice avalanches, moraines, trimlines, lake boundaries, etc. These measurements result in two- or three-dimensional, respectively, points and lines. Often, the transition between different glaciological objects is continuous or fuzzy, e.g. the transition between debris-free ice, debris-covered ice, dead ice, to morainic material or peri-glacial debris.

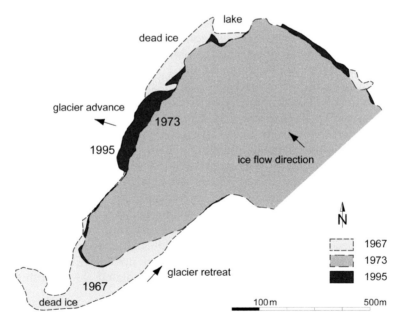

Figure 7.1 Overlay of repeat outlines of the tongue of Gruben Glacier, Switzerland, in 1967, 1973 and 1995. The outlines were derived through manual delineation within a photogrammetric stereo-model. Within the observational period, Gruben Glacier experienced retreat in a north-easterly direction, and later an advance in a westerly direction. (Cf. Figures 7.2, 7.4, 7.6, 7.8, 7.9).

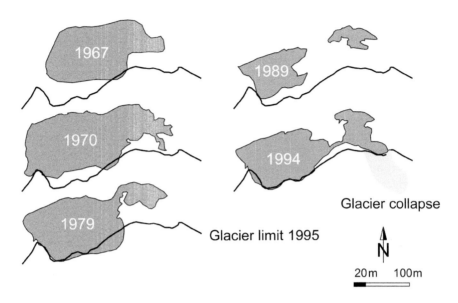

Figure 7.2 Repeat photogrammetrically-derived outlines of a glacier-marginal lake at Gruben Glacier. For location see Figure 7.1. The outlines of the glacier lake changed through an interplay between lake outbursts in 1968 and 1970, glacier changes, and melt and ice break-off processes at the calving front.

The interpretation uncertainty, therefore, often exceeds the error that is in theory achievable by the photogrammetric technique employed.

Digital image processing enables semi-automatic support of mapping work. Related techniques range from simple thresholding of images for panchromatic data to multispectral classifications for multi- or hyperspectral data.

Photogrammetric mono- or stereo-mapping has been widely applied in glacier research and in glaciology in general. Most frequently, glacier outlines and terminus positions are mapped (Figure 7.1), but also glacier lakes (Figure 7.2) and various other features and their changes over time (e.g. Kääb 1996a, Kääb 1996b, Fox and Nuttall 1997, Würländer and Eder 1998, Bacher et al. 1999, Jordan et al. 2005, Kääb 2005b).

7.3 GENERATION OF DIGITAL TERRAIN MODELS

Methods for photogrammetric generation of DEMs include analogue and analytical photogrammetry based on hardcopy photography, digital photogrammetry based on digital cameras or digitized photography (frame imagery), and digital photogrammetry based on other digital sensors (e.g. linear arrays) discussed below.

7.3.1 Analogue and analytical photogrammetry

Analogue and analytical photogrammetry are well-established tools for measuring DEMs of glaciers and rock glaciers, and have been applied for decades (e.g.

Finsterwalder 1931, Hofmann 1958, Finsterwalder and Rentsch 1980, Ebner 1987, Rentsch et al. 1990, Kersten and Meister 1993, Kääb 1996b, Fox and Nuttall 1997 Kääb and Funk 1999, Kääb 2001, Kääb 2005b). Although digital techniques will sooner or later fully replace photogrammetry that is based on analogue photography, it should be mentioned that analytical photogrammetry is still very useful for applications where human interpretation is necessary to support image measurements. This situation has arisen because of the lower resolution of digital imagery compared to photographic films, and the lower resolution and display quality of computer displays compared to optical display systems, until now. On the other hand, however, digital imagery has the powerful advantage for photogrammetry of glaciers that image processing techniques can be used to enhance the image data, for example to emphasise limited surface texture. Over snow and firn the lack of discrimination of subtle features through poor radiometric contrast is usually more a limiting factor than spatial image resolution for glacier photogrammetry.

7.3.2   Digital photogrammetry of frame imagery

In principle, the airborne methods discussed in this chapter work in most cases for different platform types such as surveying aeroplanes, small amateur aeroplanes, helicopters, or unmanned aeroplanes and helicopters. For very mobile and unstable platforms special acquisition and processing techniques might be required, which are not discussed in this chapter. In particular in terms of cost efficiency, for objects with difficult access by larger aeroplanes, or where airborne close-range techniques are required, such methods might, however, well be considered for glaciological applications.

DEMs can be automatically derived from overlapping digital images. Frame images represent central projections of the terrain (Figure 7.3):

$$\vec{X} - \vec{X}_o = \lambda \cdot \mathbf{R} \cdot (\vec{x} - \vec{x}_o) \tag{7.1}$$

where

$\vec{X} = (X, Y, Z)^T$ are the Cartesian coordinates of a terrain point in ground coordinates,
$\vec{X}_o = (X_o, Y_o, Z_o)^T$ are the ground coordinates of the projection centre o, λ is the image-to-ground scale,
$\mathbf{R} = f(\omega, \varphi, \kappa)$ is a 3×3 rotation matrix depending of the orientation angles ω, φ and κ,
$\vec{x} = (x, y, z)^T$ are image coordinates of an image point,
$\vec{x}_o = (x_o, y_o, -c)^T$ are the image coordinates of the projection centre, and c is the calibrated focal length of the camera.

Equation 7.1 represents a three-dimensional vector through the projection centre, the direction of which is given by three orientation angles, the image pixel and the projection centre. A terrain point is determined by the intersection in space of two or more vectors originating from different images (Figure 7.3).

The orientation parameters can be computed from a set of ground control points (GCPs) by solving Equation 7.1 or are measured directly from on-board differential satellite navigation (usually global position system, GPS) and inertial navigation

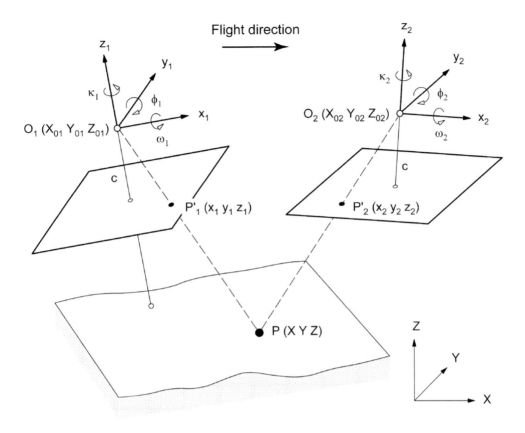

Figure 7.3 Principle of photogrammetric DEM generation: points in space are determined by the intersection of oriented rays from overlapping photography.

systems (INS). For overlapping photographs such as stereophotographs the parameters of the exterior orientation are in practice determined in a hybrid approach: the relative orientation between the images of an image block is solved from tie points, i.e. corresponding image points in two or more of the overlapping images. The absolute orientation of these three-dimensional models of two or more images to the ground space is then estimated from GCPs or GPS/INS. Tie points between overlapping images can in principle be measured automatically using image correlation techniques.

Equation 7.1 is often applied to orthoprojection by solving for X and Y of individual image points. This approach requires a DEM for introducing Z into the otherwise under-determined equation system. As a basic feature of orthoimage generation, vertical errors in terrain elevation are transformed into horizontal deviations in pixel location (see Figure 7.5).

Automatic DEMs from frame imagery can be computed from digitized analogue photographs or directly from digital imagery. Digital photogrammetric cameras can consist of one charge-coupled device (CCD) array combined with one lens system (e.g. the Emerge/Applanix Digital Sensor System) or can be composed of several CCD array cameras (e.g. the ZI Imaging Digital Mapping Camera System).

For studies involving repeat image data, i.e. change detection, multiple image sets should be oriented and adjusted as one image segment as a special measure to improve the relative deviation between multitemporal imagery and the corresponding products (Toutin 1995). For this purpose, only stable terrain points, i.e. not on glaciers, rock glaciers or other moving surfaces, must be used for ground control points and multi-temporal image tie points (Kääb and Vollmer 2000, Kääb 2002, Kääb 2005b). Finding common tie and control points in multitemporal photographs is, however, significantly hampered by different image scales and resolutions, viewing angles, illumination and terrain conditions, etc.

The methodology of subsequent automatic DEM extraction and orthophotograph generation from digital stereo imagery is well established and described (e.g. Grün and Baltsavias 1988, Baltsavias 1996, Chandler 1999, Lane et al. 2000). The computation of individual terrain heights is based on the measurement of height-parallaxes, which is the automatic assignment of corresponding terrain features in two (or more) over-lapping images. To this end, two techniques or combinations of them are used: block (or area-based) matching techniques, or feature matching techniques.

Block matching compares complete grey value arrays, i.e. image sections, to each other. Feature matching compares geometric forms such as edges or polygons previously extracted from the imagery in a preprocessing stage. Block matching techniques used for height-parallax measurement include two-dimensional cross-correlation and least-squares matching. Introduction of additional geometric constraints such as epipolar planes reduces the computing time for parallax-matching and the number and size of measurement outliers due to mismatches.

For the final measurement, the related correlation procedures focus on the comparison of image sections of only a few pixels in size. Hence, heights can best be derived where the optical contrast (also called radiometric contrast, or degree of surface texture) in the digital images is sufficient, e.g. not diminished due to fresh snow, and, secondly, where terrain sections look similar in the different photographs and are not strongly distorted. Height computation fails where parts of the terrain are occluded on one or more photographs. In summary, the error of the resulting DEM is generally larger on steep slopes or in shadow zones.

Automatic photogrammetric DEMs have been derived and investigated for Arctic glaciers and ice caps (e.g. Etzelmüller et al. 1993, Bacher et al. 1999), ice sheets (e.g. Wrobel and Schlüter 1997, Fox and Gooch 2001), and for glaciers, rock glaciers and slope instabilities in high mountains (e.g. Benson and Follet 1986, Baltsavias et al. 1996, Kaufmann and Plösch 2000, Kääb 2000, Kääb and Vollmer 2000, Weber and Herrmann 2000, Baltsavias et al. 2001, Ledwith and Lunden 2001, Julio Miranda and Delgado Granados 2003, Cox and March 2004, Ødegård et al. 2004, Jordan et al. 2005, Julio Miranda et al. 2005, Kääb 2005b, Gleitsmann and Kappas 2006).

7.3.3   Digital photogrammetry of airborne pushbroom imagery

In addition to photogrammetry of digital or digitized frame imagery, photogrammetry of imagery from digital pushbroom sensors (also called linear array CCD sensors or linear scanners), and light detection and ranging (LiDAR, laser profiling or laser scanning), play an increasingly important role in the airborne and spaceborne determination of DEMs. Laser scanning is discussed in Chapter 10 of this book.

A pushbroom array (i.e. a cross-track line of detectors) can be viewed as a two-dimensional central projection (see Equation 7.1). In contrast to frame imagery, however, the orientation parameters (position of projection centre, image rotations, scale) change for every line. These effects are usually corrected by determining them from on-board differential GPS and INS. Similar to photogrammetry of frame imagery, terrain points can be computed from along-track pushbroom arrays with different view angles (two-, three-, or multiple-line cameras) providing ray intersections at individual terrain points. Pushbroom cameras may also record multispectral information (Hauber et al. 2000, Grün and Zhang 2002, Poli 2002).

Airborne, multiple-line cameras (nadir, back-looking, and/or forward-looking push-broom lines) such as the HRSC-A (by DLR) or ADS40 (by Leica Geosystems) currently reach spatial ground resolutions of the order of 10 cm ground-projected instantaneous field of view (GIFOV) per 1000 m above ground. Corresponding DEMs have a spatial resolution between one and a few metres with a vertical error of approximately 10 cm per 1000 m above ground (Hauber et al. 2000, Grün and Zhang 2002, Zhang and Grün 2004). The general characteristics of DEMs from two- or three-line cameras are similar to those from frame imagery, where low-contrast areas or steep slopes often cause DEM errors. First tests for high-mountain terrain gave very promising results (Hauber et al. 2000, Roer et al. 2005, Otto et al. 2007).

7.4 ERRORS OF PHOTOGRAMMETRIC DEMs

7.4.1 General

The error of a DEM with respect to the real topography includes the height and positional error of individual DEM points and the representativeness of the discrete DEM points with respect to the continuous terrain. In addition, photogrammetric DEM generation delivers in principle a digital surface model (DSM), and not directly a DEM. Height-parallax measurements relate to the earth's surface including for example vegetation. If not accounted for correctly, this vertical offset between surface and terrain introduces a further error term. On glaciers and glacial environments, however, the surface will be commonly equal to the terrain.

Each DEM is composed of discrete points sampling the continuous terrain. If the DEM sampling distance, regular (grid-based) or irregular (triangular network), is significantly larger then the degree of detail in the topography measured, a vertical error between the interpolated DEM and the real topography occurs. The DEM sampling distance has thus to be chosen according to the topographic characteristics of the study site and the goal of the photogrammetric investigation. The latter defines which vertical offset between heights interpolated from measured DEM points and the real terrain is acceptable. For example, DEMs on ice surfaces, which are generally smooth, often require less dense sampling than do debris-covered glacier sections or moraines and glacier forefields.

The error in individual measured DEM points is composed of a random component (called precision) of the individual measurement and of a systematic component (called accuracy). Systematic terms are introduced, for example, by inaccurate knowledge of camera calibration, inaccurate tie points or inaccurate control points. Insufficient knowledge of camera calibration is in particular often a problem when using images from historic aerial photographs or amateur cameras. In such cases, the camera calibration in (Equation 7.1) can be introduced as a unknown and solved for in the adjustment process (self-calibrating bundle adjustment).

Random errors in automatic photogrammetric DEMs stem from errors in the height-parallax matching. These can, for example, be due to insufficient radiometric contrast (or surface texture), insufficient similarity of image sections in the overlapping images, or mismatches. Of particular concern are large measurement outliers.

7.4.2 Case study

A typical glacier surface is presented in a case study. Figure 7.4 shows the comparison between a DEM that was manually measured using analytical photogrammetry, and an automatic DEM using the BAE Systems Socet Set software. Both 25 m gridded DEMs were derived from 1:18 000 scale black-and-white imagery from 1975. The vertical error in the manual DEM is estimated as approximately ±0.5 m root mean

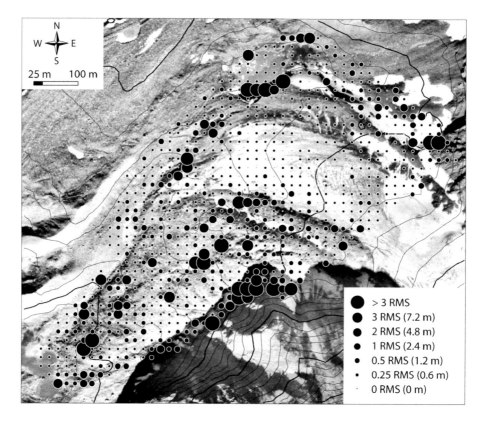

Figure 7.4 Comparison between an operator-measured analytic DEM and an automatic one for the Gruben Glacier tongue. Deviations are depicted in relation to their overall average RMS.

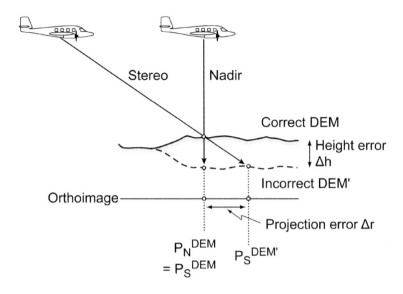

Figure 7.5 Errors in the underlying DEM lead to horizontal distortions in orthoimages that are computed using this DEM. Such a distortion in the ortho-projection depends linearly on the vertical DEM error and on the look-angle of the original image. Thus, an identical DEM error causes different distortions in different ortho-projections. This principle applies to all image data from different look angles and platform types (ground, air, space).

square (RMS) (Kääb 1996b). For the entire study site (approximately 1000 points), an root mean square error (RMSE) of 2.4 m and an error range between −20 m and +34 m was found (manual DEM minus automatic DEM). For the glacier itself an RMSE of 1.2 m and an error range between −7.5 m and +4.5 m is obtained. Thus, errors are significantly larger for debris zones where blocks of up to several metres in diameter cover the ice. The vertical deviations range from −2.2 to +3.6 m (RMSE 0.7 m) for the clean-ice zones.

Considering the above and other studies (Kääb 2005b), a vertical RMSE for automatic photogrammetric DEMs corresponding to 1−3 times the image pixel size can be achieved for moderate high-mountain topography such as glacier forefields, alpine meadows, etc., and an RMSE corresponding to 5−7 times the image pixel size for rough topography such as slopes, peaks, ridges, etc. For steep slopes with cast shadow or snow cover, maximum vertical errors of the order of 60−100 times the image pixel size may be encountered. See Baltsavias et al. (1996, 2001) for further tests on automatic photogrammetric DEMs of high-mountain terrain.

7.4.3   Error detection and DEM evaluation

For many applications, generated DEMs cannot be tested against existing reference DEMs. Approaches for evaluating automatic photogrammetric DEMs without reference data include the following: (1) Generation of DEMs with different matching parameters such as sampling distance, matching window size, or image pyramid parameters. The differences between the DEMs may hint at incorrect DEM sections where parallax matching is particularly sensitive to the parameters chosen, for example due to insufficient surface texture (Kääb 2005b). (2) Analysis of the correlation values from parallax matching. Low correlation values may point to incorrect heights. (3) Overlay of contour lines interpolated from the DEM over orthoimages. This procedure often reveals differences between the topography as visible in the images and the topography represented in the DEM. (4) Advanced digital photogrammetric software offers the possibility to view extracted DEM points three-dimensionally and overlaid over the stereo model. In this way, vertical differences between DEM points and the stereo model can be directly visualized and manually corrected by the operator using 3D vision. (5) Overlay of orthoimages from different viewing angles, as described below.

If imagery from different sensor positions is available (multi-incidence angle images), DEMs may be tested through the overlay of orthoimages generated from the image data. Such orthoimagery must be computed from the DEM to be tested and from two (or more) source images taken from different sensor positions (Baltsavias 1996, Aniello 2003, Kääb 2005a, Kääb 2005b). For a correct DEM, the two (or more) ortho-projections of corresponding image pixels overlap perfectly (Figure 7.5). Vertical DEM errors, on the other hand, translate into horizontal shifts between the orthoprojected pixels of the source images (Figure 7.5). A DEM error has no effect in the case of nadir projection vectors. The horizontal projection shifts between the orthoimages from different source imagery can be visualized by change detection techniques (e.g. ratio images) (Kääb 2005a, Kääb 2005b), or animation techniques (e.g. image flickering) (Kääb et al. 2003, Kääb 2005a). For areas showing significant differences between the orthoimages, the underlying DEM may be flagged, masked out or improved.

For known projection geometry, the resulting planimetric differences may also be measured and re-projected in order to refine the underlying DEM (Norvelle 1996, Kaufmann and Ladstädter 2002, Georgopoulos and Skarlatos 2003).

7.5 VERTICAL DEM DIFFERENCES

Terrain elevation changes over time, i.e. vertical differences between repeat DEMs, are indicators for glacial and geomorphodynamic processes such as glacier fluctuations and mass movements. In general, changes in terrain elevation are derived by subtracting repeat DEMs. If the DEMs compared represent independent measurements, the root mean square error of an individual elevation change, denoted by RMSE Δh, can be estimated from the RMSE of the repeat (here: two) DEMs (Etzelmüller 2000):

$$(\text{RMSE } \Delta h)^2 = (\text{RMSE } h_1)^2 + (\text{RMSE } h_2)^2 \qquad (7.2)$$

Special pre- and postprocessing procedures help to reduce errors in DEM differences:

Preprocessing, i.e. procedures before the DEM subtraction: As for all multitemporal analyses, correct co-registration of the multiple DEMs is a necessary prerequisite to obtain elevation changes free of global systematic errors. The co-registration of

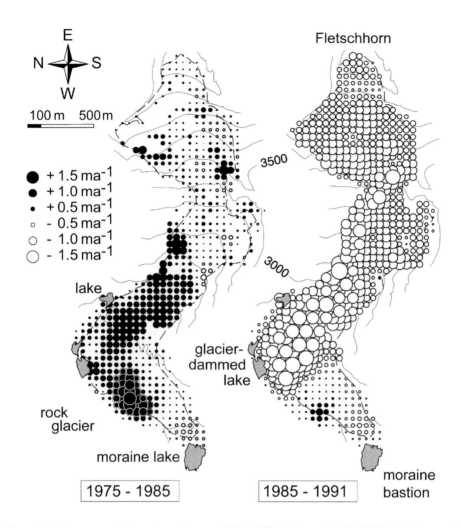

Figure 7.6 Mean annual elevation changes 1975–1985 (left) and 1985–1991 (right) over Gruben Glacier, as derived from repeat stereo aerial photographs using analytical photogrammetry (i.e. computer-aided photogrammetry based on hard-copy photographs). Over the first period the glacier predominantly gained mass, over the second period it substantially lost mass.

the DEMs (and other products) can be assured by orienting the original data as one common, multitemporal data set (see section 7.3.2). If the original sensor model and orientation are inaccessible, or the DEMs have different sources, matching between the individual DEMs to be compared is recommended (Kääb 2005b).

If the signal-to-noise ratio is insufficient to determine the spatial distribution of DEM differences at a reliable level, it might still be possible to estimate the mean elevation difference for an entire terrain section. For that purpose it has to be assumed or to be assured by adjustment measures that the error term in the DEM differences contains only a random component. Repeat DEMs might have systematic differences,

Figure 7.7 Contours of mean annual glacier elevation changes in 1975–1997 for Glaciar Chico, Southern Patagonia icefield in Argentina. The 1975 and the 1997 DEMs have both been automatically derived from digitized aerial photography. The underlying orthoimage and the 100 m contours are based on the 1997 imagery. Aerial photography provided by Andrés Rivera.

in particular if they originate from different sensors or different orientation procedures. Systematic error terms such as vertical and horizontal shifts between the DEMs have to be subtracted, or terms of higher order such as vertical and horizontal scales or rotations, can be estimated from comparing the elevation of stable terrain through matching techniques. The DEM(s) can then be corrected by, for instance, simple shifts or affine transformations, both constructed from the DEM residuals (Pilgrim 1996a, Pilgrim 1996b, Li et al. 2001, Schenk et al. 2002, Berthier et al. 2004, Kääb 2005b).

Postprocessing of the elevation differences: Once the raw differences between repeat DEMs are computed it is often necessary to filter the elevation differences obtained, because the noise in the derived differences is larger than in the original DEMs (see Equation 7.2). The task is to define a noise model adapted to the nature of the process under investigation. For instance, thickness changes of a debris-free glacier are expected to show a smooth spatial variability so that a coarse filter might be applied. Coarse filters are less suited for mass movements such as landslides with a high spatial variability of elevation changes and with many secondary local terrain movements overlain, because the filter tends to remove important "real" signals.

In general, low-pass filters exist in the spatial domain (e.g. median, mean, Gaussian, etc.) or for the spectral domain (e.g. Fourier or wavelet) (Kääb 2005b). For example through Fourier transform DEM differences or sections of them are decomposed into a sum of sine waves of different amplitudes, offsets and wavelengths. A wavelength threshold can be defined in this power spectrum of the DEM differences so that spatial variations of higher frequency are regarded as measurement noise, which is removed, and spatial variations of lower frequency are interpreted as signal, which remains after transforming the DEM differences back into the original spatial domain.

Figure 7.6 shows glacier elevation changes over an entire glacier for two consecutive periods derived from photogrammetric DEMs of three points in time (1975, 1985, 1991). Figure 7.7 shows elevation changes on a glacier tongue derived from two repeat aerophotogrammetric DEMs.

7.6   LATERAL TERRAIN DISPLACEMENTS

A change in terrain elevation (see above DEM differences) and the three-dimensional displacement of an individual particle on the surface describe different kinematic quantities (Kääb 2005b). A full examination of the kinetics of a glacier, and terrain in general, requires information on both the elevation changes and the lateral displacements. Terrain displacements can be determined by different methods. Photogrammetric methods include digital matching of repeat optical imagery, and analogue and analytical photogrammetric methods.

7.6.1   Analogue and analytical photogrammetry

Before digital photogrammetry was available, a number of analogue or analytical methods were developed for displacement measurements: (1) displacement parallax measurements from analogue terrestrial photos (e.g. Finsterwalder 1931, Melvold 1992) or aerial photographs (e.g. Hofmann 1958). (2) Manual or analogue photogrammetric point-by-point measurements (aerotriangulation, e.g. Messerli and Zurbuchen 1968, Brecher 1986, Rentsch et al. 1990). (3) Direct analogue comparison of repeat images. (4) Computer-based, point-by-point measurements (e.g. Grün and Sauermann 1977, Kaufmann 1998, Krummenacher et al. 1998). (5) Computer-based simultaneous comparison (Armenakis 1984, Knizhnikov et al. 1998, Kääb and Funk 1999) (Figures 7.8, 7.9). Here, we present digital methods in more detail.

7.6.2   Digital image matching

An efficient method for measuring terrain displacements is the digital comparison of repeat optical imagery. If original imagery is used, the displacements obtained have

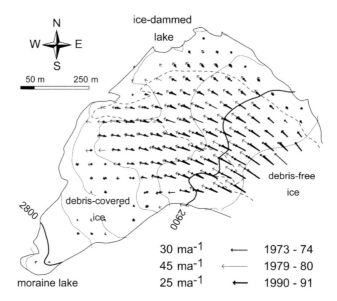

Figure 7.8 Annual horizontal velocity fields of the Gruben Glacier tongue for the years 1973/74, 1979/80 and 1991/92 indicating drastic changes in flow regime. The velocities quoted in the legend represent approximate maximum observed values for each period.

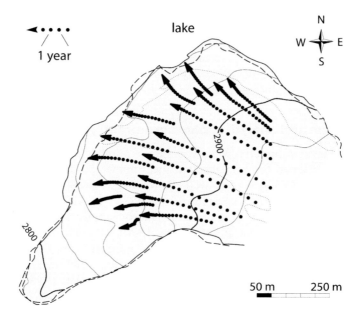

Figure 7.9 Trajectories depict the actual particle path. The trajectories on the Gruben Glacier tongue for 1973 to 1992 have been interpolated from repeated velocity fields that showed a significant change with time. Trajectories show the travel path as well as the travel time.

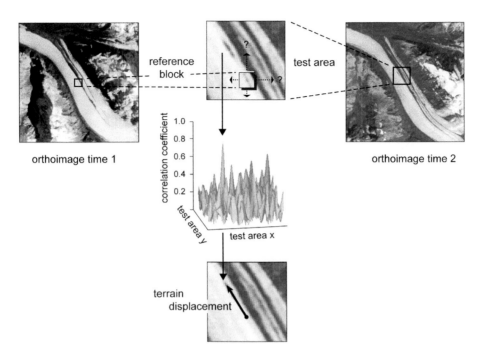

Figure 7.10 Principle of measuring horizontal terrain displacements from greyscale matching between repeat orthoimagery.

then to be rectified using the corresponding sensor model and orientation parameters (Kääb and Funk 1999, Kaufmann and Ladstädter 2002). If orthoimages are used, the image comparison directly provides the horizontal components of the displacement vector. The approach is, in principle, applicable to terrestrial, airborne and spaceborne imagery.

As for height-parallax matching, two techniques are used for digital comparison of multitemporal orthoimages: block (or area-based) matching techniques, or feature matching techniques. Block matching techniques include two-dimensional cross-correlation, least-squares matching or matching of Fourier or wavelet functions decomposed from the original image (e.g. Crippen 1992, Scambos et al. 1992, Frezzotti et al. 1998, Evans 2000, Kääb and Vollmer 2000, Van Puymbroeck et al. 2000, Kaufmann and Ladstädter 2002, Netanyahu et al. 2004, Kääb 2005b). Before performing the block matching, it might be useful to apply filters to the raw imagery such as contrast and edge enhancements, interest operators, or other global, regional and adaptive radiometric adjustments.

Here, the Correlation Image Analysis (CIAS) software (Kääb and Vollmer 2000) is used as an example. Horizontal displacements of individual terrain features are derived from multitemporal digital orthoimages. The measurement of an individual horizontal displacement vector basically follows two steps (Figure 7.10): (1) an image section (so-called reference block) with sufficient optical contrast is chosen in the orthophotograph of time 1. The ground coordinates of the central pixel are known from the

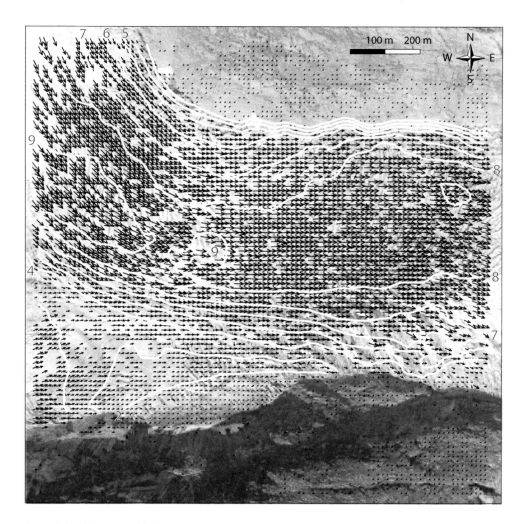

Figure 7.11 Velocity field for a section of Nigardsbreen, Norway, computed from matching between repeat airborne orthoimagery from August 19 and 29, 2001. The numbered contours give the absolute displacement in metres. Spacing between the vectors is 12 m. The applied orthoimages have 0.3 m ground resolution. The orthoimagery was provided by Bjørn Wangensteen and Trond Eiken, Department of Geosciences, University of Oslo, and was acquired within the Glaciorisk-project (http://glaciorisk.grenoble.cemagref.fr/projet_glaciorisk.htm, Wangensteen et al. 2006).

orthophotograph georeference. (2) The corresponding image section (so-called test block) is searched for in a sub-area (so-called test area) of the orthophotograph of time 2. If successfully detected, the differences in central pixel coordinates directly give the horizontal displacement vector between time 1 and 2 (Figure 7.10).

For identifying corresponding image blocks in both (or more) images, a double cross-correlation function Φ based on grey values of the images is used. The global

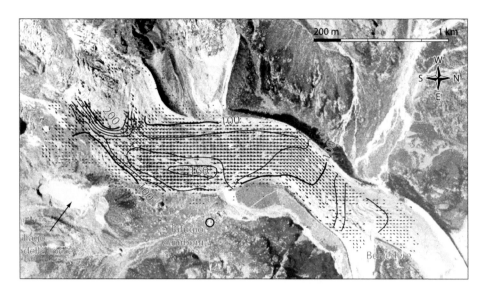

Figure 7.12 Surface velocity field on Ghiacciaio del Belvedere, Italy, between September 6 and October 11, 2001. Underlying orthoimage from 11 October 2001 (original aerial photograph acquired within the Glaciorisk-project). The black numbers and isolines indicate the glacier speeds of the autumn 2001 period. Speeds are given in metres per year. The glacier was surging during the period observed here (Kääb et al. 2004).

maximum of Φ is assumed to indicate the displaced terrain block of time 1 (Kääb 2005b).

Matching blunders are detected and eliminated from analysis of the correlation coefficients and from application of constraints, such as expected range for flow speed and direction. In the case of coherent displacement fields, additional spatial filters may be applied such as median or RMS thresholds. Glaciers usually show such coherent velocity fields due to the stress-transferring properties of ice.

Digital motion measurements from repeat airborne optical imagery have been applied for a number of glaciers (e.g. Evans 2000, Kääb and Vollmer 2001, Kääb 2002, Kääb 2005b, Wangensteen et al. 2006) (Figures 7.11, 7.12), though the method is more commonly applied to satellite images (Kääb 2005b).

In order to avoid distortions between the multitemporal products, all imagery is best adjusted as one image block connected by multitemporal tie points (see above). From comparison with ground measurements and analytical photogrammetry, and from the noise within coherent flow fields an RMSE of 0.5 to 1 times the image pixel size was found for the horizontal displacement measurements (Kääb 2005b).

7.7 CONCLUSIONS

Aerial photogrammetry offers a powerful set of methods for mapping and monitoring glaciers, glacier topography and glacier dynamics. It is one of the best established measurement methods in glaciology. Once stereoscopic aerial photographs are acquired, the information contained is very efficiently stored and can be visually interpreted at any later time, potentially using stereoscopic vision. The three-dimensional form of the terrain surface can be measured using forward intersection based on overlapping images. Today, mostly digital or digitized images and digital photogrammetric techniques are used for this purpose. One of the most severe problems for photogrammetric DEM generation is missing or insufficient radiometric contrast (i.e. optical texture) in the images, which prevents the measurement of corresponding terrain points in overlapping images. This effect typically reduces DEM quality over glacier accumulation areas and prevents photogrammetric DEM generation based on images taken in winter or after snowfall events. For optimal image and terrain conditions the vertical error of photogrammetric DEMs can be comparable to the pixel size. The main task of DEM postprocessing is to define the reliability of the results, and to find and potentially correct systematic or gross random errors. Such errors affect in particular the calculation of glacier geometry changes from repeat DEMs. Although a number of semi-automatic and manual procedures is available for finding severe DEM errors, this task is not straightforward and it usually requires a combination of different approaches and operator experience.

Because the above problems affect airborne laser scanner data less, or not at all, laser scanner techniques will become equally important, or even more so, for deriving DEMs over glacial terrain (Chapter 10). A combination of aerial photogrammetry and airborne laer scanner can provide particularly useful data for glaciological investigations, i.e. geometrically and radiometrically accurate image data combined with an accurate DEM.

If repeated aerial photographs with temporal baselines of about one year or shorter are available and if sufficiently many surface features are preserved over time, glacier movement can be measured using image correlation techniques. On debris-covered glaciers suitable surface features may be preserved longer than one year. Similar to DEM generation, the error of lateral displacements from repeat aerial imagery is roughly as low as the order of the image pixel size used.

Whilst photogrammetry from analogue film will eventually be superseded by digital sensors, analysis of archival aerial photography will still play a very important role in change detection in glaciology. In many cases aerial photographs represent the longest remotely sensed time series of glacier data and an invaluable data source for reconstructing glacier geometry during the last century. Analysing archival aerial photography and combining it with new imagery and sensors may require advanced modelling of sensors and errors.

REFERENCES

Armenakis, C. (1984). Deformation measurements from aerial photographs. *International Archives of Photogrammetry, Remote Sensing and Spatial Information Sciences* 25(A5), 39–48.

Bacher, U., S. Bludovsky, E. Dorre and U. Münzer (1999). Precision aerial survey of Vatnajökull, Iceland, by digital photogrammetry. In: M.O. Oltan and L. Gründig (eds.), *Proceedings of the 3rd Turkish-German Joint Geodetic Days – Towards a Digital Age*, Istanbul, Turkey, June 1–4, 1999, pp. 126–136.

Baltsavias, E.P. (1996). Digital ortho-images – a powerful tool for the extraction of spatial- and geo-information. *ISPRS Journal of Photogrammetry and Remote Sensing* 51(2), 63–77.

Baltsavias, E.P., H. Li, S. Mason, A. Stefanidis and M. Sinning (1996). Comparison of two digital photogrammetric systems with emphasis on DTM generation: case study glacier measurement. *International Archives of Photogrammetry, Remote Sensing and Spatial Information Sciences* 31(4), 104–109.

Baltsavias, E.P., E. Favey, A. Bauder, H. Boesch and M. Pateraki (2001). Digital surface modelling by airborne laser scanning and digital photogrammetry for glacier monitoring. *Photogrammetric Record* 17(98), 243–273.

Benson, C.S. and A.B. Follet (1986). Application of photogrammetry to the study of volcano-glacier interactions on Mount Wrangell, Alaska. *Photogrammetric Engineering and Remote Sensing* 52(6), 813–827.

Berthier, E., Y. Arnaud, D. Baratoux, C. Vincent and F. Remy (2004). Recent rapid thinning of the 'Mer de Glace' glacier derived from satellite optical images. *Geophysical Research Letters* 31(17), L17401.

Brecher, H.H. (1986). Surface velocity determination on large polar glaciers by aerial photogrammetry. *Annals of Glaciology* 8, 22–26.

Chandler, J. (1999). Effective application of automated digital photogrammetry for geomorphological research. *Earth Surface Processes and Landforms* 24(1), 51–63.

Cox, L.H. and R.S. March (2004). Comparison of geodetic and glaciological mass-balance techniques, Gulkana Glacier, Alaska, U.S.A. *Journal of Glaciology* 50(170), 363–370.

Crippen, R.E. (1992). Measuring of subresolution terrain displacements using SPOT panchromatic imagery. *Episodes* 15, 56–61.

Ebner, H. (1987). Digital terrain models for high mountains. *Mountain Research and Development* 7(4), 353–356.

Etzelmüller, B., G. Vatne, R. Ødegård and J.L. Sollid (1993). Mass balance and changes of surface slope, crevasse and flow pattern of Erikbreen, northern Spitsbergen: an application of a geographical information system (GIS). *Polar Research* 12(2), 131–146.

Etzelmüller, B. (2000). On the quantification of surface changes using grid-based digital elevation models (DEMs). *Transactions in GIS* 4(2), 129–143.

Evans, A.N. (2000). Glacier surface motion computation from digital image sequences. *IEEE Transactions on Geoscience and Remote Sensing* 38(2), 1064–1072.

Finsterwalder, R. (1931). Geschwindigkeitsmessungen an Gletschern mittels Photogrammetrie. *Zeitschrift für Gletscherkunde und Glazialgeologie* 19, 251–262.

Finsterwalder, R. and H. Rentsch (1980). Zur Höhenänderung von Ostalpengletschern im Zeitraum 1969–1979. *Zeitschrift für Gletscherkunde und Glazialgeologie* 16, 111–115.

Fox, A.J. and A.M. Nuttall (1997). Photogrammetry as a research tool for glaciology. *Photogrammetric Record* 15(89), 725–738.

Fox, A.J. and M.J. Gooch (2001). Automatic DEM generation for antarctic terrain. *Photogrammetric Record* 17(98), 275–290.

Frezzotti, M., A. Capra and L. Vittuari (1998). Comparison between glacier ice velocities inferred from GPS and sequential satellite images. *Annals of Glaciology* 27, 54–60.

Georgopoulos, A. and D. Skarlatos (2003). A novel method for automating the checking and correction of digital elevation models using orthophotographs. *Photogrammetric Record* 18(102), 156–163.

Gleitsmann, L. and M. Kappas (2006). Glacier monitoring survey flights below clouds in Alaska: oblique aerial photography utilising digital multiple-image photogrammetry to cope with adverse waether. *EARSeL eProceedings* 5(1), 42–50.

Grün, A. and H. Sauermann (1977). Photogrammetric determination of time-dependent variations of details of a glacier surface using a non-metric camera. Paper presented at the International Symposium on Dynamics of Temperate Glaciers and Related Problems, Munich, Germany, September 6–9, 1977.

Grün, A. and E.P. Baltsavias (1988). Geometrically constrained multiphoto matching. *Photogrammetric Engineering and Remote Sensing* 54(5), 633–641.

Grün, A. and L. Zhang (2002). Automatic DTM generation from three-line-scanners (TLS) images. *Proceedings of the International Symposium Photogrammetry meets Geoinformatics, Kartdagar 2002*, Jönköping, Sweden, April 17–19, 2002, 20 p.

Hauber, E., H. Slupetzky, R. Jaumann, F. Wewel, K. Gwinner and G. Neukum (2000). Digital and automated high resolution stereo mapping of the Sonnblick glacier (Austria) with HRSC-A. *Proceedings of the EARSeL-SIG-Workshop Land Ice and Snow*, Dresden, Germany, June 16–17, 2000, pp. 246–254.

Hofmann, W. (1958). Bestimmung von Gletschergeschwindigkeiten aus Luftbildern. *Bildmessung und Luftbildwesen* 3, 71–88.

Jordan, E., L. Ungerechts, B. Cáceres, A. Peñafiel and B. Francou (2005). Estimation by photogrammetry of the glacier recession on the Cotopaxi Volcano (Ecuador) between 1956 and 1997. *Hydrological Sciences Journal* 50(6), 949–961.

Julio Miranda, P. and H. Delgado Granados (2003). Fast hazard evaluation employing digital photogrammetry: Popocatepetl glaciers, Mexico. *Geofísica Internacional* 42(2), 275–283.

Julio Miranda, P., A.E. Gonzales-Huesca, H. Delgado Granados and A. Kääb (2005). Glacier melting and lahar formation during January 22, 2001 eruption, Popocatépetl volcano (Mexico). *Zeitschrift für Geomorphologie, N.F. Suppl.* 140, 93–102.

Kääb, A. (1996a). Photogrammetrische Analyse von Gletschern und Permafrost. *Vermessung Photogrammetrie Kulturtechnik* 1996(12), 639–644.

Kääb, A. (1996b). Photogrammetrische Analyse zur Früherkennung gletscher- und permafrostbedingter Naturgefahren im Hochgebirge. *Mitteilungen der Versuchsanstalt für Wasserbau, Hydrologie und Glaziologie der ETH Zürich* 145.

Kääb, A. and M. Funk (1999). Modelling mass balance using photogrammetric and geophysical data. A pilot study at Gries glacier, Swiss Alps. *Journal of Glaciology* 45(151), 575–583.

Kääb, A. (2000). Photogrammetry for early recognition of high mountain hazards: new techniques and applications. *Physics and Chemistry of the Earth, Part B: Hydrology, Oceans and Atmosphere* 25(9), 765–770.

Kääb, A. and M. Vollmer (2000). Surface geometry, thickness changes and flow fields on creeping mountain permafrost: automatic extraction by digital image analysis. *Permafrost and Periglacial Processes* 11(4), 315–326.

Kääb, A. (2001). Photogrammetric reconstruction of glacier mass balance using a kinematic ice-flow model: a 20-year time-series on Grubengletscher, Swiss Alps. *Annals of Glaciology* 31, 45–52.

Kääb, A. and M. Vollmer (2001). Digitale Photogrammetrie zur Deformationsanalyse von Massenbewegungen im Hochgebirge. *Vermessung Photogrammetrie Kulturtechnik* 99(8), 538–543.

Kääb, A. (2002). Monitoring high-mountain terrain deformation from air- and spaceborne optical data: examples using digital aerial imagery and ASTER data. *ISPRS Journal of Photogrammetry and Remote Sensing* 57(1–2), 39–52.

Kääb, A., Y. Isakowski, F. Paul, A. Neumann and R. Winter (2003). Glaziale und periglaziale Prozesse: Von der statischen zur dynamischen Visualisierung. *Kartographische Nachrichten* 53(5), 206–212.

Kääb, A., C. Huggel, S. Barbero, M. Chiarle, M. Cordola, F. Epifani, W. Haeberli, G. Mortara, P. Semino, A. Tamburini and G. Viazzo (2004). Glacier hazards at Belvedere Glacier and the Monte Rosa east face, Italian Alps: processes and mitigation. *Proceedings of the International Symposium INTERPRAEVENT 2004*, Riva del Garda, Italy, May 26, 2004, Vol. 1, pp. 67–78.

Kääb, A. (2005a). Combination of SRTM3 and repeat ASTER data for deriving alpine glacier flow velocities in the Bhutan Himalaya. *Remote Sensing of Environment* 94(4), 463–474.

Kääb, A. (2005b). Remote Sensing of Mountain Glaciers and Permafrost Creep. *Schriftenreihe Physische Geographie Glaziologie und Geomorphodynamik* 48. University of Zurich, 264 p.

Kaufmann, V. (1998). Deformation analysis of the Doesen rock glacier (Austria). Collection Nordicana 57, 551–556. Proceedings of the 7th International Permafrost Conference, Yellowknife, Canada, August 23–27, 1998.

Kaufmann, V. and R. Plösch (2000). Reconstruction and visualisation of the retreat of two small cirque glaciers in the Austrian Alps since 1850. In: M. Buchroithner (ed.), *High Mountain Cartography – Proceedings of the 2nd Symposium of the Commission on Mountain Cartography of the International Cartographic Association*, Salzburg, Austria, March 29–April 2, 2000, pp. 239–253.

Kaufmann, V. and R. Ladstädter (2002). Spatio-temporal analysis of the dynamic behaviour of the Hochebenkar rock glaciers (Oetztal Alps, Austria) by means of digital photogrammetric methods. *Grazer Schriften der Geographie und Raumforschung* 37, 119–140.

Kersten, T. and M. Meister (1993). Grosser Aletschgletscher. Photogrammetrische Auswertungen als Grundlage für glaziologische Untersuchungen. *Vermessung Photogrammetrie Kulturtechnik* 91(2), 75–80.

Knizhnikov, Y.F., R.N. Gelman, G.B. Osipova and D.G. Tsvetkov (1998). Aerophotogrammetric study of ice movement in surging glaciers. *Zeitschrift für Gletscherkunde und Glazialgeologie* 34(1), 69–84.

Krummenacher, B., K. Budmiger, D. Mihailovic and B. Blank (1998). Periglaziale Prozesse und Formen im Furggentälti, Gemmipass. *Mitteilungen des Eidgenössisches Institut für Schnee- und Lawinenforschung* (SLF), Davos, 56, 245 p.

Lane, S.N., T.D. James and M.D. Crowell (2000). Application of digital photogrammetry to complex topography for geomorphological research. *Photogrammetric Record* 16(95), 793–821.

Ledwith, M. and B. Lunden (2001). Digital photogrammetry for air-photo-based construction of a digital elevation model over snow-covered areas – Blåmannsisen, Norway. *Norwegian Journal of Geography* 55(4), 267–275.

Li, Z., Z. Xu, M. Cen and X. Ding (2001). Robust surface matching for automated detection of local deformations using least-median-of-squares estimator. *Photogrammetric Engineering and Remote Sensing* 67(11), 1283–1292.

Melvold, K. (1992). *Study of glacier motion on Kongsvegen and Kronebreen, Svalbard.* Technical Report, Department of Geography, University of Oslo.

Messerli, B. and M. Zurbuchen (1968). Blockgletscher im Weissmies und Aletsch und ihre photogrammetrische Kartierung. Die Alpen, Schweizer Alpen Club 3, 139–152.

Netanyahu, N.S., J. Le Moigne and J.G. Masek (2004). Georegistration of Landsat ata via robust matching of multiresolution features. *IEEE Transactions on Geoscience and Remote Sensing* 42(7), 1586–1600.

Norvelle, R. (1996). Using iterative orthophoto refinements to generate and correct digital elevation models (DEM's). In: L. Grewe (ed.), *Digital Photogrammetry: an Addendum to the Manual of Photogrammetry.* pp. 151–155, American Society for Photogrammetry and Remote Sensing, Falls Church, 250 p.

Ødegård, R., K. Isaksen, T. Eiken and J.L. Sollid (2004). Terrain analyses and surface velocity measurements of Hiorthfjellet rock glacier, Svalbard. *Permafrost and Periglacial Processes* 14(4), 359–365.

Otto, J.C., K. Kleinod, O. Konig, M. Krautblatter, M. Nyenhuis, I. Roer, M. Schneider, B. Schreiner and R. Dikau (2007). HRSC-A data: a new high-resolution data set with multipurpose applications in physical geography. *Progress in Physical Geography* 31(2), 179–197.

Pillewizer, W. (1938). Photogrammetrische Gletscherforschung. *Bildmessung und Luftbildwesen* 2, 66–73.

Pilgrim, L. (1996a). Robust estimation applied to surface matching. *ISPRS Journal of Photogrammetry and Remote Sensing* 51, 243–257.

Pilgrim, L.J. (1996b). Surface matching and difference detection without the aid of control points. *Survey Review* 33(259), 291–303.

Poli, D. (2002). General model for airborne and spaceborne linear array sensors. *International Archives of Photogrammetry, Remote Sensing and Spatial Information Sciences* 34(B1), 177–182.

Rentsch, H., W. Welsch, C. Heipke and M. Miller (1990). Digital terrain models as a tool for glacier studies. *Journal of Glaciology* 36(124), 273–278.

Roer, I., A. Kääb and R. Dikau (2005). Rockglacier kinematics derived from small-scale aerial photography and digital airborne pushbroom imagery. *Zeitschrift für Geomorphologie* 49(1), 73–87.

Scambos, T.A., M.J. Dutkiewicz, J.C. Wilson and R.A. Bindschadler (1992). Application of image cross-correlation to the measurement of glacier velocity using satellite image data. *Remote Sensing of Environment* 42(3), 177–186.

Schenk, T., A. Krupnik and Y. Postolov (2002). Comparative study of surface matching algorithms. *International Archives of Photogrammetry, Remote Sensing and Spatial Information Sciences* 32(B4), 518–524.

Toutin, T. (1995). Multi-source data fusion with an integrated and unified geometric modelling. *EARSeL Journal – Advances in Remote Sensing* 4(2), 118–129.

Van Puymbroeck, N., R. Michel, R. Binet, J. Avouac and J. Taboury (2000). Measuring earthquakes from optical satellite images. *Applied Optics* 39(20), 3486–3494.

Wangensteen, B., O.M. Tønsberg, A. Kääb, T. Eiken and J.O. Hagen (2006). Surface elevation change and high resolution surface velocities for advancing outlets of Jostedalsbreen. *Geografiska Annaler* 88A(1), 55–74.

Weber, D. and A. Herrmann (2000). Contribution of digital photogrammetry in spatio-temporal knowledge of unstable slopes: the example of the Super-Saute landslide (Alpes-de-Haute-Provence, France). *Bulletin de la Societe Geologique de France* 171(6), 637–648.

Wrobel, B. and M. Schlüter (1997). Digital terrain model generation in the Antarctic – A challenging task for digital photogrammetry. In: M.O. Oltan and L. Gründig (eds.), *Proceedings of the 2nd Turkish-German Joint Geodetic Days*, Berlin, Germany, May 25–29, 1997, pp. 407–416.

Würländer, R. and K. Eder (1998). Leistungsfähigkeit aktueller photogrammetrischer Auswertemethoden zum Aufbau eines digitalen Gletscherkatasters. *Zeitschrift für Gletscherkunde und Glazialgeologie* 34(2), 167–185.

Zhang, L. and A. Grün (2004). Automatic DSM generation from linear array imagery data. *International Archives of Photogrammetry, Remote Sensing and Spatial Information Sciences* 35(B3), 128–133.

Chapter 8

Optical remote sensing of glacier extent

Frank Paul
Department of Geography, University of Zurich, Switzerland

Johan Hendriks
Department of Geosciences and Geography, University of Helsinki, Finland

8.1 SPECTRAL PROPERTIES

The surface of a glacier is composed of snow, firn, ice, water and debris (e.g. rock, pebbles, dust, soot) with a highly variable fraction of each component from glacier to glacier, which significantly impacts the optical properties of the glacier (Chapter 2). In Figure 8.1 a mid-August picture of a typical small Alpine valley glacier, Oberaar Glacier in Switzerland, (46.53° N, 8.2° E) is shown. The glacier's area is 5.2 km² and the length is 4.7 km, and it is the source of the river Aare. Oberaar Glacier flows from west to east, stretches from about 2300 to 3300 m a.s.l. and is bounded by relatively steep valley walls. While the transient snow line is easily visible in the image separating the darker bare ice from the bright snow, the glacier boundary at the left side of the image is difficult to determine due to the heavy debris cover. However, human recognition is able to trace the perimeter of the glacier by the change in slope angle along the contact with the Little Ice Age lateral moraine. The debris cover to the right side is not as compact, and some bare ice is visible through it. The fractions of snow, firn, ice, water and debris of a glacier are important for measuring the spectral properties of a glacier since several different surface types may be located within one image pixel when viewing a glacier from space. With a focus on temperate glaciers as they appear at the end of the ablation season and neglecting liquid water, debris cover and other pollution, the glacier surface is composed of ice and snow. As glacier ice originates from metamorphosed snow, the optical properties of a glacier surface are largely determined by those of snow. In this chapter, only the optical properties of snow are discussed; for a thorough overview in other parts of the electromagnetic spectrum see Chapter 5 and Rees (2005). The spectral characteristics of the debris cover strongly depends on the lithology of the surrounding rock walls and varies from place to place.

The spectral properties of snow have been investigated in the field (Grenfell et al. 1981, Qunzhu et al. 1983) and modelled numerically (Dozier 1989) in several studies. Both approaches (Figure 8.2) reveal a high spectral reflectance of snow in the visible part of the electromagnetic spectrum with a large dependence on impurities and a very small dependence on grain size, a decreasing reflectance in the near infrared with decreasing dependence on impurities and an increasing dependence on grain size, and very low reflectance in the middle or shortwave infrared with little dependence

Figure 8.1 Oberaar Glacier as seen in mid-August 1993 from the barrage of the Oberaar Lake. Zones with bare ice, snow and debris are clearly visible, but the left glacier margin (orographically right) is difficult to identify. See Figure 8.4 for location of Oberaar Glacier. Photograph by Frank Paul, 1993.

on impurities and a strong dependence on grain size. Measured spectral reflectance curves for snow, firn and ice resemble the modelled curves from snow very closely (Figure 8.2), but at lower absolute values due to impurities. Spectral reflectances obtained from satellite data have frequently been used to distinguish between ice and snow zones (facies) that are different from those identified from field measurements (Hall et al. 1987, Williams et al. 1991, Winther 1993). For example, the superimposed ice zone, which belongs to the accumulation area, cannot be separated from the bare ice zone, which belongs to the ablation region, and a water saturated snow zone (slush zone) might be visible (Williams and Hall 1998). In the French Alps, satellite-derived snow lines have also been used to track glacier mass balance (Rabatel et al. 2005). However, in rugged high-mountain topography the calculation of spectral reflectance from satellite imagery requires an accurate digital elevation model (DEM) and proper orthorectification of the satellite image. Although good results had been achieved by the studies mentioned above, errors in satellite-derived snow reflectance could increase when snow conditions are more difficult to interpret. Sometimes, even visual inspection might not be able to distinguish between snow and ice in the satellite data.

For a large number of applications, such as distributed energy balance modelling and climate models, the use of albedo integrated over all wavelengths instead of spectral reflectance for discrete parts of the spectrum is more appropriate. The related conversions often assume a Lambertian reflectance characteristic of snow (100% diffuse and isotropic scattering) for simplicity. Actually, older snow exhibits a forward

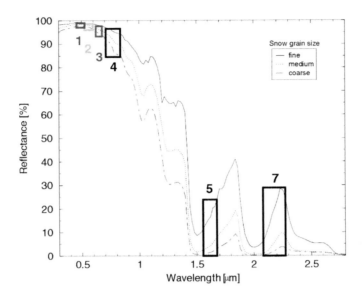

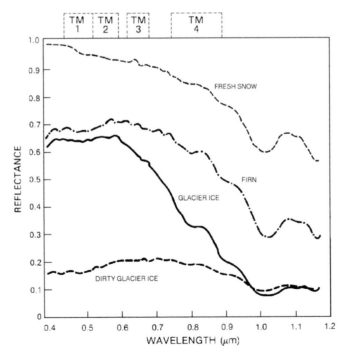

Figure 8.2 Upper panel: Modelled spectral reflectance curves of snow with different grain size and location of TM spectral bands superimposed (data taken from the ASTER spectral library at JPL, 2002). Lower panel: Measured spectral reflectances of fresh snow and firn as well as clean and dirty glacier ice (C.f. with Figure 3.2). Figure reprinted from Hall et al. (1988) with permission from Elsevier.

scattering component that increases with decreasing solar elevation (de Ruyter de Wildt et al. 2002). Hence, for accurate albedo calculations, the Bidirectional Reflectance Distribution Function (BRDF) has to be considered. The BRDF effect is most pronounced in the solar principal plane where the source of illumination, target and sensor are in one plane. Water, ice and snow are forward scattering surfaces and only freshly fallen snow resembles a Lambertian reflector. Older snow, which has gone through periods of melting and freezing or which contains dust and dirt has more anisotropic reflectance characteristics. Therefore, the reflectance anisotropy of a glacier surface varies with surface material and grain size as well as with the impurity and water content of the snow surface. As top of atmosphere (TOA) spectral reflectance of ice and snow can be measured from satellites quite accurately (Dozier 1989, Hall et al. 1988, 1990, Winther 1993) and correction for topographic and atmospheric effects are included in most digital image processing software packages, it is relatively straight forward to calculate glacier surface albedo from satellite data. However, as the satellite spectral bands typically only cover a small part of the entire electromagnetic spectrum and average the reflectance over a certain bandwidth (Table 8.2), a narrow-to-broadband conversion has to be applied. Recently, several empirical relationships for the calculation of snow, ice and glacier albedo from Landsat TM satellite data have been found through intense field work (Greuell et al. 2002) and successfully applied in distributed mass balance modelling (Paul et al. 2005). Neglecting the BRDF characteristics of snow has in general little effect on the derived albedo values, but late in the year and for low solar elevations the satellite derived albedo values might be about 0.1 smaller (Klok et al. 2003). Chapter 5 presents further discussion of snow and ice characteristics and their interaction with electromagnetic radiation.

8.2   GLACIER MAPPING AND SATELLITE SENSOR CHARACTERISTICS

The number of optical satellite sensors that could be used for mapping and monitoring of glaciers is large and still increasing. The most complete and regularly updated overview is given by Kramer (2002). However, the large number of available sensors can be reduced significantly if the focus is on glacier observation. Some important features for glacier applications are spatial resolution, region covered (swath width), number and position of spectral bands (referred to as spectral resolution), long-term availability of the data and data costs (e.g. cost per km^2). With respect to spatial resolution a compromise has to be made with the area covered, as the latter in general decreases with increasing sensor resolution. Typically, the size of individual glaciers spans about five orders of magnitude (0.01–1000 km^2) and the regions covered by glaciers often exceed thousands of square kilometres. For efficient observation, this requires sensors with swath widths larger than 50 km and a spatial resolution in the range of 10–30 m. Of course, if detailed investigations of individual glaciers are performed, sensors with very high spatial resolution (about 1 m), like Ikonos, Quickbird or SPOT 5 are the most valuable (e.g. Huggel et al. 2005). Nevertheless, the high costs of those data (>20 USD per km^2) prevent any large-scale mapping activities. The spectral resolution of the sensor is of major importance for glacier mapping purposes, at least if some degree of automated glacier classification is desired (e.g. Paul 2002a, Hendriks and Pellikka 2007).

The higher spatial resolution that can be achieved with panchromatic sensors is mainly due to the larger quantity of electromagnetic energy they receive on the same sensor size compared to the multispectral sensors. In respect to glaciers, panchromatic bands allow only manual delineation by on-screen cursor tracking, snow and cloud discrimination is difficult and sensor saturation over snow is frequent. However, after fusion with respective lower resolution multispectral bands, delineation of small scale surface features, like debris cover or trim lines, is greatly improved (Paul and Kääb 2005). Thus, automated mapping of glaciers as well as discrimination from clouds requires a band in the middle or short-wave infrared (SWIR) that is not covered by the panchromatic sensors. In combination with the required long-term availability of the data, the affordable costs and the frequent data acquisition, only very few multispectral satellite sensors remain in the selection, including Landsat TM/ETM+, SPOT HRV and ASTER on board the Terra platform. In Table 8.1 a brief overview of the satellite orbital characteristics is given and Table 8.2 summarizes major sensor features.

The costs per km^2 of a full scene are obtained by dividing the cost by the area covered by the scene (Table 8.1). From this point of view the cost of SPOT scenes

Table 8.1 Characteristics of the satellites used remote sensing of glaciers (V = visible, SWIR = shortwave infrared, TIR = thermal infrared) (Kramer 2002, NASA 2009).

Satellite	Landsat 5	Landsat 7	SPOT 4	Terra
Sensors	TM	ETM+	HRV	ASTER
Launch Date	1.3.1985	15.4.1999	22.1.1990	18.12.1999
Earth distance [km]	705	705	830	705
Temporal resolution [days]	16	16	26 (5)	16
Equator crossing time [UT]	9.30	10:00	10.31	10:30
Image size [km x km]	185 × 180	185 × 180	60 × 60	60 × 60
Spatial resolution [m]	30, 120 (TIR)	30 (V, SWIR), 60 (TIR)	20 (V, SWIR)	15 (V), 30 (SWIR), 90 (TIR)
Cost per full scene [USD]	475	600	2000	60
Cost per km^2 [USD]	0.014	0.018	0.56	0.017

Table 8.2 Spectral bandwidths of reflective bands from the different sensors (in μm). (Kramer 2002, NASA 2009).

TM band	TM	ETM+	HRV	ASTER
1 (blue)	0.45–0.52	0.45–0.52	–	–
2 (green)	0.52–0.60	0.53–0.61	0.50–0.59	0.52–0.60
3 (red)	0.63–0.69	0.63–0.69	0.61–0.68	0.63–0.69
4 (NIR)	0.76–0.90	0.75–0.90	0.79–0.89	0.76–0.86
5 (SWIR)	1.55–1.75	1.55–1.75	1.55–1.75	1.60–1.70
7 (SWIR)	2.08–2.35	2.09–2.35	–	2.15–2.43[1]
Pan	–	0.52–0.90	0.51–0.79	–

[1] *Includes five individual bands.*

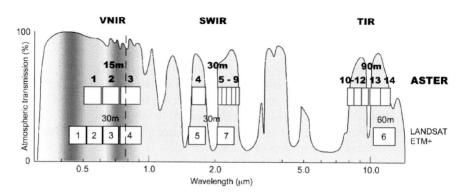

Figure 8.3 Atmospheric transmission and location of ASTER and Landsat TM spectral bands (Figure courtesy of A. Kääb).

cannot compete with TM/ETM+ and ASTER. Furthermore, considering the efforts required to orthorectify nine ASTER or SPOT scenes instead of one TM/ETM+ scene, Landsat data are the most suitable. Together with the long-term availability of the data (starting in 1982 with Landsat 4) and the similar spatial resolution of the SWIR bands (still 30 m on ASTER and 20 m for SPOT 5), Landsat data have been used widely for glacier mapping purposes (e.g. Aniya et al. 1996, Bayr et. al. 1994, Binaghi et al. 1997, Jacobs et al. 1997, Hall et al. 1988, Li et al. 1998, Paul 2002a, 2002b, Paul et al. 2002, 2003, 2004a). Therefore, the main data for glaciological studies are from Landsat 5 with its TM sensor, which provides a more or less continuous and calibrated data time-series at about 30 m spatial resolution in seven spectral bands (including a blue band) for more than 25 years now – and is still in operation. As such, this long time-series has become an invaluable treasure with respect to the monitoring of global change and in particular for documenting glacier change around the world.

In general, spatial resolution of the sensors decreases towards longer wavelengths (Table 8.1). The spectral bands can be grouped into three classes (Table 8.2), which are the visible and near infrared (VNIR) class including the bands in the blue, green, and red (0.4–0.7 μm) and the near infrared (0.7–1.4 μm), the SWIR (1.4–3 μm), and the thermal infrared (TIR) from 8–15 μm (cf. Chapter 5). The exact position of each band is strongly related to the atmospheric windows in which there is little absorption by trace gases, and therefore the spectral band ranges from sensor to sensor are fairly similar. In Figure 8.3 the spectral positions of the TM/ETM+ and ASTER bands are shown with respect to atmospheric transmission. As the Landsat TM sensor covers all of the spectral ranges that are available from other sensors, we here use the TM bands to exemplify the spectral properties of ice and snow in each band.

In Figure 8.4 the TM bands 1 to 6 acquired on 31 August 1998 are shown for comparison of the spectral characteristics of glaciers and the surrounding environment. The coverage of Figure 8.4 is about 9.5 by 9 km, and it presents the Oberaar Glacier, Switzerland, in the lower centre and some parts of the heavily debris-covered Unteraar Glacier at the top as well as numerous smaller mountain glaciers and cirques. Most of the terrain types that can be found in glacierized regions are present: glaciers, snow

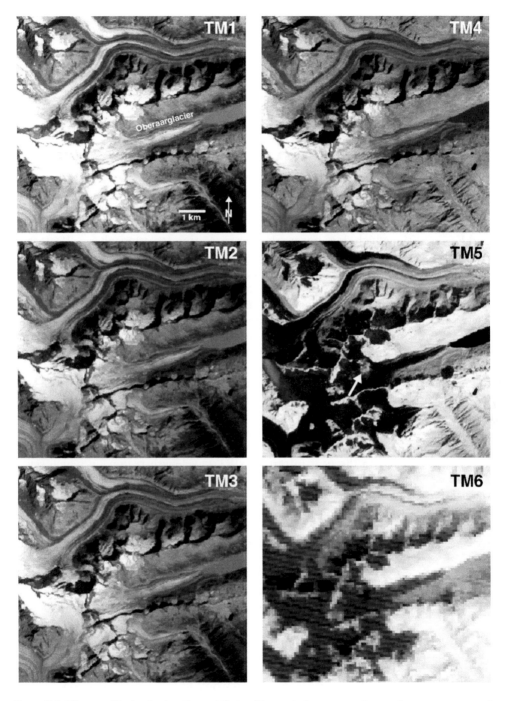

Figure 8.4 Oberaar Glacier in the Central Swiss Alps and the surrounding environment as seen in Landsat bands TM1 to TM6 after contrast stretch. Image size is 9.5 by 9 km. The white arrow in TM5 image points to a cloud invisible in the other bands.

fields, bare rock, water bodies of differing turbidity, and vegetation. There is also a small cloud above Oberaar Glacier and deep shadows cast by the steep terrain on to bare rock, ice and snow surfaces. Moreover, glacier ice exhibits a varying range of thicknesses and degree of coverage by dust or debris. This image subset therefore allows for a good demonstration of various glacier mapping algorithms.

Most of the reflectance variability in each spectral band as depicted in Figure 8.4 can be explained by the spectral curves for snow and ice (Figure 8.2). For better visibility, all images have been linearly contrast stretched and gamma corrected with the same factor. In TM band 1 (TM1), the most obvious feature is the sensor saturation over snow and firn areas. The possibility of using a low gain setting over snow covered regions for Landsat 7 and ASTER strongly reduces saturation. The high contrast between clean and dirty ice is due to the strong dependence of reflectance on contamination. Another important characteristic is the good visibility of ice and snow in cast shadow, which is due to the atmospheric attenuation (path radiance) in TM1. This contrast difference could be utilized to identify and map glaciers in this region (Paul and Kääb 2005), which does not work after correction for atmospheric effects, since it reduces the contrast in the image.

The spectral reflectances in TM bands 2 and 3 are very similar and most of the effects described above for TM1 are comparable but reduced in intensity. This implies that TM2 and TM3 have different roles in glacier mapping than TM1. For instance, the TM2 equivalent ASTER band 1 (AST1) can be used for mapping of ice and snow in cast shadow (Paul and Kääb 2005) or for a first classification of water bodies using a normalized difference water index (Huggel et al. 2002, Hendriks and Pellikka 2007). The band TM3 (together with TM4) can be utilized for mapping of vegetated areas using the normalized difference vegetation index NDVI ([TM3 − TM4] / [TM3 + TM4]). This is helpful for excluding misclassification in regions with dense vegetation cover when a TM4/TM5 ratio is used. In regions with very deep shadows (clean air and low solar elevation) a TM3/TM5 ratio has been applied successfully for glacier mapping (Paul and Kääb 2005), at the expense of misclassification of most water bodies.

TM4 is a band in the near infrared (NIR) wavelength region, in which several reflectance characteristics are important. The reflectance of snow is less in the NIR than in the green and red wavelengths of bands TM2 and TM3, which reduces the number of saturated pixels further, and the clean glacier ice is visually quite dark indicating low reflectance, which is partly due to the presence of liquid water at the surface. The strong absorption of radiation in the NIR by water is also obvious from the dark grey tone of the artificial lake in front of the Oberaar Glacier (Figure 8.4). At the same time vegetated regions have a higher reflectance being bright in the image due to mesophyll reflection, and the contrast with bare rock diminishes. In this part of the spectrum the dependence of the reflectance on snow grain size increases, while the dependence on contamination by soot or dust decreases (Hall et al. 1988, Dozier 1989, Chapter 5).

The largest changes in reflectance of ice and snow take place in the SWIR where TM5 is located (Figure 8.2). Clean glacier ice absorbs nearly all the radiation and the low digital numbers (DNs) appear almost black on a linear colour table from 0 (black) to 255 (white), while snow covered regions appear very dark grey. In Figure 8.4, the strong dependence of reflectance on snow grain size can be seen as well as the formerly saturated regions (white in TM1) now display two very distinct shades of grey, with

Table 8.3 Summary of the application fields of each band with respect to glacier mapping.

TM band	Field of application
1 (blue)	Snow/ice discrimination in cast shadow, mapping of glacier lakes
2 (green)	One part of the NDSI (snow), also snow/ice discrimination in cast shadow
3 (red)	One part of the band ratio, also for NDVI (vegetation)
4 (NIR)	One part of the band ratio, also for NDVI (vegetation) and NDWI (water)
5 (SWIR)	Mandatory band for automated classification (ratio, NDSI)
7 (SWIR)	Similar to TM5 but very noisy in shadow
6 (TIR)	Alternative for TM5 in case of a thin volcanic ash layer
Pan	Manual delineation, debris-cover on glaciers

the brighter values for regions higher up which should have smaller grains due to reduced metamorphosis. Bodies of clean water absorb nearly all radiation and appear black, while vegetated regions are very bright due to multiple scattering by the plant cell structure. And, one feature that is hardly recognizable in all the other bands is the small cloud below the snow line of Oberaar Glacier. Such clouds can be detected even better on false colour composites with TM bands 5 (SWIR, 4 (NIR) and 3 (red) as red, green and blue, respectively. Hence, satellite images must be carefully analyzed when glaciers are mapped and any cloud covered regions should be identified, at the latest during post-classification.

The thermal infrared (TIR) band TM6 is shown for comparison with the other bands in Figure 8.4. The TIR band registers the thermal emission of the surface which is closely related to temperature and could be converted to it (Richter 1990). Most obvious in the TIR band image are the larger pixel size (120 m compared to 30 m), the striping of the image, the comparably bright (i.e. warm) water surface, and the reduced thermal emission of north- compared to sun- facing slopes. Thick debris cover on the glacier has similar grey values to the mountain slopes since the temperature of the debris is not reduced much by the underlying ice. In the case of a thin debris cover the derived surface temperature could be used to estimate debris thickness (Mihalcea et al. 2008). However, the automated delineation of a debris-covered glacier tongue from its different thermal properties is challenging. When some bare ice is part of a mixed pixel, the reduction in temperature could be sufficient for a clear discrimination, but then the other glacier mapping techniques (e.g. band ratio) work as well. The TM7 band is not shown here as it is very similar to TM5 with respect to glacier-related features. Moreover, the signal to noise ratio in this far end of the reflected electromagnetic spectrum is very low and results in misclassifications in regions of cast shadow. In Table 8.3 the major applications with respect to glacier mapping for each Landsat TM band are summarized.

8.3 GLACIER MAPPING

The most prominent glaciological application of multispectral satellite data is mapping of glacier extent. However, several other interesting parameters can be obtained as well, given that image conditions and spatial resolution are sufficient. Conversion of the raw DNs to spectral reflectance or albedo has frequently been used to distinguish glacier facies and map the snow covered area (Hall et al. 1988, Rott 1994, Angelis et al. 2007) or is used as an input for distributed mass balance models (Paul et al. 2005, Machguth et al. 2006). Stereo imaging from optical sensors like ASTER (along-track) or SPOT (across-track) provides digital elevation models (DEMs) with high spatial resolution (10–30 m) and sufficient vertical accuracy on glaciers, at least in regions where image contrast is acceptable (AlRousan et al. 1997, Eckert et al. 2003, Chapter 7). Such DEMs have several important further applications, among others the proper orthorectification of the satellite images (Toutin 2004), the atmospheric correction required for albedo calculation, and the creation of a detailed glacier inventory (Paul et al. 2002). A further promising technique is the calculation of glacier surface velocity fields from feature tracking with accurately orthorectified repeat pass imagery (cf. Berthier et al. 2005, Kääb 2005).

Mapping of glacier extent is the most straightforward application of multispectral optical satellite imagery. Manual delineation of glaciers by on-screen cursor tracking is applied frequently, in particular when multispectral data are not available (e.g. using aerial photography or panchromatic satellite imagery) or when glaciers are covered by thick debris (Hall et al. 2003, Paul et al. 2004). While this method is in general the most accurate, it is extremely time-consuming for hundreds of glaciers (Lambrecht and Kuhn 2007). As such, a large number of methods have been developed and tested for automated glacier mapping from multispectral data, that all make use of the distinct spectral properties of ice and snow in the SWIR. Here, the methods are discussed based on the Landsat TM bands, but they can be applied to the corresponding bands of other sensors as well due to the similar spectral ranges covered (Table 8.2). If the blue band (TM1) is required but not available, a replacement with the green band (TM2 or ASTER band 1) is often an alternative. Most comprehensive overviews and comparisons of different mapping methods are given by Sidjak and Wheate (1999), Albert (2002), Paul et al. (2002) and Paul and Kääb (2005). They all came to the conclusion that a simple band ratio applied to the raw data is the most efficient method for glacier mapping in terms of accuracy and effort. Nevertheless, authors such as Hendriks and Pellikka (2007) or Racovitanu et al. (2008) who applied other techniques, like the normalized difference snow index (NDSI) also achieved good results. Other methods that have been applied include unsupervised (Aniya et al. 1996) and supervised classification (Li et al. 1998), fuzzy set theory (Binaghi et al. 1997) and spectral unmixing (Klein and Isacks 1999).

8.3.1 Threshold ratio images

Here, only the three most widely used methods are compared, namely thresholded ratio images (TM3/TM5 and TM4/TM5) and NDSI ((TM2 − TM5)/(TM2 + TM5)), all based on raw DNs. In general, ratio images have the advantage of strongly enhancing the image contrast for the selected surface types, while reducing the bias in illumination from the terrain at the same time. If raw DNs are corrected for atmospheric scattering (i.e. path radiance) beforehand, the spectral separation can be further improved (Crippen 1988). Such a correction can be applied by subtracting the lowest DN (e.g. obtained from a histogram) from all other DNs in the respective band (also called dark object subtraction, DOS). Otherwise illumination differences in regions of cast

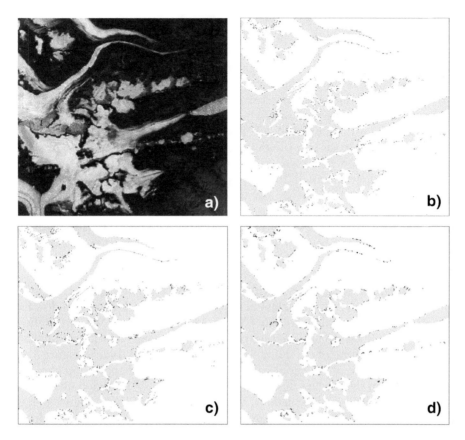

Figure 8.5 Comparison of different thresholds for glacier mapping from ratio images for the Oberaar
Glacier (see Figure 8.4). a) the TM4/TM5 ratio image after linear contrast stretch; b) three
glacier maps combined resulting from three threshold values: 1.8 (all grey shades), 1.9 (grey
and black), 2.0 (grey); c) effect of a 3 × 3 median filter: dark grey pixels are removed and
black pixels are added (shown here for the map with the threshold 1.9); d) comparison of
the three thresholds after application of the median filter (same colour scheme as in b).

shadow are strongly enhanced, especially with TM bands 1 to 3. The TM4/TM5 ratio
works without DOS, as path radiance can be neglected in both bands.

In Figure 8.5a, the TM4/TM5 ratio image is shown (values between 0.0 and 25.0
have been rescaled to 0–255), clearly illustrating the strong contrast enhancement
between ice and snow as bright grey tones and other terrain as dark tones. This is also
obvious from the grey values as depicted for each band (see Figure 8.4). When high
DNs in the VNIR over glaciers are divided by the low DNs in the SWIR, high ratio
values result. And when the high DNs over other soil, vegetation or rock, or the low
DNs over water or shadow in the NIR are divided by the corresponding high/low DN
in the SWIR, low ratio values result. Using a threshold value, glaciers can be separated
from the surrounding terrain by setting the values above the threshold to black and all
others to white. In Figure 8.5b, three such maps using different threshold values are

combined, indicating that the value is quite robust for glaciers in sunlight (i.e. little area change), while it is more sensitive in shadowed regions (more changed pixels) and should thus be optimized there. A final processing step is the application of a spatial filter such as a 3×3 median filter to the black and white glacier map for reduction of noise (Figure 8.5c). While such a filter can remove isolated snow pixels and closes gaps due to debris cover or noise in shadow, it also reduces the size of very small glaciers. For this reason and due to the pixel size of the sensor the lower limit of glacier size that can be mapped with Landsat TM is about 0.02 km^2 (Paul et al. 2003), at least if no manual corrections are applied. Finally, Figure 8.5d depicts a comparison of the median-filtered glacier maps with the same threshold values as used for Fig. 8.5b. Still, differences mainly occur along the boundaries of regions in cast shadow or with debris cover and are very small. In this respect, any of the three thresholds provided a sufficient accuracy for the bare ice, but all fail in regions of thicker debris cover and for the optically thick cloud above Oberaar Glacier. In these regions manual corrections have to be applied based on visual interpretation. A useful accuracy measure of the mapping could thus not be given as the differences to a ground truth depend on the amount of debris cover. In any case, the decision for the threshold should be optimized with respect to the required work load for post processing.

The comparison of TM4/TM5 with the TM3/TM5 ratio is depicted in Figure 8.6a. In order to separate the wrongly classified regions of bare rock in cast shadow from ice and snow, an additional threshold in TM1 was applied to the TM3/TM5 ratio image (Paul and Kääb 2005). The resulting glacier map is somewhat larger in shadowed regions and covers more of the thinly debris-covered regions (blue in Figure 8.6a). On the other hand, more of the snow patches in sunlight are mapped and some glacier-covered regions in shadow are missed (red in Figure 8.6a). The region obscured by the cloud is not mapped in either method. Visual inspection of TM6 (Figure 8.4) reveals that the cloud is also difficult to recognize in the thermal band. Overall, the differences between TM3/TM5 and TM4/TM5 are very small and there is only a slight preference for the TM3/TM5 method in this example.

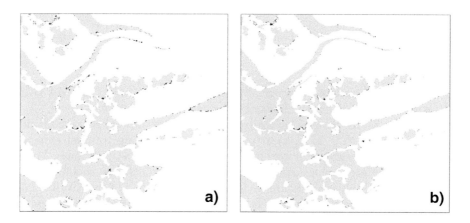

Figure 8.6 Comparison of the median filtered glacier maps for the Oberaar Glacier (see Figure 8.4). a) TM3/TM5 (grey and black) vs. TM4/TM5 (grey and dark grey); b) TM3/TM5 (grey and black) vs. NDSI (grey and dark grey).

Finally, a comparison of TM3/TM5 with the NDSI derived glacier map is depicted in Figure 8.6b, both after application of a median filter. To obtain the glacier map from the NDSI, the path radiance is first subtracted from TM2. The comparison reveals only small differences between the methods, with some more of the debris-covered regions and less of the shadowed regions being mapped from the NDSI. From this comparison it can be concluded that the NDSI performs somewhat better and is thus the best of the investigated methods. However, all three methods can be recommended for glacier mapping as they are all easy to apply and yield very similar and robust results. The threshold is most sensitive in regions of cast shadow and thin debris cover. Because manual corrections are required for debris-covered regions anyway, the threshold should be selected to minimize the required corrections in regions of cast shadow.

8.3.2 Manual corrections

The most severe corrections have to be made for wrongly classified lakes and debris-covered parts of glaciers. The latter are often not only related to obviously thick moraine, but also to the thin, but optically thick debris that is melting out along the glacier perimeter in the ablation area. For the thick debris cover, semi-automatic mapping methods have been developed that include terrain-based attributes such as slope and curvature, and object-based analysis (Bishop et al. 2001, Paul et al. 2004). However, the resulting maps from these methods must be manually corrected as well, and the work load might be smaller if the delineation is performed completely manually. In this case, higher resolution (panchromatic) imagery might help with the delineation, in particular after fusion with the multispectral bands (Paul and Kääb 2005).

The thermal band (TM6) can be used as a replacement for TM5, when the glacier surface is covered by a thin but optically thick volcanic ash layer and the glacier is large (e.g. Vatnajøkull in Iceland). In that case the lower spatial resolution of the TIR band can be neglected (Raup et al. 2007). Bilinear resampling to 30 m is a possible way to create a less blocky output. While optically thin clouds and cloud shadows are in general not problematic for the ratio mapping methods, optically thick clouds have to be corrected during post-processing. An automatic cloud detection algorithm, false colour composite using TM5, TM4 or TM3, or simply setting up a threshold for clouds using TM5 (Hendriks and Pellikka 2007) can help with identifying their position. If larger parts of a glacier are obscured by a cloud and manual correction is not possible, the glacier has to be excluded from a later statistical analysis (see Chapter 12).

8.4 CONCLUSIONS

The very contrasting spectral reflectance of ice and snow in the SWIR compared to the VNIR allows automated mapping of debris-free glaciers from all sensors that have bands in these parts of the spectrum. With respect to the required spatial resolution (10–30 m), data costs, area covered in a single scene and repeat cycle, only a few sensors (e.g. Landsat TM/ETM+, Terra ASTER) are appropriate for glacier mapping and monitoring. Combined with the available long time series since 1984 of Landsat, it has become the main data source for glacier mapping and change assessment. Other applications of optical sensors like snow line mapping, DEM generation from stereo sensors and surface velocity fields using feature tracking, greatly enlarge the variety of value-added products that can be derived. However, the scenes to be used should be excellent as well, i.e. acquired on a day without clouds, at the end of the ablation period, and in a year without snow outside of glaciers. This requires careful investigation of the archived data sets and for several regions the number of useful scenes might be quite small, e.g. due to frequent cloud cover and fresh snow. Nevertheless, with the free availability of all Landsat data in the USGS archive and the SRTM (Shuttle Radar Topography Mission) DEM a new horizon of glacier research has been opened.

REFERENCES

Albert, T. (2002). Evaluations of remote sensing techniques for ice-area classifications applied to the tropical Quelccaya Ice Cap, Peru. *Polar Geography* 26(3), 210–226.

Alrousan, N., P. Cheng, G. Petrie, T. Toutin and M.J.V. Zoej (1997). Automated DEM extraction and orthoimage generation from SPOT Level 1B imagery. *Photogrammetric Engineering and Remote Sensing* 63(8), 965–974.

Angelis, H.D., F. Rau and P. Skvarca (2007). Snow zonation on Hielo Patagonico Sur, Southern Patagonia, derived from Landsat 5 TM data. *Global and Planetary Change* 59, 149–158.

Aniya, M., H. Sato, R. Naruse, P. Skvarca and G. Casassa (1996). The use of satellite and airborne imagery to inventory outlet glaciers of the Southern Patagonia Icefield, South America. *Photogrammetric Engineering and Remote Sensing* 62(12), 1361–1369.

Bayr, K. J., D.K. Hall and W.M. Kovalick (1994). Observations on glaciers in the eastern Austrian Alps using satellite data. *International Journal of Remote Sensing* 15(9), 1733–1742.

Berthier, E., H. Vadon, D. Baratoux, Y. Arnaud, C. Vincent, K.L. Feigl, F. Rémy and B. Legrésy (2005). Surface motion of mountain glaciers derived from satellite optical imagery. *Remote Sensing of Environment* 95, 14–28.

Binaghi, E., P. Madella, M.P. Montesano and A. Rampini (1997). Fuzzy contextual classification of multisource remote sensing images. *IEEE Transactions on Geoscience and Remote Sensing* 35(2), 326–339.

Bishop, M.P., R. Bonk, U. Kamp and J.F. Shroder, Jr. (2001). Terrain analysis and data modeling for alpine glacier mapping. *Polar Geography* 25(3), 182–201.

Crippen, R.E. (1988). The dangers of underestimating the importance of data adjustments in band ratioing. *International Journal of Remote Sensing* 9(4), 767–776.

de Ruyter de Wildt, M.S., J. Oerlemans and H. Björnsson (2002). A method for monitoring glacier mass balance using satellite albedo measurements: application to Vatnajökull, Iceland. *Journal of Glaciology* 48(161), 267–278.

Dozier, J. (1989) Spectral signature of alpine snow cover from Landsat 5 TM. *Remote Sensing of Environment* 28, 9–22.

Eckert, S., T. Kellenberger and K. Itten (2003). Accuracy assessment of automatically derived digital elevation models from ASTER data in mountainous terrain. *International Journal of Remote Sensing* 26(9), 1943–1957.

Grenfell, T.C., D.K. Perovich and J.A. Ogren (1981). Spectral albedos of an alpine snow pack. *Cold Regions Science and Technology* 4, 121–127.

Greuell J.W., C.H. Reijmer and J. Oerlemans (2002). Narrowband to broadband albedo conversion for glacier ice and snow based on measurements from aircraft. *Remote Sensing of Environment* 82, 48–63.

Hall, D.K., J.P. Ormsby, R.A. Bindschadler and H. Siddalingaiah (1987). Characterization of snow and ice zones on glaciers using Landsat Thematic Mapper data. *Annals of Glaciology* 9, 104–108.

Hall, D.K., A.T.C. Chang and H. Siddalingaiah (1988). Reflectances of glaciers as calculated using Landsat 5 Thematic Mapper data. *Remote Sensing of Environment* 25(3), 311–321.

Hall, D.K., K.J. Bayr, W. Schöner, R.A. Bindschadler and J.Y.L. Chien (2003). Consideration of the errors inherent in mapping historical glacier positions in Austria from the ground and space (1893–2001). *Remote Sensing of Environment* 86(4), 566–577.

Hendriks, J.P.M. and P.K.E. Pellikka (2007). Semiautomatic glacier delineation from Landsat imagery over Hintereisferner glacier in the Austrian Alps. *Zeitschrift für Gletscherkunde und Glazialgeologie* 41, 55–75.

Huggel, C., A. Kääb, W. Haeberli, P. Teysseire and F. Paul (2002). Remote sensing based assessment of hazards from glacier lake outbursts: a case study in the Swiss Alps. *Canadian Geotechnical Journal* 39(2), 316–330.

Huggel, C., S. Zgraggen-Oswald, W. Haeberli, A. Kääb, A. Polkvoj, I. Galushkin and S.G. Evans (2005). The 2002 rock/ice avalanche at Kolka/Karmadon, Russian Caucasus: assessment of extraordinary avalanche formation and mobility, and application of QuickBird satellite imagery. *Natural Hazards and Earth System Sciences* 5(2), 173–187.

Jacobs, J.D., E.L Simms and A. Simms (1997). Recession of the southern part of Barnes Ice Cap, Baffin Island, Canada, between 1961 and 1993, determined from digital mapping of Landsat TM. *Journal of Glaciology* 43(143), 98–102.

JPL (2002). Jet Propulsion Laboratory: ASTER spectral library. URL: http://speclib.jpl.nasa.gov/

Kääb, A. (2005). Combination of SRTM3 and repeat ASTER data for deriving alpine glacier flow velocities in the Bhutan Himalaya. *Remote Sensing of Environment* 94(4), 463–474.

Klein A.G. and B.L. Isacks (1999). Spectral mixture analysis of Landsat Thematic Mapper images applied to the detection of the transient snowline on tropical Andean glaciers. *Global and Planetary Change* 22, 139–154.

Klok E.J., W. Greuell and J. Oerlemans (2003). Temporal and spatial variation of the surface albedo of Morteratschgletscher, Switzerland, as derived from 12 Landsat images. *Journal of Glaciology* 49(167), 491–502.

Kramer, H.J. (2002). *Observation of the Earth and Its Environment: Survey of Missions and Sensors.* 4th ed., Springer Verlag, Berlin, 1510 p.

Lambrecht, A. and M. Kuhn (2007). Glacier changes in the Austrian Alps during the last three decades, derived from the new Austrian glacier inventory. *Annals of Glaciology* 46, 177–184.

Li, Z., S. Wenxin and Z. Qunzhu (1998). Measurements of glacier variation in the Tibetan Plateau using Landsat data. *Remote Sensing of Environment* 63, 258–264.

Machguth, H., F. Paul, M. Hoelzle and W. Haeberli (2006). Distributed glacier mass balance modelling as an important component of modern multi-level glacier monitoring. *Annals of Glaciology* 43, 335–343.

Mihalcea, C., B.W. Brock, G. Diolaiuti, C. D'Agata, M. Citterio, M.P. Kirkbride, M.E.J. Cutler and C. Smiraglia (2008). Using ASTER satellite and ground-based surface temperature measurements to derive supraglacial debris cover and thickness patterns on Miage Glacier (Mont Blanc Massif, Italy). *Cold Regions Science and Technology* 52(3), 341–354.

NASA (2009). Landsat 7 – Science Data user handbook, URL: http://landsathandbook.gsfc.nasa.gov/handbook.html

Paul, F. (2002a). Changes in glacier area in Tyrol, Austria, between 1969 and 1992 derived from Landsat 5 TM and Austrian glacier inventory data. *International Journal of Remote Sensing*, 23(4), 787–799.

Paul, F. (2002b). Combined technologies allow rapid analysis of glacier changes. *EOS, Transactions, American Geophysical Union* 83(23), 253, 260, 261.

Paul, F. and A. Kääb (2005). Perspectives on the production of a glacier inventory from multispectral satellite data in the Canadian Arctic: Cumberland Peninsula, Baffin Island. *Annals of Glaciology* 42, 59–66.

Paul, F., A. Kääb, M. Maisch, T.W. Kellenberger and W. Haeberli (2002). The new remote sensing derived Swiss glacier inventory: I. Methods. *Annals of Glaciology* 34, 355–361.

Paul, F., C. Huggel, A. Kääb and T.W. Kellenberger (2003). Comparison of TM-derived glacier areas with higher resolution data sets. *EARSeL eProceedings* 2, 15–21. Proceedings of the EARSeL Workshop on Remote Sensing of Land Ice and Snow, Berne, Switzerland, March 11–13, 2002.

Paul, F., C. Huggel and A. Kääb (2004). Combining satellite multispectral image data and a digital elevation model for mapping of debris-covered glaciers. *Remote Sensing of Environment* 89(4), 510–518.

Paul, F., H. Machguth and A. Kääb (2005). On the impact of glacier albedo under conditions of extreme glacier melt: the summer of 2003 in the Alps. *EARSeL eProceedings* 4(2), 139–149. Proceedings of the EARSeL Workshop on Remote Sensing of Land Ice and Snow, Berne, Switzerland, February 21–23, 2005.

Qunzhu, Z., C. Meisheng, F. Xuezhi, L. Fengxian, C. Xianzhang and S. Wenkuna (1983). A study of spectral reflection characteristics for snow, ice and water in the north of China. *IAHS Publications* 145, 451–462.

Rabatel, A., J.P. Dedieu and C. Vincent (2005). The use of remote sensing data to determine equilibrium line altitude and mass balance time series, validation on three French glaciers for the 1994–2002 period. *Journal of Glaciology* 51(175), 539–546.

Racoviteanu, A.E., Y. Arnaud, M.W. Williams and J. Ordonñez (2008). Decadal changes in glacier parameters in the Cordillera Blanca, Peru, derived from remote sensing. *Journal of Glaciology* 54(186), 499–510.

Raup, B.H., A. Kääb, J.S. Kargel, M.P. Bishop, G. Hamilton, E. Lee, F. Paul, F. Rau, D. Soltesz, S.J.S. Khalsa, M. Beedle and C. Helm (2007). Remote sensing and GIS technology in the Global Land Ice Measurements from Space (GLIMS) project. *Computers and Geosciences* 33, 104–125.

Rees, W.G. (2006). *Remote Sensing of Snow and Ice*. CRC Press, Taylor & Francis Group, Boca Raton, 285 p.

Richter, R. (1990). A fast atmospheric correction algorithm applied to Landsat TM images. *International Journal of Remote Sensing* 11(1), 159–166.

Rott, H. (1994). Thematic studies in alpine areas by means of polarimetric SAR and optical imagery. *Advances in Space Research* 14(3), 217–226.

Sidjak, R.W. and R.D. Wheate (1999). Glacier mapping of the Illecillewaet icefield, British Columbia, Canada, using, Landsat TM and digital elevation data. *International Journal of Remote Sensing* 20(2), 273–284.

Toutin, T. (2004). Geometric processing of remote sensing images: models, algorithms and methods. *International Journal of Remote Sensing* 25(10), 1893–1924.

Williams, R.S. (Jr.) and D.K. Hall (1998). Use of remote-sensing techniques. In: Haeberli, W., M. Hoelzle and S. Suter (eds.), *Into the Second Century of Worldwide Glacier Monitoring: Prospects and Strategies*. pp. 97–111, UNESCO Publishing, Paris.

Williams, R.S. (Jr.), D.K. Hall and C.S. Benson (1991). Analysis of glacier facies using satellite techniques. *Journal of Glaciology* 37(125), 120–127.

Winther, J.G. (1993). Landsat TM derived and in situ summer reflectance of glaciers in Svalbard. *Polar Reserach* 12, 37–55.

Chapter 9

SAR imaging of glaciers

Kjell Arild Høgda, Rune Storvold & Tom Rune Lauknes
Norut Tromsø, Norway

9.1 INTRODUCTION

Radar (RAdio Detection And Ranging) is an active microwave system composed of a transmitter and a receiver. The radar transmits a pulse of electromagnetic energy, and waits for the echo from the target to return. Based upon the echo delay time between the transmitted and received pulse, it is possible to estimate the distance to the object being imaged. By using spaceborne imaging radars, one can image large areas on Earth frequently and independent of daylight and weather conditions. In order to obtain high spatial resolution, the principle of Synthetic Aperture Radar (SAR) has been developed. By the use of an advanced signal processing technique, a large synthetic aperture (synthetic antenna length) is created, allowing a fine resolution and signal phase preservation. The first Earth observing spaceborne SAR was the National Aeronautics and Space Administration (NASA) SEASAT satellite, launched in 1978. Despite its relatively short lifetime, SEASAT proved the SAR instrument's capability to retrieve geological and geophysical information about the surface of the earth and the oceans. Following SEASAT, a series of spaceborne SAR sensors have been launched. The most successful satellite programme, with respect to SAR applications, has been the European Space Agency (ESA) ERS-1 and ERS-2 satellites followed by Envisat. In Table 9.1 some SAR missions and their important parameters are listed.

SAR was originally used to produce two-dimensional images of radar reflectivity of an illuminated scene. These images were purely magnitude images and the phase information inherent in SAR data was discarded. However, for quite some years now interferometric SAR (InSAR) has evolved as a powerful method for many applications utilising the phase information of the reflected signal. Using the phase information derived from two or more co-registered complex SAR images, it is possible to retrieve complementary information compared to amplitude images, such as estimates of surface movement and the construction of digital elevation models (DEM).

Due to its all day, all weather imaging capabilities, large spatial coverage, and the ability to measure centimetre scale changes on the Earth's surface, SAR has many applications within glaciology. InSAR can be used to make DEMs and accordingly with knowledge of snow density estimate glacier mass balance, to measure glacier velocity (using both interferometric phase and feature/coherence tracking), and due to

Table 9.1 List of some satellite SAR missions, with their important sensor parameters.

Mission	ERS-1/2	Envisat ASAR	Radarsat-1	ALOS PALSAR	Radarsat-2	TerraSAR-X
Country/agency	ESA	ESA	Canada	Japan	Canada	Germany
Launch	1991/1995	2002	1995	2006	2007	2007
Nominal lifetime	2000/2009[a]	2010[b]	2009[c]	5 years	7 years	5 years
Frequency [GHz]	5.3	5.3	5.3	1.3	5.4	9.65
Polarization	VV	HH/VV + altpol	HH	HH, VV, HV, VH	HH, VV, HV, VH	HH, VV, HV, VH
Orbit altitude [km]	790	800	787	692	798	514
Revisit time [days]	35	35	24	46	24	11
Incidence angle [deg]	21–26	20–50	20–60	8–60	20–60	15–60
Swath width [km]	100	100–500	45–510	30–360	10–500	10–100
Antenna dimensions [m]	10 × 1	10 × 1.3	15 × 1.5	9 × 3	15 × 1.5	4.8 × 0.7
Pulse bandwidth [MHz]	15.5	16.0	11.6–30.0	14–28	11.6–100	150, 300

[a] ERS-1: 1991–2000, ERS-2: Launched in 1995, still operational in 2009.
[b] The orbit will be changed in October 2010 with planned lifetime to 2013 (InSAR not possible after 2010).
[c] Launched in 1995, still operational in 2009.

its ability to penetrate down below the glacier surface SAR can be used for detection of glacial facies changes.

In this chapter we will first discuss the basic theory and principles behind SAR. We then discuss the InSAR principle, using cross-track InSAR for DEM generation and motion estimation as examples. We also present the basics of C-band (5.6 GHz) snow and ice backscatter. Having established the necessary theoretical background, in Sections 9.5 to 9.7 we discuss the possibilities and limitations of SAR for glacier movement estimation, DEM generation and glacial facies imaging.

For interested readers, a more comprehensive description of SAR image formation theory can be found in e.g. Curlander and McDonough (1991), Franceschetti and Lanari (1999), Cumming and Wong (2005).

9.2   SAR IMAGE FORMATION

A raw SAR image has poor resolution in both range and azimuth, and in general, the unprocessed data appear similar to noise. In range, the raw slant range resolution is given by the duration of the transmitted pulse τ_p, and the speed of light c

$$\Delta R = \frac{c\tau_p}{2} \tag{9.1}$$

The factor 2 is due to two-way travel. The unprocessed range resolution for the ERS system is about 5.5 km. The pulse length τ_p enters Equation 9.1 directly. A shorter pulse length will result in a higher resolution in range direction. However, to obtain high signal to noise ratio (SNR) in the radar system, high peak powers are desirable. This conflicts with short pulse lengths, since the pulse length is limited by the instrument's ability to emit a certain amount of energy in a finite time span. The problem is solved using a linear Frequency Modulated (FM) chirp pulse in combination with matched filtering (Oliver and Quegan 1998). This allows us to use longer pulses to increase the pulse energy, leading to better SNR, while at the same time obtaining good slant range resolution. If the bandwidth of the chirp is β, the range resolution is

$$\Delta R = \frac{c}{2\beta} \tag{9.2}$$

For typical spaceborne SAR systems with pulse bandwidths from 10–40 MHz, this will result in a slant range resolution between 15–3.7 m.

The unprocessed azimuth resolution is determined by the antenna footprint size. For the radar to be able to separate two points on the ground, they can only be separated when they are not in the radar-beam at the same time. The angular spread of the antenna beam can be written as $\theta_H = \lambda/L_a$, where λ is the wavelength of the transmitted pulse, and L_a is the antenna length in the flight direction as shown in Figure 9.1. The unprocessed azimuth resolution of a SAR can be written as

$$\Delta x \approx R \cdot \theta_H = R \cdot \frac{\lambda}{L_a} \tag{9.3}$$

where R is the slant range distance. If we consider the ERS satellites, which have an antenna length L_a of 10 m (see Table 9.1), the unprocessed azimuth resolution will be about 5 km.

From Figure 9.1 we see that a longer radar antenna improves along track resolution. This, however, is a limiting factor for any spaceborne radar system. Considering

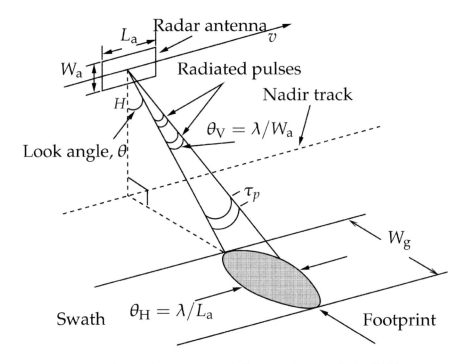

Figure 9.1 Simplified geometry of a Synthetic Aperture Radar (SAR).

the ERS-1 platform as an example, in order to have $\Delta x = 1$ km, the antenna would have to be 44 m long, which obviously is difficult to deploy in space.

Due to the motion of the radar platform, the Doppler shift creates an azimuth frequency spread when the platform passes over an object on the ground. Echoes reflected from objects in front of the moving sensor are shifted to higher frequencies than those reflected from behind. This Doppler spread is in essence a motion-induced chirp, and the same matched filter technique that is used in range can now also be employed in azimuth. This observation, which led to the development of SAR, dates from 1954 (Wiley 1954). Wiley observed that at slightly different view angles, two point targets have different frequency speeds relative to the radar platform. This means that the pulse echoes received from the two targets will be shifted according to the Doppler principle and can be resolved.

It is possible to show that the maximum azimuth resolution obtainable with a SAR is

$$\Delta x = \frac{L_a}{2} \tag{9.4}$$

independent on the distance from the radar to the target (Franceschetti and Lanari 1999).

This refinement in azimuth resolution can be considered equivalent to an increase in the size of the aperture. Since the antenna footprint, shown in Figure 9.1, is larger than the distance the radar travels between pulses, a single scatterer is illuminated by the radar in a succession of locations along the radar flight path, as shown in Figure 9.2.

With correct book-keeping of the phase history of each scatterer, a virtual aperture is synthesized by coherently combining the echoes from many pulses. Shorter antennas yield finer resolutions since a single point on the ground will be illuminated for a longer time. The limiting factor is the SNR in the radar system. What is perhaps surprising is that no other parameter, such as range, platform velocity, or wavelength, enters into the equation for theoretical azimuth resolution. This result is the fundamental aspect of SAR, and is why it has become such an important sensor, especially in spaceborne platforms.

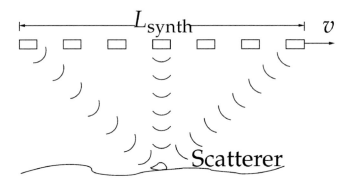

Figure 9.2 The synthetic aperture. By coherently combining echoes from different positions along the radar track, a virtual aperture much larger than the physical size of the antenna is created.

9.3 SAR INTERFEROMETRY

The SAR imaging process creates a focused two-dimensional image of the complex radar reflectivity. Often, only the amplitude signal is used. The inherent phase information in the signal is not utilized except for image formation purposes. InSAR uses the phase difference between two or more images acquired at slightly different positions in space to characterize the scene. For an in-depth discussion concerning theory and applications of SAR interferometry, we recommend Gens and Genderen (1996), Bamler and Hartl (1998), Massonnet and Feigl (1998), Rosen et al. (2000) and Hanssen (2001).

The concept of radar interferometry was first demonstrated for surface mapping during the exploration of Venus and the Moon (Rogers and Ingalls 1969, Zisk 1972a, 1972b, Shorthill et al. 1972). The first example of using InSAR for topographic mapping was shown by Graham (1974), while the first practical results of observations performed with a dual antenna side-looking airborne radar was reported by Zebker and Goldstein (1986).

There are basically three different concepts of SAR interferometry: 1) single-pass along-track interferometry, 2) single-pass across-track interferometry, and 3) repeat-pass interferometry.

The single-pass concepts illuminate the surface simultaneously in time with two spatially separated antennas. For the along-track case, the antennas are separated in the along-track direction. Systems using the along-track InSAR technique are mainly airborne, but the technique has also been demonstrated on the Space Shuttle. The along-track configuration is mainly used to study ocean surface currents and wave properties.

In the across-track configuration, the antennas are separated in the across-track direction. Such systems are operated onboard aircrafts and has also been flown on a Space Shuttle mission. The single-pass across-track mode is mainly used for digital elevation mapping. As an example, the NASA Shuttle Radar Topography Mission (SRTM) in 1999 was able to produce a global digital topography dataset of the Earth's surface covering all land areas between 60° north and 54° south latitude.

The last concept, and the one relevant for most satellite systems, is the repeat-pass interferometry configuration. This mode utilizes the fact that the repeat cycle of the satellite is not perfect, resulting in an across-track shift of orbits of typically a few hundred metres. This means that the repeat-pass data are acquired at different times compared with the single-pass modes. This provides the possibility to study phenomena related to surface changes such as e.g. ice dynamics, land deformation, and land cover changes occurring between the satellite acquisitions.

9.3.1 InSAR for DEM generation

SAR data can be used with interferometric techniques for DEM generation. Both repeat pass and across track systems have been operated (Henderson and Lewis 1998). However, the accuracy of DEMs generated by repeat-pass satellite radar interferometric techniques is lowered by the loss of coherence and glacier movement in-between passes. Accordingly, across-track techniques are best applied (Høgda et al. 2007).

Let us consider the across-track (or repeat pass processed at zero-doppler) InSAR configuration as shown in Figure 9.3. A point target P is located in the plane perpendicular to the flight direction. The distances from the two antennas to the point P are R_1 and R_2, for antennas S_1 and S_2, respectively. The two SAR antennas are separated by a baseline vector B. Each radar independently measures the time delay for the radar pulse to reach the point on the ground and return to the antenna. A radar is said to be coherent if the phase of each received echo is proportional to the time delay. The path difference between the two signals can then be determined to within a fraction of the wavelength by looking at the phase difference of the received echoes.

The complex signals received by the two antennas S_1 and S_2, can be written as

$$u_1 = |s_1| \exp(j\phi_1)$$
$$u_2 = |s_2| \exp(j\phi_2)$$

(9.5)

where $|s_1|$, $|s_2|$, ϕ_1 and ϕ_2 are the amplitudes and phases for the SAR signal, respectively. The observed phase values from the two antennas are

$$\phi_1 = -\frac{4\pi}{\lambda} R_1 + \psi_1$$
$$\phi_2 = -\frac{4\pi}{\lambda} R_2 + \psi_2$$

(9.6)

where λ is the radar wavelength, R_1 and R_2 are the geometric slant range distances, and ψ_1 and ψ_2 are the scattering mechanism contributions to the observed phases.

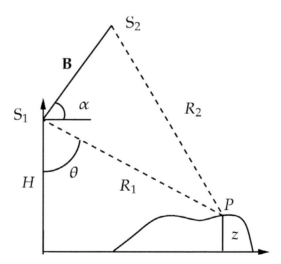

Figure 9.3 The geometry of a dual pass InSAR configuration. Flight paths are perpendicular into the plane. The InSAR configuration has two antennas, separated by a baseline, B. The phase difference of the two received echoes is proportional to the difference in path length R_2–R_1. The path difference depends on the baseline distance B, baseline angle α, look angle θ, range R_1, spacecraft altitude H, and the altitude of the point viewed z.

From Equation 9.5 we generate the complex interferogram

$$u_1 u_2^* = |s_1| |s_2| \exp[\, j \, (\phi_1 - \phi_2)] = |s_1| |s_2| \exp[\, j \delta \phi] \tag{9.7}$$

where $*$ denotes the complex conjugate. If we assume that the scattering mechanism on the ground does not change between the acquisitions, the measured phase difference is then only dependent on the path length difference

$$\delta \phi = \phi_1 - \phi_2 = -\frac{4\pi}{\lambda} (R_1 - R_2) = -\frac{4\pi}{\lambda} \Delta R \tag{9.8}$$

By exploiting the geometry in Figure 9.3, and applying the cosine rule we get

$$R_2^2 = R_1^2 + B^2 - 2R_1 B \cos(90° - \theta + \alpha) \tag{9.9}$$

which can be rewritten as

$$\sin(\theta - \alpha) = \frac{R_2^2 - R_1^2 - B^2}{2R_1 B} \tag{9.10}$$

Geometrically, by applying the far-field or parallel-ray approximation ($B \ll R$), we can write the path length difference ΔR as (Zebker and Goldstein 1986)

$$\Delta R = B \sin(\theta - \alpha) \tag{9.11}$$

The interferometric phase can now be written as

$$\delta \phi \approx \frac{-4\pi}{\lambda} B \sin(\theta - \alpha) \tag{9.12}$$

Equation 9.12 relates the interferometric phase $\delta \phi$ to the cylindrical coordinate θ of the imaged point P in Figure 9.3. This equation actually solves the full three-dimensional (3D) location of the point since all its three coordinates, slant range R_1, azimuth x, and look angle θ, are determined.

9.3.2 InSAR for surface displacement measurement

InSAR is also a powerful technique for detection of surface changes. A ground target displacement ΔR_d in the line of sight (LOS) direction of the radar occurring between two SAR passes will create an additional interferometric displacement phase term $\delta \phi_d$. Typically, for spaceborne geometries, $B < 1$ km, $R_1 \approx 800$ km, and $\Delta R_d < 1$ m. This justifies the parallel-ray approximation that ($R_1 \parallel \Delta R_d$) (Rosen et al. 2000). The displacement component of the interferometric phase in the LOS direction from the radar can then be written as

$$\delta \phi_d \approx \frac{4\pi}{\lambda} \Delta R_d \tag{9.13}$$

SAR interferometry is only able to measure the projection of the displacement vector in the radar LOS direction. In order to reconstruct the full 3D vector displacement, several observations have to be made from different viewing angles. The sensitivity of $\delta\phi$ to topography changes and surface displacement, respectively, is significantly different in magnitude. For spaceborne interferometers, the slant range distance R_1 is much larger than the baseline B. It is then clear that the phase is more sensitive to changes in displacement than to changes in topography.

Using a numerical example with a nominal ERS-1 configuration ($B = 100$ m, $\alpha = 45°$, $R_1 = 800$ km, $\lambda = 5.66$ cm), a topography change of $z = 1$ m gives an interferometric phase signature of $\delta\phi = 3.8°$, while a displacement of $\Delta R_d = 1$ m in the LOS of the radar gives a phase signature of $\delta\phi = 12700°$. In this example, the sensitivity is about 3000 times greater for measuring surface displacement than for measuring topography change. Even a modest 1 cm displacement in the LOS direction creates a phase signature of $\delta\phi = 127°$, which is easily measured by the ERS type satellites.

The time interval over which the displacement is measured must match the time scale of the geophysical quantity of interest. For example, if we want to measure ocean currents, we need a temporal baseline of the order of a fraction of a second for the backscattered phase not to decorrelate. If we want to measure long-term crustal displacements, we will need temporal baselines of several years to be able to detect any changes. For glacier velocity measurements we will need temporal baselines of hours to a few weeks.

9.3.3   Error contributions in InSAR observed surface displacements

Up to now we have only considered very idealized conditions, and we will now discuss some limitations for SAR interferometry, which is important in order to interpret the measurements quantitatively. In practice, due to the complex SAR imaging process, the interferometric phase may only be measured on the restricted interval $[-\pi, \pi]$ (wrapped phase). A retrieval operation must therefore be carried out on the measured data in order to relate it to any geophysical phenomenon. This operation is referred to as phase unwrapping, the process of restoring the correct multiple of 2π to each point of the interferometric phase image.

Phase unwrapping is one of the most challenging aspects in the successful application of SAR interferometry. The absolute unwrapped interferometric phase is directly proportional to the difference in path lengths for the SAR image pair. Many different phase unwrapping algorithms exist, and the topic of two-dimensional (2D) phase unwrapping is still one of the most challenging in InSAR. Some well-used phase unwrapping algorithms are discussed by Bamler and Hartl (1998), Franceschetti and Lanari (1999), Hanssen (2001), Chen and Zebker (2001), among others.

In the first studies of InSAR for analysis of geophysical phenomena, the work was highly concentrated on showing the applicability of the new technique, often overlooking the measurement reliability issue. During the last years, work has been done in order to understand and quantify the different error sources contained in interferometric phase measurements. We will discuss the most important error sources here and for a more in-depth discussion on error sources in SAR interferometry, please refer to Hanssen (2001).

Most of the errors in measurements by using InSAR are not due to our ability to estimate the interferometric phase correctly. Rather, they are connected with our ability to separate and analyze the different phase contributions coming from surface topography, deformation, atmospheric delay, and refractivity changes in the scattering object. Sensor parameters such as the radar frequency, spatial resolution and incidence angle also have a strong impact on the feasibility of InSAR for surface deformation mapping.

9.3.4 Phase noise estimation

An interferometric radar relies on the coherence of the two signals; incoherent signals can not be used for interferometry. The phase accuracy in SAR interferograms is mainly affected by phase noise and decorrelation. One quality measure of the interferometric phase is the complex correlation coefficient, or complex coherence. It is defined as (Bamler and Hartl 1998)

$$\gamma = \frac{E[u_1 u_2^*]}{\sqrt{E[|u_1|^2]E[|u_2|^2]}} \qquad (9.14)$$

where $E[\cdot]$ is the expectation operator, u_1 and u_2 are the two complex SAR image values. In order to estimate the quality of the interferometric phase in a pixel, the correlation value is computed on a small sample of pixels around every location. Assuming ergodicity, the ensemble averages in Equation 9.14 can then be replaced with spatial averages around the pixel of interest. If we assume Gaussian image statistics, the maximum likelihood estimate of the coherence $|\gamma|$ is (Bamler and Hartl 1998)

$$|\gamma|_{ML} = \frac{\left| \sum_{n=1}^{L} u_1(n) u_2^*(n) \right|}{\sqrt{\sum_{n=1}^{L} |u_1(n)|^2 \sum_{n=1}^{L} |u_2(n)|^2}} \qquad (9.15)$$

summing over L independent samples. Note that often only the magnitude value of the complex coherence is used, referred to as coherence. From Equation 9.15 and by using the Cauchy-Schwarz inequality, we see that the values of $|\gamma|$ are between 0 and 1. A correlation value of 1 corresponds to perfect phase coherence between the two measurements. Coherence values less than unity correspond to reduced phase coherence, resulting in noisy phase measurements.

The estimate in Equation 9.15 is always biased towards higher values (Joughin and Winebrenner 1994, Touzi and Lopes 1996). Therefore, in practice, the theoretical lower limit of 0 is often not reached. The trade-offs between window size and estimation accuracy is a common problem in coherence estimation. The standard deviation of the phase noise, σ_ϕ, is a function of the coherence, and can, for a large number of sample points L, be reached asymptotically as (Rodriguez and Martin 1992)

$$\sigma_\phi = \frac{1}{2L} \frac{\sqrt{1 - |\gamma|^2}}{|\gamma|} \qquad (9.16)$$

9.3.5   Decorrelation sources

The different sources of decorrelation in the SAR interferogram contribute multiplicatively (Zebker and Villasenor 1992, Hanssen 2001). The most important decorrelation sources can be written as

$$\gamma = \gamma^{\text{spatial}} \gamma^{\text{temporal}} \gamma^{\text{thermal}} \gamma^{\text{DopplerCentroid}} \gamma^{\text{volumetric}} \tag{9.17}$$

Spatial baseline decorrelation occurs when the interferometric baseline is not exactly zero. Since the radar receives the coherent sum of all independent scatterers within the resolution cell, these contributions are added slightly differently due to the different viewing geometry. The baseline decorrelation is related to the different look angles of the two SAR acquisitions, and leads to a critical baseline, above which the interferometric phase is pure noise. For ERS, assuming flat terrain, the critical baseline is about 1150 m (Bamler and Hartl 1998).

Temporal decorrelation is the most problematic to characterize theoretically. It is due to geometrical or electrical changes in the properties of the surface, as a function of time between the acquisitions. These changes may be caused by randomly moving scatterers. Terrain containing a variable amount of liquid water, such as e.g. areas covered with wet snow, will also have different scattering properties from one observation to the next (Guneriussen et al. 2001). Since temporal decorrelation is due to changes of the surface mainly at the scale of the radar wavelength, temporal decorrelation is highly dependent on the operating frequency of the radar (Zebker and Villasenor 1992). InSAR data acquired with longer wavelengths, for example L-band (23 cm wavelength), exhibits a longer temporal decorrelation time than C-band (5.7 cm wavelength) (Strozzi et al. 2003). Accordingly, the launch of the Japanese ALOS satellite with an L-band SAR in January 2006 is beneficial for glacier velocity studies.

Thermal decorrelation is due to system noise, and can be related to the SNR of the radar system as (Zebker and Villasenor 1992)

$$\gamma^{\text{thermal}} = \frac{1}{1 + SNR^{-1}} \tag{9.18}$$

which for the ERS sensors is close to unity (Hanssen 2001).

Doppler centroid decorrelation is caused by the differences in doppler centroids for the two acquisitions. It is the azimuth equivalent of the spatial decorrelation component. Volumetric decorrelation is caused by penetration of the radar wave through the scattering medium. It depends highly on the radar wavelength and the dielectric properties of the scattering medium. The decorrelation terms for the InSAR processing have not been included, since it normally introduces only small amounts of decorrelation (Hanssen 2001).

9.4   SAR BACKSCATTER FROM SNOW AND ICE

One of the main advantages of SAR for glacier monitoring is its ability to penetrate into the snow and ice, giving us information about the internal structure of the glacier. Penetration depths up to tens of metres are reported from dry snow conditions over Greenland (Rignot et al. 2001).

In Section 9.7 we describe how SAR can be used to monitor impacts of climate change on glaciers. This technique uses the different scattering properties of the upper layer of the glacier. To properly identify the different glacier facies we need to decouple the different scattering mechanisms so we can retrieve the snow and ice properties of this upper layer. To perform the decoupling we need to be able to model backscatter from the different glacier facies. In Figure 9.4 we show a schematic of the different facies and pictures of snow and ice from cores taken at the Kongsvegen glacier on Svalbard.

The backscattering cross-section, σ_{0pq}, measured by the SAR instrument is defined as

$$\sigma_{0pq}(\theta_{0i}) = \cos\theta_{0i}\frac{I_{rp}}{I_{tq}} \tag{9.19}$$

Here p and q represents transmitted (t) and received (r) polarizations (horizontal, vertical or cross), θ_{0i} is the radar incidence angle relative to the surface normal and I_{tq} and I_{rp} transmitted and received intensity by the antenna. The received intensity comes from scattering by the snow grains and air bubbles in the snow and ice (volume

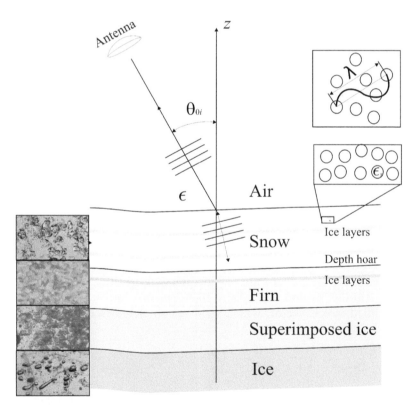

Figure 9.4 Model geometry for active remote sensing of glacier facies properties. The snow layer consists of densely packed discrete spherical snow grains with dielectric constant ϵ_s, and radii a. Pictures from cores at Kongsvegen glacier at Svalbard.

scattering), and from scattering and reflections due to rough snow and ice interfaces (surface scattering). For further reading, see Chapter 5. The relative scattering contributions from the volume and the surfaces depend on the frequency, polarization, and incidence angle of the radar and the surface, snow and ice properties. The surface is described by its dielectric contrast and roughness, usually parameterized by the root mean square (RMS) height variation and correlation length. The volume is described by the dielectric constant of the medium and of the inclusions (snow grains or air bubbles), the bulk density, size and shape of the inclusions. For firn (Chapter 2) one would also need to include the effect of "sticky" particles as the snow grains here have partially melted and refrozen. For more details, see e.g. Ulaby et al. (1986), Ulaby and Elachi (1990), Wen et al. (1990), Tsang et al. (2001).

9.4.1 Backscatter modelling

Conventional radiative transfer (CRT) theory is based on a stochastic approach built on probabilities and summation of intensity contributions, which implies that contributions from the different scatterers must be incoherent. This assumption is called the independent scatterer approximation. If the distances between the scatterers are of the order of a wavelength or less or if the scatterers positions are correlated, like in a grating, this approximation is not valid, and one has to use dense medium radiative transfer theory (DMRT). The radiative transfer equation for DMRT takes the same form as in the case of CRT, and for active remote sensing the radiative transfer equation can be formulated as

$$\cos\theta \frac{\partial \underline{I}(z;\theta,\phi)}{\partial z} = -\kappa_e \underline{I}(z;\theta,\phi) + \int_0^{2\pi} \int_0^{\pi} \sin\theta' \underline{\underline{P}}(z;\theta,\phi;\theta',\phi')\underline{I}(z;\theta',\phi')d\theta'\,d\phi'$$

(9.20)

Equation 9.20 gives the change in the intensity in a given direction of interest described by polar coordinates (θ, φ) with vertical distance (z). It equals the loss of the intensity in the incoming field (first term on the right-hand side) by the rate given by the extinction coefficient κ_e, plus a gain in intensity due to energy scattered into this direction of interest (second term on the right-hand side). The extinction is the sum of loss due to scattering and absorption. The double integral sums the contributions from all directions scattered into the direction of interest. $\underline{\underline{P}}$ is a 4 × 4 phase matrix describing the probability that intensity incoming in the direction (θ', φ') is scattered into the direction (θ, φ) and depends on the extinction and scattering cross sections. $\underline{I}$ is the modified Stokes vector, which yields a complete description of the radiation field. It can be written as

$$\underline{I} = \begin{bmatrix} I_V & I_H & U & V \end{bmatrix}^T$$

(9.21)

where I_V and I_H are the intensities of the vertical and horizontal polarizations, and U and V are the degree of linear and circular polarizations, respectively. The extinction rate, albedo, and phase matrix are computed based on the physical parameters of the dense medium and EM theory. For details see Wen et al. (1990).

9.4.2 First order solution

For the backscatter problem, as long as the wavelength is much larger than the scatterers, one can use a first order (single scattering) approximation with good results. This reduces the DMRT equation from the integro-differential form above to an ordinary first order differential equation. The solution for multiple layers can be written on the form

$$\sigma_0^{pq}(\theta_0) = \sigma_{ss1}^{pq}(\theta_0) + T_{ss1}^2(\theta_0)[\sigma_{sv1}^{pq}(\theta_1) + T_{sv1}^2(\theta_1)[\sigma_{ss2}^{pq}(\theta_1) + T_{ss2}^2(\theta_1)[\cdots]]] \quad (9.22)$$

where the snow-covered surface cross section, σ_0^{pq}, is a superposition of the contributions from the layers and surfaces in the glacier. σ_{ssn}^{pq} is the contributions from scattering at the rough surface n between layers $n-1$ and n, which has different optical properties. σ_{svn}^{pq} is the backscattering from the snow volume in layer n. $T_{ssn}^{pq}(\theta_n)$ is the transmissivity through interfaces, $T_{svn}^{pq}(\theta_n)$ is the one-way propagation loss due to extinction by the layer, θ_0 is the incidence angle at the top surface, and θ_n is the refracted angle in layer n provided by Snell's law.

The surface transmittance and backscatter cross-section for the rough surfaces is a combination of coherent and incoherent scattering where the coherent contribution can be expressed using the Fresnel reflection and transmission coefficients. The incoherent component depends on the roughness of the surfaces usually described by a RMS height variation and a correlation length. The method used to calculate these contributions will depend on the scale of the roughness compared to the wavelength. The surface scattering can be expressed by

$$\sigma_{ssn}^{pq} = cR_{n-1,n}^{pq}(\theta_{0i}; \theta_{0s})\, \delta\theta_i, \theta_s + (1-f)\mathcal{R}_{n-1,n}^{pq}(\theta_n, \phi_n; \pi - \theta_n, \phi_n) \quad (9.23)$$

where $R_{n-1,n}^{pq}$ is the Fresnel reflection coefficient and $\mathcal{R}_{n-1,n}^{pq}$ the incoherent surface reflection coefficient for the surface between layer $(n-1)$ and (n). In the backscattering case the coherent part only contributes if the incident beam is normal to the parts of surface large enough to allow for specular reflections. f denotes the fraction of coherent to incoherent scattering by the surface.

The volume scattering and extinction (layer transmittivity) is computed using the phase matrix, which can be calculated from Rayleigh or Mie theory in case of spherical scattering particles as

$$\sigma_{svn}^{pq} = P_n^{pq}(\theta_n, \phi_n; \pi - \theta_n, \phi_n)\left(1 - \exp\left(\frac{2\kappa_e \Delta z}{\cos\theta_n}\right)\right) \quad (9.24)$$

To decouple the two scattering mechanisms to retrieve surface and bulk properties of the glacier, one can use the result that the angular dependence is stronger for surface scattering than for volume scattering. In addition we have the possibility of using up to three measurements of different polarization combinations; vertical-vertical (*VV*), *horizontal-horizontal* (*HH*), vertical-horizontal (*VH*) or horiozontal-vertical (*HV*). The first letter indicates transmit and the last letter indicates receive polarization state.

For a more detailed description of microwave radar backscatter, we refer to e.g. Ulaby et al. (1986), Fung (1994), Tsang et al. (2001).

9.5   SAR GLACIER FLOW VELOCITY MEASUREMENTS

9.5.1   InSAR velocity

Three SAR based methods are commonly used for retrieval of glacier surface velocity parameters. These are InSAR, feature tracking and speckle/coherence tracking. The potential of SAR interferometry to map surface displacements at centimetre resolution and its application to glacier surface velocity estimations have been demonstrated by many, see e.g. Goldstein et al. (1993), Joughin et al. (1996a, 2001), Rignot et al. (1997, 2001, 2004), Rignot and Kanagaratnam 2006, and Mohr et al. (1998).

In Section 9.3 the sensitivity of the interferometric phase to surface topography was explained. However, the interferometric phase is as noted sensitive to both surface topography and coherent displacement of scatterers along the radar look direction in the time between the acquisitions of the interferometric image pairs. However, the two contributions cannot be separated from a single interferometric image pair alone.

There are in principle three possible ways of retrieving glacier velocity information from SAR interferograms:

1   If we have zero baseline it is observed from Equation 9.11 that the sensitivity to topography is zero. The problem is that current and planned InSAR systems very seldom satisfy this condition.
2   The topographic contribution to the interferogram can be simulated and removed if an independent and accurate DEM exists. This is however seldom the case over glaciers with time varying elevation and flow velocity.
3   If the underlying ice movement remains constant and we have three coherent SAR images with equal temporal baselines, it is possible to use the technique of 3-pass differential interferometry. Two different interferograms are produced, one with mixed topography/motion phase and one with topography only phase. Subtracting the topography-only phase from the mixed phase then gives the motion contribution.

The successful use of differential SAR interferometry is limited by the phase noise, usually characterized by the coherence. For glacier surfaces, the coherence is affected by both meteorological conditions and glacier flow velocity, and generally decreases rapidly with increasing time between the acquisitions of the two SAR images. Meteorological causes of decorrelation include ice and snow melting, snowfall and wind redistribution of snow and ice. The ERS-1/2 1-day tandem mission has been shown to be very good for glacier applications, but it was only available for 9 months in 1995–1996 and is planned for Radarsat-2/3.

The differential use of two interferograms with similar displacements allows the removal of the topographic phase from the interferogram to derive a displacement map from which the line-of-sight velocity vector can be measured (Joughin et al. 1996b, 1996c, Wegmüller et al. 1998). Following Kwok and Fahnestock (1996), the phase from two different interferograms computed from three SAR acquisitions (passes 1, 2 and 3) is

$$\Delta\phi_{12} = \frac{4\pi}{\lambda} B_{12} \sin(\theta - \alpha_{12}) + \frac{4\pi}{\lambda} \Delta R_d \qquad (9.25)$$

and

$$\Delta\phi_{23} = \frac{4\pi}{\lambda} B_{23} \sin(\theta - \alpha_{23}) + \frac{4\pi}{\lambda}\Delta R_d \tag{9.26}$$

where subscripts refer to passes 1 through 3. Assuming that the line of sight displacement, ΔR_d, is the same in between the passes, subtracting Equation 9.26 from Equation 9.25 gives

$$\Delta\phi'_{13} = \frac{4\pi}{\lambda} B_{12} \sin(\theta - \alpha_{12}) - \frac{4\pi}{\lambda} B_{23} \sin(\theta - \alpha_{23}) \tag{9.27}$$

We observe that the contribution from the glacier movement is removed in Equation 9.27. We then resolve the two different baselines in Equation 9.27 into vertical and horizontal components, getting

$$\Delta\phi'_{13} = \frac{4\pi}{\lambda}((B_{12}^y - B_{23}^y)\sin(\theta) - (B_{12}^z - B_{23}^z)\cos(\theta)) \tag{9.28}$$

$$= \frac{4\pi}{\lambda}(B_{13}^{'y}\sin(\theta) - B_{13}^{'z}\cos(\theta)) \tag{9.29}$$

$$= \frac{4\pi}{\lambda} B_{13}^{'y} \sin(\theta - \alpha'_{13}) \tag{9.30}$$

which is the same as the topography-only interferogram with baseline between passes 1 and 3.

If we then compute the mixed interferogram between the first and third passes, $\Delta\phi_{13}$, and subtract the topography only interferogram $\Delta\phi'_{13}$ from Equation 9.30, after phase unwrapping we get

$$\Delta\phi_{13} - k\Delta\phi'_{13} = \frac{4\pi}{\lambda}\Delta R_d \tag{9.31}$$

where k is a scaling factor scaling the B'_{13} fringe pattern (phase shift pattern) into the B_{13} interferogram (Kwok and Fahnestock 1996). The scaling factor is necessary since the sensitivity to topographic variations in the two interferometric combinations will be different if the baselines are not the same. It is to be noted, that the phase differences, typically displayed as fringes with a 2π periodicity, must be unwrapped to generate a continuous field of motion. Also to be noted, is that the velocity measured is a relative component in the SAR line of sight. Fixed points (rocks etc.) or the known velocity of one point at the surface is needed to control the velocity field. By combining two different imaging geometries, such as ascending and descending orbits, together with a DEM, it is possible to estimate the full three-dimensional velocity field (Joughin et al. 1998, Mohr et al. 1998).

9.5.2 Feature tracking velocity

As discussed, shortcomings of InSAR are temporal decorrelation and the ability to measure the line of sight velocity component only. Feature tracking determines the ice-surface velocity field by detecting and tracking features such as crevasses that are visible in images (Fahnestock et al. 1993, Luckman et al. 2006). Doing this, it is necessary to use high resolution and co-registered sequential pairs of images. Optical feature tracking has been most widely used, but the SAR penetrating below the glacier surface is likely to locate more tracking features such as internal inhomogeneities and crevasses (Massom and Lubin 2006).

Feature tracking, or intensity cross correlation between SAR images obtained at different times, has been used to derive two dimensional ice-surface velocity fields from tandem (1-day), 3-day and 35-day repeat pass ERS-1/2 SAR data. Luckman et al. (2003) have shown, covering a 500×500 km^2 area of East Greenland, that 35-day tracking yields only a slightly lower density of velocity measurements than 1-day tracking. Results were in agreement with the spatial pattern of ice velocity except at the glacier termini, where tidal effects may dominate. Luckman et al. (2003) concluded that SAR feature tracking can be used to routinely monitor ice-discharge velocities on a regional basis and thereby inform studies of regional mass balance. In Figure 9.5 an example of glacier velocity measurements by SAR feature tracking is illustrated. The result is from the fast flowing Kronebreen in Svalbard using ERS-1/ERS-2 3-day Tandem mission data obtained on March 8 and 11, 1994.

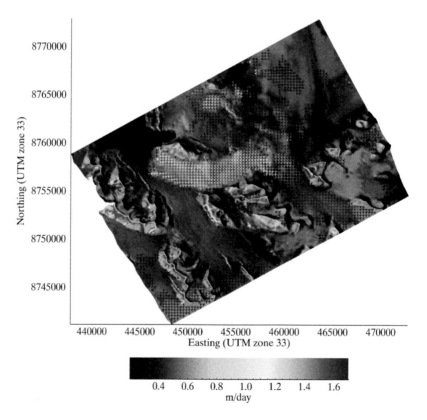

Figure 9.5 Feature tracking glacier velocity from Kronebreen on Svalbard. Data from ERS-1/2 3-day tandem mission March 8, 1994 and March 11, 1994. Image processed by Yngvar Larsen, Norut Tromsø. (See colour plate section).

9.5.3 Speckle/coherence tracking velocity

When coherence is retained between the images considered, the speckle pattern of the two images is correlated, and intensity tracking with small image patches can be performed with high accuracy (Strozzi et al. 2002). After initial processing, ice flow speed and direction are determined to sub-pixel accuracy by cross correlation of small sub-areas across the SAR images. Compared with feature tracking, which relies on visible features such as crevasses, the accuracy achieved in image co-registering can be an order of magnitude better, down to 1/10 of a pixel (Gray et al. 1998).

Coherence tracking is comparable with speckle tracking (Gray et al. 1998, 2001). The difference is that it uses the phase information in the complex SAR image rather than the speckle information in the real-valued amplitude image. Small data patches are selected, a series of small interferograms with changing offset is constructed, and the coherence is estimated. The location of the coherence maximum is determined at sub-pixel accuracy and, in theory, the obtainable accuracy of the velocity is very high.

9.5.4 SAR glacier velocity summary

Strozzi et al. (2002) concluded that InSAR is the most accurate method to measure ice movement in the slant-range direction. The limiting factors are the feasibility of phase unwrapping and the loss of coherence between the available passes. Single areas of high coherence can often not be used because of the lack of reference points. Coherence tracking can be used in areas with high phase coherence. A major advantage is that it measures absolute motion in two horizontal directions. Its disadvantage compared to InSAR is that its accuracy is less; it is also more computational demanding. Intensity tracking is a source of information for the study of fast-flowing glaciers and may be used in the analysis of large surging glaciers. It can also be used when we have long acquisition time intervals between the two SAR images. In some cases, intensity/feature tracking may be the only technique that can be applied for SAR data acquired with more than a 24-day time interval, which is the case for all the current and in near-future planned satellite missions. A combination of the different methods is probably the most efficient mean of measuring glacier motion.

9.6 SAR GLACIER DEM

In Section 9.5 we saw how repeat-pass InSAR can be used to simultaneously determine ice sheet topography and flow velocity. However, the accuracy of DEMs generated by repeat-pass satellite radar interferometric techniques is decreased by the loss of coherence and varying glacier movement in-between acquisitions. Achievable accuracy is shown to be between 5 to 40 m in elevation with a pixel size of 25–50 m, elevation accuracy decreasing on rough and inclined surfaces and at higher glacier velocity (Joughin et al. 1996b, Forsberg et al. 2000, Rignot et al. 2001). This will in many cases not be sufficient for mass balance studies (Chapters 2, 3).

Across-track InSAR can be used for accurate DEM generation over glaciers (Dall et al. 2001, Høgda et al. 2007). The main problem is that there is currently no available operational satellite across-track InSAR system. However, there are several airborne

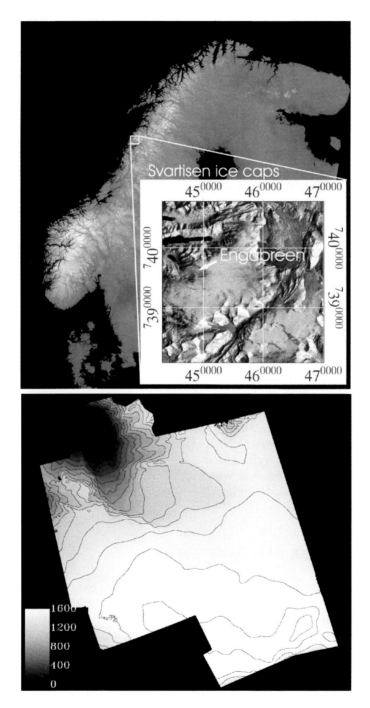

Figure 9.6 DEM from airborne X-band SAR over Engabreen, outlet of Svartisen ice caps, northern Norway. Greyscale in right panel indicate meters above sea level.

systems that can and have been used for glacier DEM purposes, for example the German DLR E-SAR and the Danish EMISAR.

A DEM of the Geikie ice cap in East Greenland was generated from interferometric C-band InSAR data acquired with the EMISAR system (Dall et al. 2001). They used GPS surveyed radar reflectors and an airborne laser altimeter as a part of the experiment. The accuracy of the SAR DEM was shown to be about 1.5 m. The mean difference between the surface elevation recorded using laser and the SAR differed from 0 m in the soaked zone to a maximum of 13 m in the percolation zone. The difference was explained by the fact that the snow in the soaked zone contained liquid water, which attenuated the radar signals, while the transparency of the firn in the percolation zone made volume scattering dominant at the higher elevations.

In Figure 9.6 we see an example of an airborne X-band across-track SAR DEM from Engabreen, an outlet from the Svartisen ice caps (13°59' E; 66°40' N) in northern Norway, which was generated as a part of OMEGA project (Pellikka 2007). For this DEM the E-SAR onboard a DLR Dornier DO 228 aircraft was used for the InSAR acquisition (Schreiber et al. 1999). The sensor operates in 4 frequency bands, X-, C-, L- and P-band, hence it covers a range of wavelengths from 3 to 70 cm. The polarization of the radar signal is selectable, horizontal as well as vertical. In polarimetric mode the polarization is switched from pulse to pulse. The measurement modes include single channel operation, i.e. one wavelength and polarization at a time, and the modes of SAR interferometry and SAR polarimetry. The X-band across track interferometric mode was used for this DEM.

In the Engabreen study the snow was wet within the whole study area during the SAR over-flights, ensuring low penetration into the snow pack from the X-band radar, but lowering the SNR. Four 70 cm dihedral corner reflectors were also deployed as ground control points to calibrate the InSAR elevation measurements. The accuracy of the InSAR DEM compared with laser scanning DEM and simultaneous obtained in-situ GPS profiling was 0.3 m $\pm$ 3.6 m (Høgda et al. 2007).

In general, we would expect InSAR DEMs to be affected by the SAR penetration depth on glaciers, as penetration depth as previously discussed can be of the order of tens of meters in dry snow conditions. Data supporting this was obtained by Langley et al. (2007) studying C-band backscatter on the Kongsvegean glacier on Svalbard, using a ground penetrating radar (GPR). They reported that the depth of observed scattering sources was greatest in the superimposed ice zone, where layers were clearly seen to a depth of approximately 14 m. The high scattering properties of the firn layer precluded layers deeper than approximately 6 m from being imaged at C-band in the firn area.

9.7 SAR GLACIER FACIES DETECTION

As a result of climate differences with elevation and distance from the ocean, arctic glaciers have a series of distinctive altitudinal snow zones and snow facies (Chapters 2, 5). At high altitude is the dry snow zone where little surface melt occurs. The percolation zone is characterised by episodic melting in the summer. The melt-water percolates downwards through the firn where it refreezes and forms ice lenses and layers within the firn. The superimposed ice zone is a zone of such high melt and refreezing that ice lenses have merged to a continuous mass. The wet snow zone is where melting saturates all the depth of annual accumulation. The ablation or ice zone is at the lowest elevations, with a large part of the yearly accumulation melting. Ablation and accumulation zones are separated by the equilibrium line, where the net mass balance is zero. The firn line delineates the upper limit of the ablation zone (Chapter 3).

As outlined in Section 9.4, the strength of the SAR backscatter signal reflected by glacier ice and snow depends on wavelength, polarization and incidence angle, and is affected by the dielectric and geometric properties of the glacier. The geometric properties include layer surface roughness characteristics, morphology, grain size and shape, for further information, see Chapter 5. Also important is the presence of internal reflectors such as ice lenses and crevasses. The interpretation of SAR data over glaciers is complicated by the penetration into the glacier snow and ice. The penetration depth at C-band (5.6 GHz) is shown to be of the order of 10 metres in dry snow conditions (Bingham and Drinkwater 2000, Dall et al. 2001). At L-band (1.27 GHz) as the PALSAR (The phased array type L-band synthetic aperture radar) on the recently launched Japanese ALOS satellite, the SAR penetration depth will be substantially greater (Rignot et al. 2001).

Accordingly, satellite SAR data can be used to distinguish areas of contrasting backscatter on glaciers and relate these to the glacier facies. Rau et al. (2000) define the following radar glacier zones that can be classified by backscatter characteristics and their elevation positions relative to each other: a dry snow radar zone, a frozen percolation radar zone, a wet snow radar zone and a bare ice radar zone. However, these zones do not necessarily coincide with classical glaciological zones. In general, dry snow zones (temperature never above 0°C) are characterized by a moderately layered snowpack, having small snow crystals and no significant internal ice layers. High penetration depths and dominating volume scattering results in low radar backscatter (-14 to -20 dB) (Rau et al. 2000). In the percolation zone, refreezing of melt-water results in ice lenses and pipes, giving very high radar backscatter (up to 0 dB while in the frozen state) (Rau et al. 2000, Langley et al. 2008). Detection of melt events is easily performed with microwave radar, as the inclusion of liquid water within the snow pack will result in high absorption and a penetration depth of few centimetres, resulting in lowered radar backscatter (Malnes et al. 2006). Accordingly, depending on surface roughness, scattering from a melting snow cover (wet snow radar zone) will be low (-15 to -25 dB) (Rau et al. 2000). At the lowest elevation, the ablation zone with its bare ice can again give higher backscatter due to enhanced surface scatter compared to the wet snow radar zone (-6 to -14 dB) (Rau et al. 2000, Massom and Lubin 2006).

König et al. (2002) used SAR to detect superimposed ice on Kongsvegen glacier (78°48′N, 13°E) in Svalbard. Superimposed ice contains small air bubbles that act as scattering centres at high frequencies as in C-band. König et al. (2001, 2004), and Engeset et al. (2002) all show that the firn-line can be detected using space-born C-band radar. König et al. (2001) used airborne multi-polarization SAR images to study equilibrium and firn line detection on Austre Okstindbreen, Norway. They concluded that the snow line could not be seen on the SAR images (König et al. 2001, 2002).

However, König et al. (2004), when classifying the SAR backscatter in Kongsvegen glacier with clustering methodology, found a relation between the areas of different classified zones and the last years mass balance. It is however not clear what caused this correlation. In Figure 9.7 an Envisat ASAR image of the Kongsvegen glacier is shown. The image is an average of 156 scenes acquired in spring 2004 and 2005. The image is composed of both ascending and descending orbits, VV, VH, HH, and HV polarizations, and incidence angles from ca 17° to ca 44°. We can clearly identify different facies (ice, superimposed ice and firn) on the glacier. As these zones are sensitive to the weather and climate, long-term trends can be viewed as indicators of climate change and variability. Accordingly, analysing time-series of SAR images, it is possible to measure trends within glacier facies, e.g. caused by a warming climate. Important is also that is possible to use relative low resolution Scan-SAR images, covering large areas.

Recently, fully polarimetric satellite SAR systems as Radarsat-2 (C-band), PAL-SAR and TerraSAR-X (X-band) have been launched. These systems will measure the amplitude and phase in HH, VV, HV and VH polarizations and will also support Pol-InSAR. These new instruments will increase the detection and monitoring capabilities of glacier facies as the polarimetric face will contain information on the structural characteristics of the scatterers and will enable us to better discriminate between volume and surface scattering. It is believed that cross-polarization (VH or HV) data will give information based largely on volume scattering, while co-polarised (HH or VV) signals will largely be contributions from surface scattering (Massom and Lubin 2006).

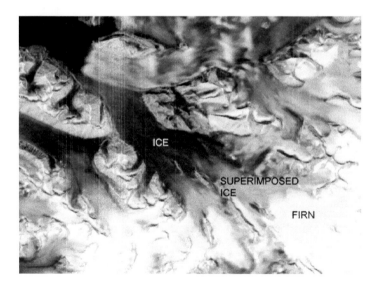

Figure 9.7 Envisat ASAR image of Kongsvegen glacier on Svalbard. Average of 156 scenes (ascending, descending, S1–S7, VV, VH, HH, and HV) acquired in spring 2004 and 2005 (mostly dry snow conditions). We can clearly identify the different glacier facies (ice in centre and towards upper right corner of image, superimposed ice in centre and towards lower left corner and firn in lower right corner) within the image. The blurred parts of the image are caused by glacier movement.

Pol-InSAR will also be an emerging new field with the increased availability of fully polarimetric SAR systems. It will give us information on the physical properties of the ice-sheet, and could potentially more accurately separate volume and surface scattering contributions (Dall et al. 2004). The basic idea is that the interferometric properties of polarimetric data can give information about where the scattering signature is originating from. However, it must be stressed that the utilisation of Pol-InSAR data from glaciers is still at an early stage of research.

9.8 SUMMARY

Data from SAR instruments can be obtained both day and night in all weather conditions and provides both surface and sub-surface information of glaciers, making SAR well suited for long time monitoring purposes. This is especially important for monitoring effects of climate change on remote glaciers difficult to access and investigate by traditional glaciological methods. However, quantitatively relating changes on glaciers observed by SAR to climate change is not trivial due to the large natural variability in the polar regions. This problem emphasizes the need for long observational time series that enable us to detect and understand the natural variability and infer trends.

One core variable to monitor on glaciers would be the yearly mass balance. However, this is a variable that is difficult to measure directly by SAR and at present accurate computation of mass balance requires substantial in-situ field measurement efforts as presented in Chapter 2. DEMs can be obtained from InSAR data. However, the main problem is the need for short temporal baselines (tandem-missions) in order to have sufficient coherence between the data acquisitions. This problem is today best-solved applying airborne SAR systems with across-track interferometric capabilities.

Another way of monitoring climate change impacts on glaciers is to measure long-term trends in parameters, such as e.g. firn-line, wet snow areas and melting season intensity. Studying backscatter properties of SAR images will give information on the different facies on the glacier, and time series will give information about trends in them. Glacial backscattering properties has also been related to mass balance (König et al. 2004), though the absolute value of the mass balance estimate can not be derived by SAR data alone. Wet snow is readily separated form dry snow in SAR backscatter and the long-term evolution of the wet snow season duration and extent, will give information about effects of climatic changes and natural climate variability as well.

SAR interferometry allows monitoring of ice movement. The main problem is again the need for short temporal baselines in order to have sufficient coherence. However, in lack of tandem missions, feature tracking using SAR imagery can be used to track glacier velocities. This technique can also be used on archived data back to 1991 (ERS-1) to generate historical data series to compare with present measurements and eventually deduce changes.

To conclude, SAR is an instrument giving unique and complementary information compared to optical sensors, especially because of the all day all weather capabilities, penetration depth and techniques as InSAR not possible with other types of sensors. With the steady increase in the number and capabilities of satellite SAR systems, SAR will obviously become more and more important for the monitoring of remote glaciers in the years to come.

REFERENCES

Bamler, R. and P. Hartl (1998). Synthetic aperture radar interferometry. *Inverse Problems* 14, R1–R54.

Bingham, A.W. and M.R. Drinkwater (2000). Recent changes in the microwave scattering properties of the Antarctic Ice Sheet. *IEEE Transactions on Geoscience and Remote Sensing* 38(4), 1810–1820.

Chen, C.W. and H.A. Zebker (2001). Two-dimensional phase unwrapping with use of stastistical models for cost functions in nonlinear optimization. *Journal of the Optical Society of America* 18(2), 338–351.

Cumming, I.G. and F.H. Wong (2005). *Digital Processing of Synthetic Aperture Radar Data: Algorithms and Implementation* Artech House, Norwood, 630 p.

Curlander, J.C. and R.N. McDonough (1991). *Synthetic Aperture Radar: Systems and Signal Processing.* John Wiley & Sons, New York, 672 p.

Dall, J., S. Nørvang Madsen, K. Keller and R. Forsberg (2001). Topography and penetration of the Greenland ice sheet measured with airborne SAR interferometry. *Geophysical Research Letters* 28(9), 1703–1706.

Dall, J., K.P. Papathanassiou and H. Skriver (2004). Polarimetric SAR interferometry applied to land ice: Modelling. *Proceedings of the EUSAR 2004, 5th European Conference on Synthetic Aperture Radar*, Ulm, Germany, May 25–27, 2004. VDE Verlag, pp. 247–250.

Engeset, R.V., J. Kohler, K. Melvold and B. Lundén (2002). Change detection and monitoring of glacier mass balance and facies using ERS SAR winter images over Svalbard. *International Journal of Remote Sensing* 23(10), 2023–2050.

Fahnestock, M.A., R.A. Bindschadler, R.A. Kwok and K. Jezek (1993). Greenland ice sheet surface properties and ice dynamics from ERS-1 SAR imagery. *Science* 262, 1530–1534.

Forsberg, R., K. Keller, C.S. Nielsen, N. Gundestrup, C.C. Tscherning, S. Nørvang Madsen and J. Dall (2000). Elevation change measurements of the Greenland Ice Sheet. *Earth, Planets and Space* 52, 1049–1053.

Franceschetti, G. and R. Lanari (1999). *Synthetic Aperture Radar Processing.* CRC Press, Boca Raton, 307 p.

Fung, A.K. (1994). *Microwave scattering and emission models and their applications.* Artech House, Boston, 573 p.

Gens, R. and J.L.V. Genderen (1996). SAR interferometry-issues, techniques, applications. *International Journal of Remote Sensing* 17(10), 1803–1835.

Goldstein, R.M., H. Engelhardt, B. Kamb and R.M. Frolich (1993). Satellite radar interferometry for monitoring ice sheet motion: Application to an Antarctic ice stream. *Science* 262, 1525–1530.

Graham, L.C. (1974). Synthetic interferometer radar for topographic mapping. *Proceedings of the IEEE* 62(6), 763–768.

Gray, A.L., K.E. Mattar and P.W. Vachon (1998). InSAR results from the RADARSAT Antarctic mapping mission: Estimation of glacier motion using a simple registration procedure. *Proceedings of IGARSS'98*, Seattle, Washington, U.S.A., July 6–10, 1998, pp. 1638–1640.

Gray, A.L., N.H. Short, K.E. Mattar and K.C. Jezek (2001). Velocities and ice flux of the Filchner Ice Shelf and its tributaries determined from speckle tracking interferometry. *Canadian Journal of Remote Sensing* 27(3), 193–206.

Guneriussen, T., K.A. Høgda, H. Johnsen and I. Lauknes (2001). InSAR for estimation of changes in snow water equivalent of dry snow. *IEEE Transactions on Geoscience and Remote Sensing* 39(10), 2101–2108.

Hanssen, R.F. (2001). *Radar Interferometry – Data Interpretation and Error Analysis.* Kluwer Academic Publishers, Dordrecht, 308 p.

Henderson, F.M. and A.J. Lewis (eds.) (1998). *Principles and applications of imaging radar.* Manual of Remote Sensing, American Society of Photogrammetry and Remote Sensing. 3rd ed., Vol. 2. Wiley and Sons, Somerset, 800 p.

Høgda, K.A., T. Geist, M. Jackson, H. Elvehøy, J. Stötter and I. Lauknes (2007). Comparison of digital elevation models from airborne SAR technology and airborne laser scanner technology at Engabreen, Svartisen, Norway. *Zeitschrift fur Gletscherkunde und Glazialgeologie 41,* 205–226.

Joughin, I. and D.P. Winebrenner (1994). Effective number of looks for a multilook interferometric phase distribution. *Proceedings of IGARSS'94,* Pasadena, CA, U.S.A., August 8–12, 1994, pp. 2276–2278.

Joughin, I., R. Kwok and M. Fahnestock (1996a). Estimation of ice sheet motion using satellite radar interferometry: Method and error analysis with application to the Humboldt glacier, Greenland. *Journal of Glaciology 42(142),* 564–575.

Joughin, I., D. Winebrenner, M. Fahnestock, R. Kwok and W. Krabill (1996b). Measurement of ice-sheet topography using satellite radar interferometry. *Journal of Glaciology 42(149),* 10–22.

Joughin, I., S. Tulaczyk, M. Fahnestock and R. Kwok (1996c). A mini-surge on the Ryder Glacier, Greenland observed via satellite radar interferometry. *Science 274,* 228–230.

Joughin, I., R. Kwok and M. Fahnestock (1998). Interferometric estimation of the three dimensional ice-flow velocity vector using ascending and descending passes. *IEEE Transactions on Geoscience and Remote Sensing 36(1),* 25–37.

Joughin, I., M. Fahnestock, D. MacAyeal, J.L. Bamber and P. Gogineni (2001). Observation and analysis of ice flow in the largest Greenland ice stream. *Journal of Geophysical Research 106(D24),* 34021–34034.

König, M., J.G. Winther, N.T. Knudsen and T. Guneriussen (2001). Equilibrium and firn line detection on Austre Okstindbreen, Norway, with airborne multi polarisation SAR. *Journal of Glaciology 47(157),* 251–257.

König, M., J. Wadham, J.G. Winther, J. Kohler and A.-M. Nuttal (2002). Detection of superimposed ice on the glaciers Kongsvegen and Midre Lovenbreen, Svalbard, using SAR satellite imagery. *Annals of Glaciology 34,* 335–342.

König, M., J-G. Winther, J. Kohler and F. König (2004). Two methods for firn area and mass balance monitoring of Svalbard glaciers with Synthetic Aperture Radar (SAR) satellite images. *Journal of Glaciology 50(168),* 116–128.

Kwok, R. and M.A. Fahnestock (1996). Ice sheet motion and topography from radar interferometry. *IEEE Transactions on Geoscience and Remote Sensing 34(1),* 189–200.

Langley, K., S.-E. Hamran, K.A. Høgda, R. Storvold, O. Brandt, J.O. Hagen and J. Kohler (2007). Use of C-band ground penetrating radar to determine backscatter sources within glaciers. *IEEE Transactions on Geoscience and Remote Sensing 45(5),* 1237–1245.

Langley, K., S.-E. Hamran, K.A. Høgda, R. Storvold, O. Brandt, J. Kohler, and J.O. Hagen (2008). From glacier facies to SAR backscatter zones via GPR. *IEEE Transactions on Geoscience and Remote Sensing 46(9),* 2506–2516.

Luckman, A., T. Murray, H. Jiskoot, H. Pritchard and T. Strozzi (2003). ERS SAR feature-tracking measurement of outlet glacier velocities on a regional scale in East Greenland. *Annals of Glaciology 36,* 129–134.

Luckman, A., T. Murray, R. de Lange and E. Hanna (2006). Rapid and synchronous ice-dynamic changes in East Greenland. *Geophysical Research Letters 33,* L03503.

Malnes, E., R. Storvold, I. Lauknes and S. Pettinato (2006). Multi-polarisation measurements of snow signatures with air- and satelliteborne SAR. *EARSeL eProceedings 5(1),* 111–119.

Massom, R. and D. Lubin (2006). *Polar Remote Sensing, Volume II: Ice Sheets.* Springer Praxis Books, Chichester, 426 p.

Massonnet, D. and K.L. Feigl (1998). Radar interferometry and its application to changes in the earth's surface. *Reviews of Geophysics 36(4),* 441–500.

Mohr, J.J., N. Reeh and S.N. Madsen (1998). Three-dimensional glacial flow and surface elevation measured with radar interferometry. *Nature* 391, 273–276.

Oliver, C. and S. Quegan (1998). *Understanding Synthetic Aperture Radar Images*. Artech House, Norwood, 479 p.

Pellikka, P. (2007). Monitoring glacier changes within the OMEGA project. *Zeitschrift für Gletscherkunde und Glazialgeologie* 41, 3–5.

Rau, F., M. Braun, M. Friedrich, F. Weber and H. Goßmann (2000). Radar glacier zones and their boundaries as indicators of glacier mass balance and climatic variability. *Proceedings of the EARSeL-SIG-Workshop Land Ice and Snow*, Dresden, Germany, June 16–17, 2000, pp. 317–327.

Rignot, E.J., S.P. Gogineni, W.B. Krabill and S. Ekholm (1997). North and Northeast Greenland ice discharge from satellite radar interferometry. *Science* 276(5314), 934–937.

Rignot, E., K. Echelmeyer and W. Krabill (2001). Penetration depth of interferometric synthetic-aperture radar signals in snow and ice. *Geophysical Research Letters* 28(18), 3501–3504.

Rignot, E., G. Casassa, P. Gogineni, W. Krabill, A. Rivera and R. Thomas (2004). Accelerated ice discharge from the Antarctic Peninsula following the collapse of Larsen B ice shelf. *Geophysical Research Letters* 31, L18401.

Rignot, E. and P. Kanagaratnam (2006). Changes in the velocity structure of the Greenland ice sheet. *Science* 311(5763), 986–990.

Rodriguez E. and J.M. Martin (1992). Theory and design of interferometric synthetic aperture radars. *IEEE Proceedings, Part F: Radar and Signal Processing* 139(2), 147–159.

Rogers, A.E.E. and R.P. Ingalls (1969). Venus: mapping the surface reflectivity by radar interferometry. *Science* 165, 797–799.

Rosen, P.A., S. Hensley, I.R. Joughin, F.K. Li, S.N. Madsen, E. Rodriguez and R.M. Goldstein (2000). Synthetic aperture radar interferometry. *Proceedings of the IEEE* 88(3), 333–382.

Scheiber, R., A. Reigber, A. Ulbricht, K.P Papathanassiou, R. Horn, St. Buckreuß and A. Moreira (1999). Overview of interferometric data acquisition and processing modes of the experimental airborne SAR system of DLR. *Proceedings of IGARSS'99*, Hamburg, Germany, 28 June–2 July, 1999, pp. 35–37.

Shorthill, R.W., T.W. Thompson and S.H. Zisk (1972). Infrared and radar maps of the lunar equatorial region. *Earth, Moon and Planets* 4(3–4), 442–446.

Strozzi, T., A. Luckman, T. Murray, U. Wegmüller and C. L. Werner (2002). Glacier motion estimation using SAR offset-tracking procedures. *IEEE Transactions on Geoscience and Remote Sensing* 40(11), 2384–2391.

Strozzi, T., U. Wegmüller, C.L. Werner, A. Wiesmann and V. Spreckels (2003). JERS SAR interferometry for land subsidence monitoring. *IEEE Transactions on Geoscience and Remote Sensing* 41(7), 1702–1708.

Touzi, R. and A. Lopes (1996). Statistics of the Stokes parameters and of the complex coherence parameters in one-look and multilook speckle fields. *IEEE Transactions on Geoscience and Remote Sensing* 34(2), 519–539.

Tsang, L., J.A. Kong, K.-H. Ding and C.O. Ao (2001). *Scattering of Electromagnetic Waves; Numerical Simulations*. Wiley Series in Remote Sensing, New York, 736 p.

Ulaby, F.T., Moore R.K. and Fung A.K. (1986). *Microwave Remote Sensing: Active and Passive, Vol. III – Volume Scattering and Emission Theory, Advanced Systems and Applications*. Artech House, Dedham, 1100 p.

Ulaby, F.T. and C. Elachi (eds.) (1990). *Radar Polarimetry for Geoscience Applications*. Artech House, Norwood, 376 p.

Wegmüller, U., T. Strozzi and C. Werner (1998). Characterization of differential interferometry approaches. Proceedings of the EUSAR'98, 2nd European Conference on Synthetic Aperture Radar, Friedrichshafen, Germany, 25–27 May 1998, pp. 237–240.

Wen, B., L. Tsang, D.P. Winebrenner and A. Ishimura (1990). Dense medium radiative transfer theory: Comparison with experiment and application to microwave remote sensing and polarimetry. *IEEE Transactions on Geoscience and Remote Sensing* 28, 46–59.

Wiley, C.A. (1954). *Pulsed Doppler Radar Methods and Apparatus*. Technical report, United States Patent, No. 3196436.

Zebker, H.A and R.M. Goldstein (1986). Topographic mapping from interferometric synthetic aperture radar observations. *Journal of Geophysical Research* 91(B5), 4993–4999.

Zebker, H.A. and J. Villasenor (1992). Decorrelation in interferometry radar echoes. *IEEE Transactions on Geoscience and Remote Sensing* 30(5), 950–959.

Zisk, S.H. (1972a). A new earth-based radar technique for the measurement of lunar topography. *Moon* 4, 296–306.

Zisk, S.H. (1972b). Lunar topography: First radar-interferometer measurements of the Alphonsus-Ptolemaeus-Arzachel region. *Science* 178, 977–980.

Chapter 10

Airborne laser scanning in glacier studies

Thomas Geist
FFG – Austrian Research Promotion Agency, Vienna, Austria
Institute of Geography, University of Innsbruck, Austria

Johann Stötter
Institute of Geography, University of Innsbruck, Austria

10.1 MEASUREMENT PRINCIPLES AND RESULTING DATA SETS

Laser scanning is an active remote sensing technology operated from airborne and terrestrial platforms for directly measuring three-dimensional coordinates of points on surfaces, including the terrain and objects thereupon (e.g. trees or houses). In literature often the term airborne *LiDAR* (= Light Detection And Ranging) is used synonymously. Airborne laser scanning is becoming the standard method for acquiring area-wide topographic data replacing image-based photogrammetry as it has numerous advantages. First of all, airborne laser scanning operates with high degree of automation, from data acquisition to digital terrain model generation, which has a positive effect on data processing costs. The measurements can be made with high density (1 point/m^2 or higher as a standard value) and with a vertical accuracy (ca. ± 10 cm), which allows a very detailed terrain representation. Airborne laser scanner signal has also the ability of penetrating through vegetation cover and thus recording the ground surface also in wooded areas, which obviously is not relevant to glacier studies. As laser scanner system is an active system, the direct measurement allows data acquisition during night or over areas with very limited texture (e.g. snow surfaces). The ability to acquire information from snow surfaces is probably the most important advantage in glacier studies. The original idea of using airborne laser scanning technology in OMEGA was to apply and evaluate airborne laser scanning as a stand-alone method for different tasks in glaciological research. A further intention was to provide a benchmark topographic dataset against which datasets derived from other remote sensing technologies could be tested. In particular, additional interest was laid in investigating further possible applications of airborne laser scanner data which goes beyond the simple provision of high resolution digital elevation models.

An airborne laser scanning system is a multi-sensor measurement system that incorporates the following time-synchronized components presented in Figure 10.1. The laser scanner itself consists of the laser range finder measuring the distance from the sensor on the airborne platform to an observed, reflecting surface, and a device, that deflects the laser beam perpendicular to the flight direction ($+/- 20°$ as common value). The laser range finder operates by measuring the two-way travel time required

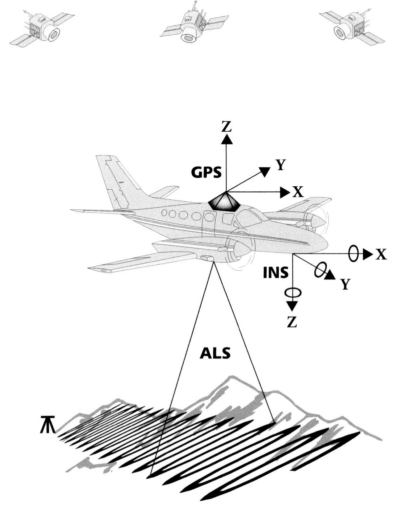

Figure 10.1 Schematic illustration of the airborne laser scanning system components and the data acquisition geometry.

for a pulse of laser light (typically in the near infrared part of the electromagnetic spectrum) to travel to the location of reflection, and back to the receiver. The system also has a global positioning system (GPS) receiver for determining the absolute position (x, y, z) of the sensor platform in a differential mode using ground reference stations and an inertial measurement unit (IMU) for determining the orientation (angular attitude) of the platform, which are described as roll, pitch and heading. The flight path of the platform is calculated from combined analysis of the GPS and IMU data. From the ground, data from another GPS is used as well as a flat calibration target, which is measured on the ground and also by the airborne laser scanner.

The measurements of the position and orientation of the platform by GPS and IMU are performed with a certain frequency, which are typically 1 Hz for GPS and 1 kHz for IMU. Current systems are capable of operating up to 5000 m above ground level. Range measurements provide pulse repetition rates up to 150 kHz. Comprehensive overviews on the basic technical aspects of airborne laser scanning are given by e.g. Wehr and Lohr (1999) and Baltsavias (1999).

The range is measured via a pulsed signal with certain length, which is emitted by the laser source. The laser beam has a specified divergence thus causing a foot-print with a specific diameter on the scanned surface (e.g. a beam divergence of 0.25 mrad causes a 0.25 m footprint diameter, when the system is 1000 m above ground level). The signal interacts with the surface in a combination of absorption and reflectance. According to Rees (2001) three forms of interaction are possible. In *absorption* the energy is absorbed by the target and no energy is reflected at the object, in *specular (mirror-like) reflection* of the non-absorbed part, where it is impossible to measure the range as the angle of the reflected ray is equal to the angle of incidence thus, and in *diffuse reflection* (Lambertian scattering) of the non-absorbed part, where the energy is diffusely reflected into all directions and thus can be used for measuring the distance.

The backscattered energy (the echo) is preferably detected only at the wavelength of the emitted signal. This makes the detection relative insensitive to background radiation. The laser scanner system records polar co-ordinates, i.e. directions and distances. The data pre-processing comprises the determination of the absolute position and orientation of the laser scanner during the flight by analysis of the time-synchronised differential GPS and INS data, calculation of the relative coordinates, system calibration and finally calculation of the coordinates in WGS84 format (WGS = World Geodetic System). For the transformation of laser scanner points from WGS84 to the local coordinate system (usually with an orthometric height system) detailed information on geoid undulation is required.

So far, standard laser scanning systems are able to detect two reflections per shot which are referred to as *first echo* and *last echo* and result in e.g. different surface representations in multilayered areas (e.g. forest canopy surface and forest ground surface). Recently, multi-pulse and so-called full waveform laser scanning systems have gained more importance. As they record the strength of the returning echoes as a function of time, this gives information on the vertical distribution of reflecting objects within the footprint (Wagner et al. 2006).

The result of a laser scanning mission is a set of points in a defined coordinate system (so called point cloud). The point density depends on the measurement frequency, the flying speed and height, the swath width, the distance between the flight lines, and the resulting degree of swath overlapping. In addition to the coordinate acquisition, also the intensity of the backscattered pulse is recorded thus providing a complementary data source to be used for visual interpretation or for more refined data analysis, e.g. surface classification.

One of the most important applications of airborne laser scanning is the very detailed (high resolution in planimetry) and accurate (small errors in vertical and horizontal direction) digital representation of a surface in a DEM (Digital Elevation Model). Therefore the original points are transformed into a regular raster where elevation is stored as a value for each raster cell. An increasing number of mathematical algorithms are available for filtering the point cloud and interpolating to raster models.

Due to very high point density the vertical distance between the resulting raster DEM calculated with two different algorithms is probably smaller than the vertical accuracy of the single point measurement, and thus the choice of the point to raster method is of less importance than it was when the surface information was related to a limited number of points. While precision is inevitably lost when processing the points in the raster domain, the advantage is that tools for digital image processing can be readily applied.

The most common types of DEMs generated from the point cloud are DSM (Digital Surface Model) and DTM (Digital Terrain Model). A DSM consists of the uppermost surfaces including objects as trees or houses. A DTM represents the terrain surface with objects removed. On glaciers and on every other open terrain, the DSM is equal to the DTM. For computing the DTM the ground points and off-terrain points have to be separated by different filtering methods (see Sithole and Vosselman 2004).

The agreement between the z-values of the measured point and that of the *true* terrain surface provide a quantifiable measure for the accuracy respective the quality. Apart from random errors, gross and systematic errors in the z-values of data points have to be considered. Random errors depend on the measurement system. The points measured on the terrain surface have a more or less random distribution with respect to the true terrain surface. In general, this random error distribution is characterized by the standard deviation of the measurement system and should be considered in the digital model generation process. Gross errors can occur in all datasets. In the case of airborne laser scanning these errors can be split in *real* measurement errors (e.g. caused by multi-path effects) and errors caused by reflecting surfaces above the terrain (e.g. on vegetation or houses). The system accuracy depends primarily upon the contributing error budgets from the subsystems (see Figure 10.1). See Baltsavias (1999) for a discussion and examples of how each system parameter contributes to overall laser scanning system accuracy. Favey (2001) split up the overall accuracy of a single measurement into the contributing parts of the different sub-systems, and Kennett and Eiken (1997) documented that the accuracy of the measurement is affected principally by uncertainties in laser range (about 0.07 m vertically) and GPS position (about 0.1 m vertically). Errors in scanner and platform orientation angles will also have an effect, although this has a minor impact on the result for relatively flat glacier surfaces. The accuracy can be improved when systematic errors, i.e in the overlapping zones of neighbouring strips, are removed. Approaches for the consideration of systematic errors in laser scanner data can be found in Filin (2001).

10.2   PREVIOUS USE OF AIRBORNE LASER SCANNING IN GLACIOLOGICAL STUDIES

One of the fundamental requirements for the creation of accurate models of glacier mass balance using the geodetic method (Chapter 2) is the provision of accurate topographic data in the form of digital representations of the glacier surface. Normally, such data has been derived from either maps, often with a photogrammetric origin and with limited spatial resolution, or, more recently, from satellite – based remote sensing data. The last two decades have seen increasing interest in the use of systems for monitoring the surface form, elevation and dynamics of both ice sheets and mountain glaciers, partly driven by major improvements in the capabilities of such systems.

Since the 1990s airborne laser scanning has been used to assess the ice thickness change of polar ice caps. There are several studies in Greenland revealing elevation changes of the Greenland ice sheet and identifying areas of significant surface elevation change (Garvin and Williams 1993, Krabill et al. 1995, 1999, 2000, Thomas et al. 1995, 2003, Csatho et al. 1996, Christensen et al. 2000, Abdalati et al. 2001, 2002), the results of which contribute to e.g. global sea-level rise studies. The survey net covers the entire Greenland ice sheet with a maximum distance of ca. 20 km from each point on the ice sheet surface to the nearest flight line. One specific result of the surveys was the fact that outlet glaciers show far more spatial and temporal variability in elevation change as expected and consequently require individual and more detailed attention in order to determine its balance. The role of airborne laser elevation surveys in the future is likely to focus more specifically on the study of these outlet glaciers. Also in Antarctica multi-temporal airborne laser scannning was used to derive surface elevation changes, mainly on parts of the West Antarctic ice sheet (Spikes et al. 1999, 2003a, 2003b).

As on contrary to the profound experience in Greenland, there have been only few applications on smaller ice caps or mountain glaciers. The first documented attempt of using airborne laser altimetry for glacier monitoring was carried out in Alaska, where the elevation data acquired by a laser profiler were compared with photogrammetrically derived maps in terms of elevation change along the profile lines and consequently in the estimation of glacier volume change (Echelmeyer et al. 1996, Aðalgeirdottir et al. 1998, Sapiano et al. 1998). Further investigations with laser scanning systems were conducted at the Unteraar Glacier in Switzerland (Favey et al. 1999, Baltsavias et al. 2001) and the Hardangerjökulen in Norway (Kennett and Eiken 1997). More recent research activities include glacier change studies in Svalbard with flight campaigns in summer 2003 and summer 2005 (e.g. Arnold et al. 2006), with specific investigations at the Austfonna ice cap where repeated airborne laser profiles indicate a clear thickening of the upper central part of the ice cap and a peripheral thinning (Bamber et al. 2004).

Given that one of the key advantages of laser scanning data is the option to provide high resolution DEMs of glaciated areas and, through multi-temporal surveys, to calculate changes of the topography, the accuracy of these data needs to be assessed. It is particularly important that the degree of measurement accuracy is both high and spatially consistent. Furthermore, high levels of accuracy must be repeatable in subsequent surveys to enable the quantitative comparison of results necessary for change analysis. Most studies on the accuracy of airborne laser scanning are restricted to analyzing the vertical accuracy of the measured points or a derived DEM. The accuracy of a DEM can be much higher than the accuracy of one single point measurement due to the (partly) elimination of random errors. It has been shown that accuracy depends on the object surface measured. The elevation of asphalt surfaces, for example, can be measured more precisely than the height of grass land, which is less well defined, too. Likewise, accuracy gets worse with increasing terrain slope and decreasing point density (see Kraus 2004).

Some of the studies mentioned above include investigations on the accuracy of the collected data on the glacier surface (Table 10.1). All authors tend to highlight the vertical accuracy of point measurements which range from ca. 0.1 m to 0.3 m in glaciated environments.

For some decades aerial photography has been the most common remote sensing resource for glacier studies. Comparative studies (Baltsavias et al. 2001)

Table 10.1 Reported vertical accuracy of airborne laser scanner measurements on glacier surfaces.

Author	Region	Vertical accuracy
Krabill et al. (2002)	Greenland	0.085 m
Echelmeyer et al. (1996)	Alaska	0.3 m
Spikes et al. (2003a)	West Antarctica	0.09–0.22 m
Arnold et al. (2006)	Svalbard	0.05–0.15 m

show that if airborne laser scanning reaches the expected accuracy of 0.1–0.3 m photogrammetry can compete only by increasing the image scale to about 1:4000. Other studies compare laser scanning with airborne SAR (Bindschadler et al. 1999) and satellite radar altimetry (Bamber et al. 1998). In this book, Chapter 13 gives evidence of a comparison of DEMs from different data sources derived in the OMEGA project.

10.3 THE AIRBORNE LASER SCANNER DATA SETS IN THE OMEGA PROJECT

The primary goal of using airborne laser scanning in the OMEGA (Pellikka 2007) project was to investigate and evaluate the possibilities and limitations of laser scanning as an independent remote sensing method in research on mountain glaciers, and to provide calibration data for other remote sensing techniques (Geist et al. 2003). An overall number of 14 airborne laser scanning campaigns were carried out within the glaciological years 2001/2002 and 2002/2003, out of which ten were carried out in Austria covering Hintereisferner, Kesselwandferner and neighbouring areas (40 km^2 in total), and four in Norway covering Engabreen and its forefield (62 km^2 in total). In addition, other single data sets were acquired for Vernagtferner in Austria and Svartisheibreen in Norway (Geist et al. 2005, Geist and Stötter 2007).

Laser scanner data acquisition campaigns were carried out with Optech ALTM laser scanning systems. In Austria, the data of two permanent GPS receiving stations were used for the differential correction of the position of the aircraft. In Norway, data of one permanent receiving station was used, accompanied by data collection on geodetic points closer to the study area. At both sites football fields were surveyed as calibration areas for the laser scanning system. More details on the technical aspects of data acquisition in Austria and Norway can be found in Geist and Stötter (2007) and Geist et al. (2005).

The result of the airborne laser scanning surveys were point clouds with x, y and z coordinates representing the scanned surface in UTM/WGS84. By using standard routines for interpolation, the high point density allowed for the generation of high resolution raster DEMs having cell size of one metre. These DEM's were the input for all further investigations. From these DEMs digital standard products such as shaded reliefs (Figure 10.2) or contour lines were derived for visualisation.

The quality of airborne laser scanning data has been assessed by applying several procedures. The data point density and data point distribution was evaluated visually and the sensor position accuracy was estimated from GPS reference data. The z-values of the DEMs were compared with independent GPS measurements on

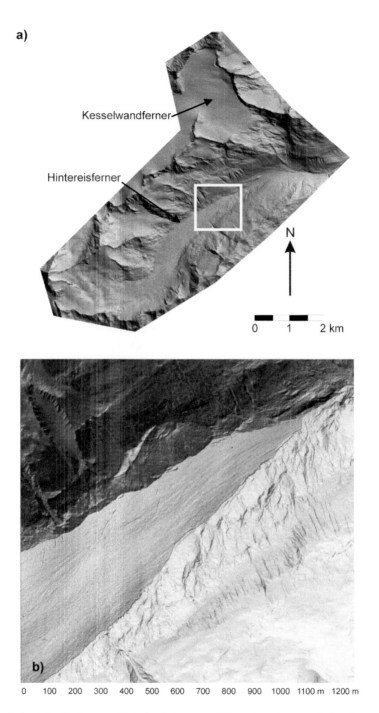

Figure 10.2 a) shaded relief image of Hintereisferner and Kesselwandferner derived from laser scanning data from September 9, 2008; b) detail from the snout of Hintereisferner revealing meltwater channels and crevasses.

the glacier surface. For the assessment of the vertical accuracy, the z-values of point measurements within the football fields (control area) were compared, and for the assessment of the horizontal accuracy, images derived from laser scanning DEMs with shapes of clearly defined objects, e.g. crevasses and mountain huts were compared. The results show that 99,99% of all points lie within a vertical distance of 30 cm to the calibration surface while the horizontal accuracy can be characterized as in the sub-meter range (Geist et al. 2005, Geist and Stötter 2007).

The spatial resolution of the surface information represented in the DEMs reveals abundant details (Figure 10.2). Linear features such as meltwater channels and crevasses on the glacier surface are clearly visible. The relatively smooth surface of the glacier itself can easily be distinguished from the rough unglaciated terrain.

10.4 APPLICATION OF AIRBORNE LASER SCANNING DATA IN GLACIER STUDIES

10.4.1 Surface elevation change

Glacier surface elevation change integrates effects due to both mass accumulation/mass ablation and glacier dynamics. An analysis comprising the magnitude of vertical surface changes and the spatial patterns of these changes give a better picture on short-term changes in glacier mass balance and climate than measuring glacier length variations. In any case, surface elevation measurements between two determined dates can only provide of positive or negative surface elevation change. Geist and Stötter (2007) present the surface elevation changes of the Hintereisferner and Kesselwandferner glaciers in Austria for different time scales, such as whole observation period, glaciological years, ablation and accumulation seasons, and discuss the observations. Figure 10.3 shows a map of surface elevation changes at Hintereisferner for the glaciological year 2001/2002.

The link between surface elevation change, volume change and mass balance change is one of the challenges in ongoing research in utilising lser scanning for glaciological applications (see Abdalati et al. 2002). As a first step in this direction, Bauder (2001) describes the implementation and validation of an indirect method for the determination of glacier mass balance. The relation, which is expressed by the kinematic boundary condition at the glacier surface, is used to connect the mass balance rate at a specified point with the horizontal surface velocity, the surface elevation change with time and the surface slope. The necessary information is generated with remote sensing data and numeric flow modelling in high resolution without in-situ measurements. With multi-temporal airborne laser scanning, the surface elevation change with time and the horizontal flow velocity can be determined, the 3D numeric flow model calculates the missing vertical velocity component along the surface. The crucial point in this approach is the profound knowledge on the spatial distribution of the vertical velocity component. Geist et al. (2005) discuss the derivation of the vertical velocity component from repeated laser scanner data for Engabreen glacier outlet of West Svartisen in winter 2001/2002 and compare it with independent reference data. Due to the comparatively inert adaptation of glacier dynamics on changing mass balance conditions, mass-balance changes for single years can be assessed, at least in their relative magnitude. This is supported by Arnold et al. (2006) who state that the

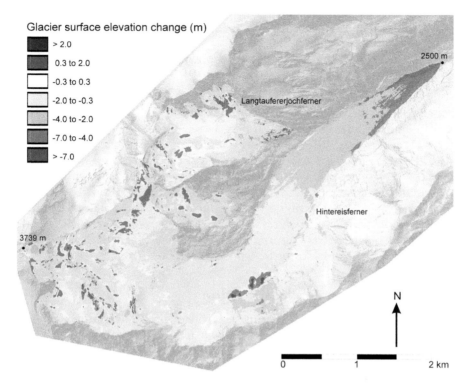

Glacier surface elevation change (m)

- > 2.0
- 0.3 to 2.0
- -0.3 to 0.3
- -2.0 to -0.3
- -4.0 to -2.0
- -7.0 to -4.0
- > -7.0

Figure 10.3 Surface elevation change of Hintereisferner for the glaciological year 2001/2002, derived from airborne laser scanner data of October 10, 2001 and September 18, 2002. Some areas increased in elevation due to avalanches like in the lower right, for example, but in general glacier was thinning. The overall elevation change of Hintereisferner was −1.3 m. (See colour plate section).

resolution of airborne laser scanner data is suitable for mass-balance estimation using repeated data on an annual or semi-annual basis.

10.4.2 Derivation of glacier velocities

Within small-scale studies, airborne laser scanning can contribute to the derivation of values of glacier surface velocity from multi-temporal datasets. Abdalati and Krabill (1999) illustrated for parts of the Greenland ice sheet how airborne laser scanner data could be used to track features and estimate surface velocities. Glacier velocity can be determined with sequential digital elevation models by observing the displacement of surface features (e.g. crevasses) in time (Bucher et al. 2006). In comparison to the aerial photographs, laser scanning has the advantage not only to have the same illumination in each data set, but by changing the illumination they offer additional possibilities for identifying traceable features. The ability to observe quite small-scale objects increases the possibility that particular features may survive from year to year, and their motion down-glacier provides information to estimate surface flow rates. Bucher et al. (2006) used shaded reliefs of DEMs with a spatial resolution of 1 m, as input to the open-source software IMCORR. The movement of the glacier was determined by cross-correlating selected features in the shaded relief. The method works well in the lower snow-free parts of the glacier.

10.4.3   Surface roughness values as input for energy balance modelling

Topographic shading, slope and aspect, and correction of the surface albedo for high solar zenith angles, are understood to play a crucial role for determining the spatial patterns of surface energy balance. The surface roughness of a glacier (or more general, small-scale topographic variations) is a major controlling factor of the turbulent heat fluxes at the ice-atmosphere interface, and hence in controlling glacier melt rates, and ultimately the mass balance. Chasmer and Hopkinson (2001) showed how airborne laser scanning data can contribute to determination of those parameters. Van der Veen et al. (1998) have used high resolution airborne laser scanner data to determine the small-scale surface relief in Central Greenland and estimate the contribution from spatial noise to stratigraphic records.

10.4.4   Glacier surface classification

Accumulation and ablation zones can be divided into glacier zones (Chapter 3), these being distinct zones with characteristic features in the surface layers of a glacier. Optical sensors have been widely used for classifying ice and snow surfaces, thus there is comprehensive knowledge on reflectance properties of snow and ice. As most laser scanning systems are operating in the near infrared part of the electromagnetic spectrum (e.g. 1064 nm, as the Optech ALTM systems), individual reflectance behaviour of different surface types (e.g. snow, firn, ice) can be expected. From evaluation of the point density distribution, it is obvious, that no or only very limited reflectance is occurring on water bodies or wet surfaces and consequently no coordinate measurement. Figure 10.4 shows the backscattered laser intensity measurements acquired with an airborne laser scanner from about 1000 m above ground. The distinct differences between ice, snow, and moraine which can be observed in the intensity data seems to indicate that multitemporal imaging may be used to track snowline retreat during melting seasons. For instance, in the middle of the Figure 10.4b, which represents the firn area of Hintereisferner, the snow and firn patches seem to have other reflectance properties than ice or rock surface. In the Figure 10.4a, the areas discovered under the retreating Hintereisferner and Kesselwandferner since the Little Ice Age seem to have different refelectance properties than the vegetated areas surrounding them.

Lutz et al. (2003) discuss some factors influencing the signal intensity (e.g. range, scanning angle) and present a glacier surface classification for Svartisheibreen based on statistical analysis. Hopkinson and Demuth (2006) state that intensity is a good indicator for glacial surfaces. Höfle et al. (2007) introduce a comprehensive method for glacier surface segmentation and classification using spatial and intensity information of the unstructured laser scanning point cloud. The intensity is used to compute a value (the *corrected intensity*) proportional to the surface property reflectance by applying the laser range equation. Ice, firn, snow and surface irregularities (mainly crevasses) show a good differentiation in terms of geometry and reflectance.

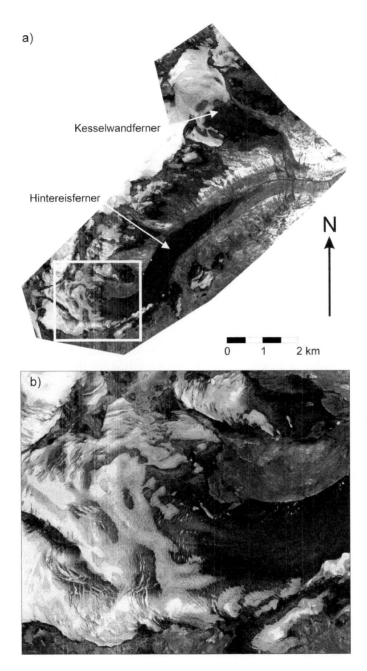

a)

Kesselwandferner

Hintereisferner

N

0 1 2 km

b)

Figure 10.4 a) airborne laser scanner intensity image of Hintereisferner and Kesselwandferner of August 12, 2003; b) detail image of Hintereisferner at the equilibrium line altitude around 3000 m a.s.l. A clear distinction between snow, firn, ice and moraine/bedrock can be made. The impact of the acquisition geometry (e.g. the height above ground) on the intensity is visible.

10.4.5 Automatic glacier delineation and crevasse detection

With view to the increasing availability of laser scanner data in high mountain areas, the development of tools and algorithms to derive objective and comparable information for glacier inventories on both regional and national scales gains growing importance. Using laser scanner data from Hintereisferner, Kodde et al. (2007) developed an approach for the automatic delineation of glaciers and detection of crevasses based on the DEM with a delineation accuracy of the pixel size. For classifying between glacier and non-glacier surface a variance based smoothness classification is applied, where smoothness stands for a minimum of irregularities and roughness. The method takes the implicit assumption that a glacier surface can be recognised by a comparable smooth surface. Other delineation criteria are the set of hydrological constraints, the connectivity of all pixels that belong to a glacier and the use of the intensity values. The resulting raster map can be converted to a vector representing the outline of the glacier. First results fit well with manually derived delineations. Further problems to be solved are debris-covered parts of glaciers which show a similar roughness as non-glacier areas.

Crevasses within the glacier are detected by assuming that they are deviations from a regular glacier surface. Such a surface can be calculated from the DEM by applying algorithms of mathematical morphology by using a *closing filter*, for example. The result is a surface without crevasses, which can be compared with the original surface. The difference will reveal the location, shape and minimum depth of crevasses. Additionally, Kodde et al. (2007) developed a method for the minimum volume estimation of single crevasses based on the laser point cloud data, assuming that crevasses have a V-like shape. For the location, delineation and reconstruction of crevasses the upcoming full-waveform laser scanner systems will be of advantage as they are able to capture the edge of crevasses very precisely.

10.5 CONCLUSIONS

The results show that airborne laser scanning is not only just another remote sensing technique for delivering DEMs for glaciological purposes, but it is accurate enough for investigating short term geometry changes (area, surface elevation, volume). This can be done in detailed spatial resolution as both relative vertical and relative horizontal accuracy are in the decimetre range. Furthermore, it offers new possibilities for the derivation of additional parameters relevant for glaciological applications (e.g. glacier velocity, surface characteristics). One striking point is that airborne laser scanning can give reliable information for the entire glacier area, due to the fact that it is an active remote sensing technology which works on both snow and ice surfaces. Multi-temporal data on annual or semi-annual basis seem to be accurate enough for mass-balance estimation for glaciers with ablation and accumulation rates of at least 1 m a^{-1} (Arnold et al. 2006).

As the measurement frequency (or data collection rate) and the position accuracy of laser scanning systems continue to increase, and new methods for data acquisition, data management and data analysis will be developed, the potential for glacier monitoring becomes ever greater. As a consequence the time lapse between multi-temporal

Figure 10.5 3D-visualisation of Hintereisferner showing general trend of cumulative decrease in elevation between the first and last laser scanning data acquisitions of the OMEGA project; October 10, 2001 and September 26, 2003. Glacier was thinning many metres more in the lower parts of the glacier than in the accumulation areas. (See colour plate section).

DEMs can be shortened considerably thus enabling to quantify slower processes over larger areas. Objective and comparable multi-temporal data sets might enable the construction of a new generation of area-wide glacier inventories. In addition to the scientific use of airborne laser scanning, there is an immense potential for using static or dynamic visualisation products of repeated laser scanning for communicating results to a wider public and for education purposes (Figure 10.7). And obviously there is a high synergy potential in fusing laser scanner data with other modern remote sensing data, such as digital camera data. Summarizing all these findings leads to the final conclusion that airborne laser scanning meets many requirements for operational glacier monitoring and seems to be a well fitting technology especially for investigating fast retreating and downwasting glaciers with accelerated global warming.

REFERENCES

Abdalati, W. and W. Krabill (1999). Calculation of ice velocities in the Jakobshavn Isbrae area using airborne laser altimetry. *Remote Sensing of Environment* 67, 194–204.

Abdalati, W., W. Krabill, E. Frederick, S. Manizade, C. Martin, J. Sonntag, R. Swift, R. Thomas, W. Wright and J. Yungel (2001). Outlet glacier and margin elevation changes: near-coastal thinning of the Greenland ice sheet. *Journal of Geophysical Research* 106(D24), 33729–33741.

Abdalati, W., W. Krabill, E. Frederick, S. Manizade, C. Martin, J. Sonntag, R. Swift, R. Thomas, W. Wright and J. Yungel (2002). Airborne laser altimetry mapping of the Greenland ice sheet: application to mass balance assessment. *Journal of Geodynamics* 34, 391–403.

Aðalgeirsdóttir, G., K.A. Echelmeyer and W.D. Harrison (1998). Elevation and volume changes on the Harding Icefield, Alaska. *Journal of Glaciology* 44(148), 570–582.

Arnold, N.S., W.G. Rees, B.J. Devereux and G.S. Amable (2006). Evaluating the potential of high resolution airborne LiDAR data in glaciology. *International Journal of Remote Sensing* (27) 1233–1251.

Baltsavias, E. (1999). Airborne laser scanning: basic relations and formulas. *ISPRS Journal of Photogrammetry and Remote Sensing* 54, 199–214.

Baltsavias, E., E. Favey, A. Bauder, H. Bösch and M. Pateraki (2001). Digital surface modelling by airborne laser scanning and digital photogrammetry for glacier monitoring. *Photogrammetric Record* 17(98), 243–273.

Bamber, J.L., S. Ekholm and W.B. Krabill (1998). The accuracy of satellite radar altimeter data over the Greenland ice sheet determined from airborne laser data. *Geophysical Research Letters* 26(16), 3177–3180.

Bamber, J.L., W. Krabill, V. Raper and J. Dowdeswell (2004). Anomalous recent growth of part of a large Arctic ice cap: Austfonna, Svalbard. *Geophysical Research Letters* 31(12), L12402. doi: 10.1029/2004GL019667.

Bauder, A. (2001). Bestimmung der Massenbilanz von Gletschern mit Fernerkundungsmethoden und Fließmodellierungen. *Mitteilungen der Versuchsanstalt für Wasserbau, Hydrologie und Glaziologie an der Eidgenossischen Technischen Hochschule Zürich* 169, 171 p.

Bindschadler, R., M. Fahnestock and A. Sigmund (1999). Comparison of Greenland ice sheet topography measured by TOPSAR and airborne laser altimetry. *IEEE Transactions on Geoscience and Remote Sensing* 37(5), 2530–2535.

Bucher, K., T. Geist and J. Stötter (2006). Ableitung der horizontalen Gletscherbewegung aus multitemporalen Laserscanning-Daten Fallbeispiel: Hintereisferner/Ötztaler Alpen. In: J. Strobl et al. (eds.), *Angewandte Geoinformatik 2006 – Beiträge zum AGIT-Symposium Salzburg* 277–286.

Chasmer, L. and C. Hopkinson (2001). Using airborne laser altimetry and GIS to assess scale-induced radiation loading errors in a glacierised basin. *Proceedings of the 58th Eastern Snow Conference*. Ottawa, Canada, May 14–17, 2001, 11 p.

Christensen, E.L., N. Reeh, R. Forsberg, J.H. Jörgensen, N. Skou and K. Woelders (2000). A low-cost glacier-mapping system. *Journal of Glaciology* 46(154), 531–537.

Csatho, B., T. Schenk, R. Thomas and W. Krabill (1996). Remote sensing of polar regions using laser altimetry. *International Archives of the Photogrammetry, Remote Sensing and Spatial Information Science* 31(B1), 42–47.

Echelmeyer, K.A., W.D. Harrison, C.F. Larsen, J. Sapiano, J.E. Mitchell, J. Demallie, B. Rabus, G. Aðalgeirsdottir and L. Sombardier (1996). Airborne surface elevation measurements of glaciers: a case study in Alaska. *Journal of Glaciology* 42(142), 538–547.

Favey, E., A. Geiger, G.H. Guðmundsson and A. Wehr (1999). Evaluating the potential of an airborne laser-scanning system for measuring volume changes of glaciers. *Geografiska Annaler* 81A (4), 555–561.

Favey, E. (2001). Investigation and improvement of airborne laser scanning technique for monitoring surface elevation changes of glaciers. *Mitteilungen – Institut fur Geodasie und Photogrammetrie an der Eidgenossischen Technischen Hochschule Zürich* 72, 152 p.

Filin, S. (2001). Recovery of systematic biases in laser altimeters using natural surfaces. *International Archives of the Photogrammetry, Remote Sensing and Spatial Information Sciences* 34(3/W4), 85–91.

Garvin, J.B. and R.S. Williams (1993). Geodetic airborne laser altimetry of Breiðamerkurjökull and Skeiðarárjökull, Iceland and Jakobshavns Isbræ, West Greenland. *Annals of Glaciology* 17, 379–385.

Geist, T., E. Lutz and J. Stötter (2003). Airborne laser scanning technology and its potential for applications in glaciology. *International Archives of Photogrammetry, Remote Sensing and Spatial Information Science* 34(3/W13), 101–106.

Geist, T., H. Elvehøy, M. Jackson and J. Stötter (2005). Investigations on intra-annual elevation changes using multi-temporal airborne laser scanning data: case study Engabreen, Norway. *Annals of Glaciology* 42, 195–201.

Geist T. and J. Stötter (2007). Documentation of glacier surface elevation change with multi-temporal airborne laser scanner data – case study: Hintereisferner and Kesselwandferner, Tyrol, Austria. *Zeitschrift für Gletscherkunde und Glazialgeologie* 41, 77–106.

Höfle, B., T. Geist, M. Rutzinger and N. Pfeifer (2007). Glacier surface segmentation using airborne laser scanning point cloud and intensity data. *International Archives of the Photogrammetry, Remote Sensing and Spatial Information Sciences* 36(3/W52), 195–200.

Hopkinson, C. and M.N. Demuth (2006). Using airborne lidar to assess the influence of glacier downwasting on water resources in the Canadian Rocky Mountains. *Canadian Journal of Remote Sensing* 32(2), 212–222.

Kennett, M. and T. Eiken (1997). Airborne measurement of glacier surface elevation by scanning laser altimeter. *Annals of Glaciology* 24, 293–296.

Kodde, M.P., N. Pfeifer, B.G.H. Korte, T. Geist and B. Höfle (2007). Automatic glacier surface analysis from airborne laser scanning. *International Archives of the Photogrammetry, Remote Sensing and Spatial Information Sciences* 36(3/W52), 221–226.

Krabill, W., R. Thomas, K. Jezek, K. Kuivinen and S. Manizade (1995). Greenland ice sheet thickness changes measured by laser altimetry. *Geophysical Research Letters* 22(17), 2341–2344.

Krabill, W., E. Frederick, S. Manizade, C. Martin, J. Sonntag, R. Swift, R. Thomas, W. Wright and J. Yungel (1999). Rapid thinning of parts of the southern Greenland ice sheet. *Science* 283(5407), 1522–1524.

Krabill, W., W. Abdalati, E. Frederick, S. Manizade, C. Martin, J. Sonntag, R. Swift, R. Thomas, W. Wright and J. Yungel (2000). Greenland ice sheet: high elevation balance and peripheral thinning. *Science* 289(5478), 428–430.

Kraus, K. (2004). *Photogrammetrie, Band 1: Geometrische Informationen aus Photographien und Laserscanneraufnahmen.* 7th ed., De Gruyter, Berlin, 516 p.

Lutz, E., T. Geist and J. Stötter (2003). Investigations of airborne laser scanning signal intensity on glacial surfaces – utilising comprehensive laser geometry modelling and orthophoto surface modelling. *International Archives of Photogrammetry, Remote Sensing and Spatial Information Science* 34(3/W13), 143–148.

Pellikka, P. (2007). Monitoring glacier changes within the OMEGA project. *Zeitschrift für Gletscherkunde und Glazialgeologie* 41, 3–5.

Sapiano, J., W. Harrison and K. Echelmeyer (1998). Elevation, volume and terminus changes of nine glaciers in North America. *Journal of Glaciology* 44(146), 119–135.

Sithole, G. and G. Vosselman (2004). Experimental comparison of filter algorithms for bare-Earth extraction from airborne laser scanning point clouds. *ISPRS Journal of Photogrammetry & Remote Sensing* 59(1–2), 85–101.

Spikes, B., B. Csatho and I. Whillans (1999). Airborne laser profiling of Antarctic ice streams for change detection. *International Archives of Photogrammetry, Remote Sensing and Spatial Information Science* 32(3/W14), 169–175.

Spikes, V.B., B.M. Csatho and I.M. Whillans (2003a). Laser Profiling over Antarctic ice streams: methods and accuracy. *Journal of Glaciology* 49 (165), 315–322.

Spikes, V.B., B.M. Csatho, G.S. Hamilton and I.M. Whillans (2003b). Thickness changes on Whillans Ice Stream and Ice Stream C, West Antarctica, derived from laser altimeter measurements. *Journal of Glaciology* 49 (165), 223–230.

Thomas, R., W. Krabill, E. Frederick and K. Jezek (1995). Thickening of Jakobshavns Isbrae, West Greenland, measured by airborne laser altimetry. *Annals of Glaciology* 21, 259–262.

Thomas, R., W. Abdalati, E. Frederick, W. Krabill, S. Manizade and K. Steffen (2003). Investigation of surface melting and dynamic thinning on Jakobshaven Isbrae, Greenland. *Journal of Glaciology* 49 (165), 231–239.

Van der Veen, C.J., W. Krabill, B. Csatho and J.F. Bolzan (1998). Surface roughness on the Greenland ice sheet from laser altimetry. *Geophysical Research Letters* 25(20), 3887–3890.

Wagner, W., A. Ullrich, V. Ducic, T. Melzer and N. Studnicka (2006). Gaussian decomposition and calibration of a novel small-footprint full-waveform digitising airborne laser scanner. *ISPRS Journal of Photogrammetry and Remote Sensing* 60(2), 100–112.

Wehr, A. and U. Lohr (1999). Airborne laser scanning – an introduction and overview. *ISPRS Journal of Photogrammetry and Remote Sensing* 54, 68–82.

Chapter 11

Ground-penetrating radar in glaciological applications

Francisco Navarro
Departamento de Matemática Aplicada, ETSI de Telecomunicación,
Universidad Politécnica de Madrid, Spain

Olaf Eisen
Alfred Wegener Institute, Bremerhaven, Germany

11.1 INTRODUCTION

Ice-radar is a very useful tool for the study of glaciers and ice sheets. Whereas the radar applications discussed in other chapters, like satellite-borne SAR (Chapter 9), mostly focus on the surface of the ice, applications of ground-penetrating radar (GPR) try to peer into and through the ice. This is achieved by investigating not just a single pulse reflected from the medium, but a whole time series of reflections.

This chapter is not intended to present a complete overview of GPR, but just to provide the reader with the necessary background to the subject in order to clearly understand the main focus of the chapter, which is the measurement of ice thickness as a tool to determine the ice volume stored by glaciers and their temporal variations. We will start by presenting the fundamentals of radio-wave propagation in ice, followed by a description of the main radar systems used in glaciology and the key elements of a typical radar system. We will then explain how such systems are used in fieldwork and outline the main steps of radar data processing. Next we will give an overview of the applications of GPR in glaciology, and finally, we will discuss in more detail how ice volume and bed topography are determined from radar-measured ice thickness, paying particular attention to the error estimates.

We will split the applications of radar in glaciology into two main types: those that are principally concerned with the internal properties of the ice body, i.e. the full snow/firn/ice column from glacier surface to its base, and those more directed at basal properties, for which one focuses on the interface between the ice body and the underlying bed. We will pay more attention to internal layering and density/water content. The former, because of its use as a tool to determine accumulation rates, and the latter because of their influence on radio-wave velocity and hence on ice thickness and volume estimates, and also for the role of density in the volume to mass conversion.

11.2 RADIO-WAVE PROPAGATION IN GLACIER ICE

The physics of ground-penetrating radar is the same as that for other electromagnetic sensors, as outlined in Chapters 5 and 9. The propagation of electromagnetic waves for ground-penetrating radar applications is likewise determined by the Maxwell equations. Within any medium, including ice, the dielectric properties determine the specific behaviour of the wave. The dielectric properties can be expressed by the dielectric constant or complex-valued relative permittivity

$$\varepsilon = \varepsilon' - i\varepsilon'' = \varepsilon' - \frac{i\sigma}{\varepsilon_0 \omega} \tag{11.1}$$

where the real part ε' is the ordinary relative permittivity (dielectric constant) of the medium and the imaginary part ε'' is the dielectric loss factor. The latter can be expressed as a function of conductivity σ, angular frequency $\omega = 2\pi f$, f being frequency, and permittivity of vacuum ε_0. The loss factor is often expressed relative to the real part, which yields the loss tangent $\tan\delta = \varepsilon''/\varepsilon'$. The permittivity determines the wave speed with which the wave travels. As impurities in glacier ice are usually low, it can be treated as a low loss medium (Ulaby et al. 1982). For the wave speed, the contribution of the real part of the permittivity dominates over the imaginary part. Neglecting the imaginary part yields the simplified expression for the wave speed

$$c = \frac{c_0}{(\varepsilon')^{1/2}} \tag{11.2}$$

with c_0 the wave speed in vacuum. During wave propagation, different processes change the amount of energy transported by the wave: scattering, attenuation and spreading. The imaginary part ε'' of the permittivity determines the attenuation. It depends on the type and amount of impurities and also on temperature (Dowdeswell and Evans 2004). Whenever the complex permittivity in the medium changes, either its real or imaginary part, or both, a fraction of the propagating wave is scattered or reflected. The reflected energy recorded at the surface constitutes the signal in which we are interested. For most applications of GPR in glaciology, it is sufficient to use the Fresnel power reflection coefficient R for coherent backscattering at the interface of two media with properties ε_1 and ε_2. To quantify the fraction of reflected to incident power, we can use, for normal incidence,

$$R = \left(\frac{\varepsilon_1^{1/2} - \varepsilon_2^{1/2}}{\varepsilon_1^{1/2} + \varepsilon_2^{1/2}} \right)^2 \tag{11.3}$$

Three factors are known to change the dielectric constant in firn or ice, and are dominant over certain depth ranges, for which we will mention typical values for polar ice sheets. 1) Changes in the real part, the permittivity, are mostly related to density; they dominate radar reflections in the upper hundreds of metres. 2) Variations in the imaginary part are proportional to conductivity, related to acidity, and depend on frequency; they are the governing cause of reflections from deeper ice, usually below the firn-ice transition. 3) The third mechanism involves dielectric anisotropy of the crystal fabric, but it becomes significant and has been observed only at deeper levels (>500–1000 m) of ice sheets, where changes in anisotropic crystal fabrics could develop as a result of ice dynamics. For each of these cases simplifications for the reflection coefficient can be derived (Paren 1981). For instance, a small change in the real part $\Delta\varepsilon' = \varepsilon_1 - \varepsilon_2$, yields

$$R = \left(\frac{1}{4}\frac{\Delta\varepsilon'}{\varepsilon'}\right)^2 \tag{11.4}$$

whereas a change in the imaginary part $\Delta\varepsilon''$, or loss tangent $\Delta(\tan\delta) = \Delta\varepsilon''/\varepsilon'$, can be approximated by

$$R = \left(\frac{1}{4}\Delta(\tan\delta)\right)^2 \tag{11.5}$$

Finally, the energy per unit area of the propagating wave decreases with distance from the transmitter, because the area covered by the wave increases. This is known as geometric spreading. The combined effect of all these different influences on the received power P_r in relation to the transmitted power P_t can be combined in the radar equation (Dowdeswell and Evans 2004). For simplifying assumptions of reflection from a smooth plane interface (like the bedrock) and antennae located directly on the glacier surface and close to each other, the radar equation can be approximated by

$$P_r = \frac{P_t G^2 \lambda^2 \varepsilon'}{(4\pi)^2 (2z)^2}\frac{R_r}{L} \tag{11.6}$$

The first quotient includes the system characteristics, where G is the gain of the receiving and transmitting antennae (assumed to be equal) and λ is the wavelength. The factor ε' stands for the mean permittivity of ice. It arises from the distinctive stratification and density distribution in firn, resulting in refraction of oblique rays towards the normal, causing the wave to become more focused. The factor z^2 in the denominator, z being the depth of the interface, results from geometric spreading. The second quotient considers, through the power reflection coefficient R_r, the reflection loss at the interface. The loss L includes attenuation caused by impurities and reflection losses from inhomogeneities along the propagation path between the surface and the reflecting interface.

Different properties contribute to the permittivity of glacier ice. Most important for the real part of permittivity, and thus for wave speed, is the density ρ. Different mixing formulae were developed to relate ε' and ρ. A widely used version for dry snow, firn, and ice is the empirical formula published by Robin et al. (1969), which was later improved by Kovacs et al. (1995) by comparing field measurements of ε' versus density. Their study leads to the relation

$$\varepsilon' = (1 + 0.000845\,\rho)^2 \tag{11.7}$$

where ρ is given in kg m^{-3}, with a standard error of ± 0.031 for ε' (or 1% for ε'_{ice}). Typical values of ρ range from 300 kg m^{-3} ($\varepsilon' = 1.6$) for dry snow zones to 917 kg m^{-3} ($\varepsilon'_{ice} \approx 3.2$) for solid glacier ice. The usage of mixing formulae becomes increasingly complicated once ice is at the pressure melting point and thus contains a small, but significant amount of liquid water, which has a much higher permittivity of $\varepsilon' \approx 80$ (see Section 11.6.2, and also Chapters 5 and 9).

11.3   RADAR SYSTEMS

11.3.1   An overview of radar systems used in glaciology

The origins of the ice-radar or radio-echo sounding (RES) technique can be traced back to before World War II, when pilots flying over Greenland and Antarctica reported that their radar altimeters were giving unreliable data. In particular, in 1933 at Admiral Byrd's Little America base, in Antarctica, the first indications that snow and ice are transparent to high frequency electromagnetic radiation (VHF and UHF bands) were pointed out. Based on this, investigation by US Army researchers lead Amory Waite and others to demonstrate in 1957 that a radar altimeter could be used to measure the ice thickness (Waite and Schmidt 1961). This observation can be considered as the birth of the ice-radar concept, and soon led to the development of the RES technique.

We will present here only a brief outline of the most prominent ice-radar developments, aimed at presenting the different types of radar systems. The interested reader is referred to the reviews by e.g. Gogineni et al. (1998) and Plewes and Hubbard (2001). Before proceeding, however, we should point out that there is a wide scatter in the ice-radar terminology. It is quite easy to find the same term used for different concepts in the literature. The first of several VHF radar systems specifically designed for radioglaciological applications was that developed by Evans in 1963 at the Scott Polar Research Institute (SPRI) of the University of Cambridge (Evans 1967). Several radar equipments soon followed, developed by different institutions from the U.S.A., Soviet Union, UK, Denmark, Canada and Australia. These earlier RES systems were all *pulsed systems*, i.e. radars sending out signals in short (tens to hundreds of nanoseconds in these earlier systems) but powerful bursts or pulses. This is opposed to continuous-wave radars (either Doppler or frequency-modulated), which send out a continuous signal instead of short bursts. The earlier pulsed systems used in glaciology operated at frequencies between 30 and 600 MHz, meaning by this the carrier frequency modulated in amplitude by the envelope making up the "burst" of pulses. These systems were successfully employed for determining the ice thickness and internal layering of the ice sheets of Greenland and Antarctica, as well as ice caps and glaciers in polar and subpolar regions. Further developments included e.g. an increase in radar frequency (up to 930 MHz) for improving system directivity, of interest to reduce the reflections from the side walls in valley glaciers. The advent of emerging technologies in digital data acquisition and recording, and signal processing, led to the availability of coherent RES systems, capable of measuring both amplitude and phase of the received signal (e.g. Walford et al. 1977) and RES systems using synthetic aperture radar (SAR) techniques (described in Chapter 9 of this book) (e.g. Musil and Doake 1987, Raju et al. 1990). The latter was the first RES system based completely on solid-state, computerized components, and employed pulse compression to reduce transmitter power requirements and coherent signal processing to improve the signal-to-noise ratio (SNR) and along-track resolution. SAR techniques are also used in the radar sounders such as MARSIS and SHARAD on board the spacecraft orbiting Mars, which have mapped the polar layered deposits of that planet (e.g. Safaeinili et al. 2008).

The use of pulsed radars on temperate glaciers posed serious problems because of the limited radio-wave penetration due to scattering and absorption losses. The former is mainly due to the presence of water-filled channels and cavities of different shapes and sizes, while the higher absorption losses are attributed to the higher ice temperature and also the presence of liquid water. These limitations led to the application in glaciology of *impulse radar* systems.

An impulse radar is one whose waveform is a short individual pulse (typically, a single-cycle sine wave or a Ricker wavelet). Its most distinctive characteristic is its very wide bandwidth related to the short pulse, as opposed to the very narrow bandwidth of conventional pulsed radar systems. Other radar systems achieve a large bandwidth by other means, e.g. by stepping through multiple frequencies (frequency modulation or chirping) to obtain more information about a scene. In the case of impulse (or ultrawide-band) radars, individual pulses are emitted that contain energy over a very wide range of frequencies. The shorter the pulse, the wider the band, thereby generating even greater information about reflected objects. Because the pulse is so short, very little energy is needed to generate the signal. The drawback of using short, low-power pulses is that less energy can be measured on the radar returns. This problem is circumvented by piling up (stacking) many pulses rapidly and averaging all returns, a process which also contributes to enhance the SNR. Impulse radars typically use resistively loaded dipole antennae (Wu and King 1965) coupled with a device generating high-voltage pulses, at a rate given by the pulse repetition frequency (PRF), and transmit approximately one cycle at the centre frequency determined by the dipole antenna length. It should be noticed that the signal fed to the antenna resembles a spike (i.e. an impulse signal), with a certain pulse length dependent on the antenna size, and the dipole antenna acts as a differentiatior, which radiates to the medium a short pulse with length inverse to frequency (typically, a full cycle of sine wave or a Ricker wavelet). Several such impulse systems were developed in the seventies and early eighties, with frequencies of 1–32 MHz, both for ground-based measurements (e.g. Watts and England 1976, Sverrisson et al. 1980) and airborne measurements, though the latter impose quite complex antenna design and layout (e.g. Watts and Wright 1981), given the low frequencies involved. These systems achieve depth penetrations of the order of 1 km in ice, while exploiting the improved resolution of the mono-cycle wavelet shape. The main drawback is the large antenna size required. In the nineties, and up to present, the Narod and Clarke (1994) miniature impulse transmitter, which can operate in a range of frequencies from 1 to 200 MHz, has become quite popular in glaciology, as it is commercially available and can easily be used in conjunction with a digital oscilloscope as receiver, with output to a laptop or palmtop computer. Such a receiving-recording system has, however, some shortcomings, with the implication that special-purpose receivers combined with digital recording systems are still in use and being developed for impulse radars (e.g. Vasilenko et al. 2002, 2008).

The fact that impulse radars are mono-cyclic (i.e. transmit a single cycle) has implied that some authors in the glaciological literature have used the term monopulse radars for them. We do not recommend such use, as the term monopulse radar has a different meaning (referring to an adaptation of conical scanning) in the more general radar literature. We should also point out that impulse radars could be considered as a particular case of pulsed radar, in the sense that they are not continuous-wave radars.

The use of commercial ground-penetrating radars has also become extremely popular in glaciology during recent years. Although commercially available from the early seventies, it was the use of fibre optics and real-time digitization that led to a rapid expansion of the use of GPR in geophysical, archaeological, and civil engineering applications, among others. Complete GPR systems, including dedicated control and analysis software, are commercially available from various suppliers and are widely used in glaciological studies. GPR uses essentially the same technique as impulse radars. Perhaps the main difference is that GPRs achieve higher frequencies than classical impulse radars, as present GPRs cover a range of frequencies between 12.5 MHz and 2.3 GHz. Such a wide range of frequencies allows the use of GPRs for e.g. high-resolution (but shallow penetration) snow stratigraphy studies, mid-penetration (at medium resolution) investigation of firn layer thickness, or deep-penetration (but lower resolution) ice thickness measurements. The improved resolution given by the mono-cycle wavelet shape, together with the increased performance of present GPR systems, also allows phase changes (or polarity sequence) of reflected radar signals to be resolved, and thus assists the interpretation of the permittivity contrast between the host material and the reflector (e.g. Arcone 1995).

Other radar techniques have also been applied in glaciology, such as the Frequency Modulated Continuous Wave (FMCW) radar. This is a radar system where a monochromatic radio wave is modulated in frequency by a triangular modulation signal, so that it varies gradually, and then mixes with the signal reflected from a target object with this transmit signal to produce a beat signal. The very high frequencies involved (between 1 and 18 GHz) imply correspondingly high vertical resolutions, up to 3 cm, as compared to the 17 cm achieved by the commercial GPRs with highest frequency. This makes the FMCW radars ideal for high resolution (but low penetration) snow studies (e.g. Marshall et al. 2005). Nevertheless, one of these systems, with a bandwidth centred around 1.4 GHz, has been able to reproduce englacial internal horizons down to 60 m, and has even identified isolated events down to 150 m (Arcone and Yankielun 2000).

Historically, airborne radar applications are often termed radio-echo sounding, or RES, though many authors use this term for any kind of ice-radar application. Similarly, some authors restrict the term ground-penetrating radar, or GPR, to surface-based radar, while others use the term for both surface-base and airborne systems. Referred to radar type, some authors restrict the term RES to pulsed radar systems, while others use this term for any kind of radar. Similarly, most authors restrict the term GPR to impulse radar systems, while some others make a wider use of the term, applying it to any kind of radar. The reason for some of the above usages of terms is clear. The earliest radar systems were mostly airborne pulsed systems, while impulse radar systems – which were introduced slightly later in glaciological applications – are mostly, but not only, ground-based systems. In any case, it is clear that there is not a consensus in the ice-radar terminology, and this calls for an effort of the ice-radar community in order to reach it.

11.3.2 Radar system elements

All radar systems, no matter in what particular technique they are based, consist of a transmitter and a receiver, with corresponding antennae (a bistatic arrangement, i.e. separate transmitting and receiving antennae, is commonly used), plus a control unit and recording system (digital, for present systems). Some form of synchronisation system between transmitter and receiver is also needed. Figure 11.1 shows a typical equipment set-up for fieldwork measurements.

The key element of the transmitter is the pulse generator. The transmitters of classical pulsed radars add some complexity to those of impulse radars, as they also require a modulator. Coherent signal processing adds further requirements to the system design. The pulse generators of impulse radars have evolved along two lines: those based on avalanche transistors, and those based on thyristors.

Receivers usually require a front end consisting of a band-pass filter to eliminate spurious signals and a limiter to protect the receiver from strong signals (this is particularly important for the direct signal through the air). This is followed by amplifiers, usually with variable gain. Again, coherent detection adds further complication to the receiver electronics. Bandwidth and stability are critical properties of receivers.

Antennae of pulsed radar systems (usually antenna arrays) have a large variety of implementations, often constrained by physical limitations imposed by the aircraft structure. The most commonly used antennae for impulse radars (including GPRs) are resistively loaded dipoles of the Wu and King (1965) type. The resistive load is aimed at preventing antenna ringing. Dipole antennae are highly directional, radiating most of their energy in the direction orthogonal to the antenna wires. Although the radiation lobes of such an antenna placed in a free space would have axial symmetry, when placed over a half-space of ice (the upper half-space being air), due to the permittivity

Figure 11.1 Typical equipment set-up for on-glacier GPR measurements illustrated by an image of a 25 MHz GPR profiling on Livingston Island, Antarctica. Transmitter (Tx) and receiver (Rx), with their corresponding dipole antennas, are placed on separate sledges. The inset shows the contents of the white box on the first sledge: recording laptop, GPR control unit and GPS receiver. The optical fibre link shown in the image provides the transmission channel for the triggering signal from control unit to transmitter. The radar signal detected by the receiver is also sent to the control unit using optical fibre (thin cable joining top of receiver with white box). Photograph by Andreas Ahlstrøm, 2003.

contrast of ice and air most of the energy is radiated in a conical beam in the downward direction. Because of this strong directionality, when working on valley glaciers caution has to be taken in the choice of antenna layout and design of radar profiles in order to avoid spurious reflections from valley walls.

The transmitter and receiver need to communicate with each other for triggering of pulses and optimizing memory usage by minimizing the data recording window to the period when the reflected signal is expected. This is accomplished by means of some kind of synchronisation system, implemented either by a dedicated radio-channel or by an optical fibre cable.

Digital recording of data of course implies analog-to-digital conversion, usually followed by addition of a certain number of digitized waveforms (stacking) to enhance the SNR, data formatting and digital storage. The control specification of variable system parameters, as well as the control of the system as a whole, is done using some type of control unit.

Different implementations of the above units are possible such as the digital oscilloscope and laptop computer combination mentioned earlier or special-purpose digital recording systems which also perform control unit functions. Two implementations are common in commercial GPR systems: either they have a dedicated unit that combines the control and recording functions, or they have a control unit, from which radar data is downloaded to a laptop computer, where dedicated software for system control and data recording is installed. (Sometimes a separate unit is provided, which replaces the functions of the laptop computer and constitutes a more robust system under rough conditions).

Following each transmitted pulse, the received signal is sampled and recorded, with a given sampling frequency, during a fixed time window. Each of these records is referred to as a trace. The time-window size limits the depth of ice that can be sampled. Some radar systems allow the user to select the sampling frequency and recording time window, whose values depend inversely on each other.

11.3.3 Detection and resolution

In signal theory, *detection* refers to a means to quantify the ability to discern between signal and noise. In our context, this translates into the ability of a radar system to be sensitive to the energy reflected by an object or interface, despite the losses produced in the propagation medium. This depends on many factors, such as the depth, size, shape and orientation of the object, its permittivity contrast with the host medium, the signal losses due to dissipation, scattering and attenuation and the radar system performance.

In contrast, *resolution* refers to the ability to interpret accurately the shape and size of a detected reflector, which in practice translates into the smallest scale at which two adjacent reflectors can be distinguished from each other.

Vertical resolution is controlled by the wavelength of the radar signal in the propagation medium. The theoretical vertical resolution is about one quarter of the wavelength (Sheriff and Geldart 1995), i.e. $\lambda/4$. In practice, however, this may be worsened to $\lambda/2$ due, among other factors, to the progressive filtering by the medium of the higher frequency elements of the wave-train, which attenuate faster. Wavelength is directly proportional to the period of the wave (and inversely proportional to frequency) though the relation $\lambda = cT = c/f$. However, in the case of pulsed systems what is relevant to resolution is not the carrier frequency but the time duration of the actual wavelet (i.e. the envelope of the train of pulses). This is why the resolution of the old but powerful pulsed radio-echo sounding systems was so poor, not because of their carrier frequency, but because of their burst length.

Data spacing is a crucial factor in controlling horizontal resolution. In order to avoid spatial aliasing, radar traces should be collected less than one quarter wavelength apart. Where spatial aliasing has been avoided, horizontal resolution becomes a function of both the transmitted signal wavelength and the depth of the reflector. However, according to studies by Welch et al. (1998), based on earlier assertions by Yilmaz (1987), while the previous statement holds for unmigrated radargrams (see migration in Section 11.5), for properly migrated radar sections the horizontal resolution becomes $\lambda/2$, independent of the reflector depth. Pre-migration horizontal resolution is controlled by the illumination area of the radar beam (footprint), which defines the first Fresnel zone around the object of interest. If the object is larger than the first Fresnel zone, then it will become faithfully reproduced in the radargram; otherwise, its reflection will contain diffraction patterns which will obscure the interpretation of the objects's image and, moreover, several objects located within a single Fresnel zone cannot be distinguished from each other. The radius of the fist Fresnel zone is given by (Robin et al. 1969)

$$F_r = \sqrt{\frac{\lambda z}{2} + \frac{\lambda^2}{16}}$$ (11.8)

and is therefore dependent on both wavelength and reflector depth z. The above equation describes the area of the reflecting surface that contributes to a single reflection of a single radar trace. However, by using data from adjacent traces, migration has a focusing effect on the transmitted signal, so that, according to the results of Welch's experiment, the resolution of migrated radar profiles becomes $\lambda/2$, as opposed to the depth-dependent Fresnel zone used to define the resolution of a single unmigrated trace.

11.4 OPERATING RADARS ON GLACIERS

Operation of GPR for imaging the snow, firn, and ice column usually utilize a transmitter and receiver moved at a fixed distance from each other across the surface along the survey profile. This setup is called common offset (CO), referring to the constant distance between transmitter and receiver. See the scheme in Figure 11.2. To convert travel time to depth, the velocity-depth function has to be known. This can be obtained by different approaches. The most usual is the common-midpoint method (CMP), which is a special case of the general radar wide-angle refraction and reflection measurements (e.g. Macheret et al. 1993). The CMP technique makes use of a special linear geometry set-up, such that the points of reflection at a certain depth remain the same for all transmitter-receiver offsets. The velocity-depth function can be inferred from the increase of travel time with offset, assuming near-horizontal reflectors. The down-hole radar technique (e.g. Murray et al. 2000, Eisen et al. 2002) makes use of

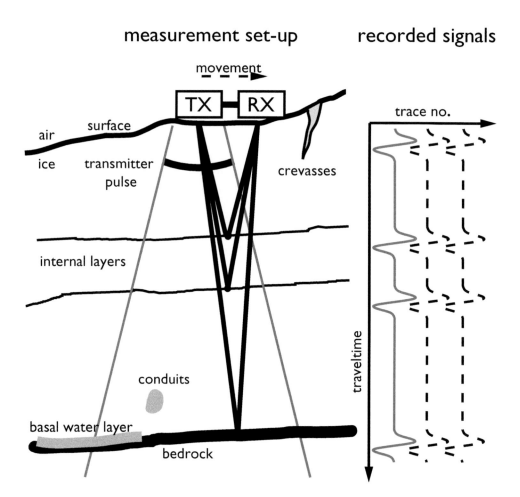

Figure 11.2 Scheme of a GPR common-offset measurement set-up, where transmitter (TX) and receiver (RX) are moved at a fixed distance across the glacier's surface. Within the illuminated area, the transmitter pulse travels downwards, is partially reflected at discontinuities and travels back to the surface, where the signals are detected by the receiver. We schematically show here three selected ray paths, two of them to internal layers and another to the glacier-bedrock interface. For each measurement a trace is recorded as a function of travel time, and plotted next to each other to provide a picture of the sub-surface.

a borehole to record travel times as a function of depth of a reflecting target. Interval velocities can then be derived from the transmitter-target-receiver travel time as a function of depth. For large distances of the surface unit from the borehole unit, or if two boreholes or snow pits are available, the radar measurements provide a tomographic view of the subsurface (e.g. Musil et al. 2006).

GPR systems can be operated directly on the glacier surface or from a helicopter or aeroplane. Either method has advantages and disadvantages, so the choice depends on the focus of the measurements and the available equipment. Surface-based radars have the advantage that the radar wave directly couples into the ice, reducing the power loss of some −5 dB caused by the reflection at the air-firn/ice interface. Moreover, ground-based measurements allow basically the operation of every desired frequency, from 1 MHz to several GHz, as well as a large suite of different survey geometries. Ground-based radar is possible almost everywhere, apart from heavily crevassed areas, where operation is too dangerous. For ground-based measurements, the GPR system can be towed by hand on foot (in ablation areas) or skis (in the presence of snow cover). An example is shown in the right-hand side picture of Figure 11.3. This operation mode is time consuming. On glaciers, ice caps and ice sheets with a rather smooth and flat surface, GPR systems are towed by either snowmobiles or other vehicles (as shown in the left-hand side picture of Figure 11.3), which allows larger distances to be covered. However, the distance between consecutive traces is usually larger, decreasing the along-profile resolution because of the trace repetition limit of the GPR system.

To cover larger areas, airborne radars have been in use since the 1950s (Bogorodsky et al. 1985). To obtain large penetration depth, these radars usually operate with relatively long pulses of several hundred nanoseconds to microseconds, thereby limiting the resolution of the reflection characteristics. Application of high resolution airborne radar, capable of mapping annual accumulation, is still rare and mostly limited to fixed-wing aircraft, although the utilisation of multi-beam antennae and application of processing techniques comparable to satellite SAR nowadays also enable metre resolution. To cope with the difficulties imposed by measurements in

Figure 11.3 Left: GPR survey on Vernagtferner, Ötztal Alps, Austria. The image shows the survey setup of a self-assembled 40-MHz GPR to determine ice thickness towed by a snowmobile. The transmitter is the grey box feeding a bow-tie antenna coupled to crossed dipoles mounted on a pair of cross-country skis. The receiver antenna (in front of the snowmobile) has the same set-up. The receiver signals are transferred to a storage oscilloscope operated from the snowmobile. Photograph by Olaf Eisen, 2007. Right: 200 MHz GPR survey on Las Palmas glacier, Hurd Peninsula, Livingston Island, Antarctica. In this case, the small size of the antennas allows to place the full radar system on a single pulka sledge towed by a skier. Transmitter and receiver can be seen in the image. The control and recording unit is under the plastic cover at the back of the sledge. A GPS receiver and an odometer can also be seen at the right hand side. Photograph by Francisco Navarro, 2008.

mountainous terrain and valley glaciers, airborne radar is most suitable. Apart from quasi-airborne measurements, such as from aerial tramways, helicopter-borne radar provides the most versatile platform and has been applied for glacier thickness measurements (e.g. Macheret and Zhuravlev 1980, Thorning and Hansen 1987, Damm 2004), determination of intraglacial features and physical properties, and accumulation measurements (Arcone and Yankielun 2000, Arcone 2002, Machguth et al. 2006). Apart from the already mentioned power loss caused by airborne operation, another shortcoming is the limitation of operation frequency because of antenna size. On polar ice sheets and polythermal glaciers, operation at frequencies above 100 MHz is rewarding because of the low absorption of electromagnetic energy in cold firn and ice. On temperate glaciers, however, application aiming at depths greater than 100 m is usually restricted to sounding of ice thicknesses with low frequencies (<100 MHz). Although successful attempts have been made by dragging low-frequency antennae behind aeroplanes, this operation mode can still be considered to be in the prototype phase.

Each radar survey is performed along a profile consisting of quasi-continuous measurements. The spatial trace interval depends on the system characteristics, the target (e.g. ice thickness or internal features) and the required needs for data analysis. Whether the mode of operation is airborne or ground-based, the geometry of the survey is usually decided at the office in the planning phase. Typical geometries for valley glaciers often consist of several profiles across the glacier to obtain the cross section at different distances from the terminus. These can be connected by a profile parallel to flow direction, but especially for narrow glaciers the interpretation of the latter is more difficult because of multiple reflections from the glacier sides. The latter comment refers to the case in which transmitting and receiving antennae are arranged coaxially along the profiling direction, because the radiation lobes of dipole antennae are perpendicular to the antenna axis. This is probably the best antennae setup, as it implies a lower coupling between antennae. However, when high frequency antennae are used, quite often they are arranged parallel to each other for ease of transportation (despite the larger coupling between antennae implied). In such a case, if they are arranged perpendicular to the profiling direction, the converse comment on profile design would apply. For studies concerned with the spatial distribution of shallow internal features, such as reflections from buried annual layers in the firn, profiles are often oriented along flow, thus covering a large region, with several profiles parallel to each other. To get a full spatial coverage of the region of interest, profiles are set up in an orthogonal grid-like geometry. Due to logistic limitations, often only one direction of such a grid is surveyed. Surveying both grid dimensions, however, results in a large number of cross-over points, which allows a reliable analysis of the accuracy of observed features. The increment between survey lines, again, depends on the time available. To avoid aliasing effects from spatial undersampling, the distance between parallel profiles should be less than a quarter of the expected roughness of the feature of interest.

Recently, it has also become possible to detect internal features in ice bodies from airborne radar systems designed for satellite-borne operation. A prominent example is the ASIRAS system, whose observation concept is based on the SIRAL system to be operated on board the ESA satellite CryoSat-2. Current applications mainly focus on the distribution of accumulation in the dry snow and percolation zones (Hawley et al. 2006, Helm et al. 2007). Developments are also underway to allow the detection of ice thickness of large ice bodies, e.g. the ice sheets, from satellite platforms operating at P-band (435 MHz).

To allow accurate processing of radar data and their interpretation in the context of other data obtained at the same site, accurate knowledge of the position is needed. In the early 1970s, inertial navigation systems were available for airborne RES, while systems employing microwave signals and transponders at known geographical locations were used for ground-based measurements, the latter with accuracies of the order of 10 m. In the late seventies, LORAN and satellite navigation became available, providing affordable, though less accurate, positioning. Doppler navigation systems were used for airborne measurements during the 1980s. At present, GPS positioning is customarily collected together with radar data, providing accuracies of a few metres, which can be further reduced using differential GPS techniques. For operation of the latter, a static (that is, not moving) base GPS station is set-up near the region of interest and logs data for the whole measurement period. Along with the radar system, a so-called GPS rover is operated. When working in real-time kinematic mode, the base unit radio-transmits to the rover unit the differential corrections. If running in post-processing mode (e.g. because of absence of a good radio-link between base and rover), both units record the information separately, which is later post-processed to obtain accurate coordinates for the rover unit. If standalone – rather than differential – GPS is used during profiling, it is convenient to have at least some control points (e.g. the profile endpoints) measured with a higher accuracy (e.g. differential GPS or total station). Some radar systems allow simultaneous logging of both radar and GPS data, minimizing post-processing needs. Apart from geolocating the radar traces with GPS, another advantage is that one simultaneously obtains a digital elevation model of the surveyed area, which is important for data interpretation.

11.5 PROCESSING TECHNIQUES

We sketch in this section the main steps of radar data processing, focusing on data from common-offset surveys, which are by far the most usual type. Processing techniques are strongly based upon those of seismic reflection data. A detailed description of the latter is given e.g. in Yilmaz (1987, 2001). Focused on radar data, a suitable reference may be Chapter 7 of Daniels (2004).

Processing is preceded by a series of data handling procedures, referred to as preprocessing, aimed at preparing the radar data in a format suitable for processing. This includes issues such as data file merging, data header or background information updates, static corrections needed to assign a common start time to all traces or repositioning/interpolation of traces. The latter may be needed to obtain equidistant trace spacing, which can be a requirement for certain processing procedures. The original position of traces is determined by the shooting of radar pulses, which can be triggered either at regular time intervals – which would generate uniformly spaced traces only in the case of constant velocity of the vehicle transporting the radar – or at constant distance as determined e.g. by an odometer – though actual distance is often far from constant because of improper functioning of the odometer on irregular snow/ice surfaces. Pre-processing also includes trace-editing tasks such as correction of polarity reversals or deletion of noisy traces, traces with transient glitches or monofrequent signals, and subsequent interpolation of deleted traces from neighbouring ones. Though many of theses tasks are quite trivial, pre-processing is often the most time-consuming operation.

Following post-processing, the initial processing step is to remove very low frequency components from the radar signal. Because of the large energy input from the direct waves through the air and the ice, as well as reflections from near-surface reflectors, and due to the limitations inherent to the dynamic range of the receiver, this may become saturated and unable to adjust fast enough to the large variations in signal amplitude. This induces a low-frequency, slowly decaying "wow" in the radar signal. The filtering to remove such effects is frequently referred to as de-wowing the data.

The next step is usually to select a time gain for the data set. A gain-recovery function is applied to the data to correct for the amplitude effects of wavefront (spherical) divergence. This amounts to applying a geometric spreading function, which depends on travel time, and an average velocity (or velocity function). Additionally, an exponential gain function may be used to compensate for attenuation losses. Generally the first step in selecting gain is to examine the amplitude versus time fall-off of the data. The choice of the gain function depends on the purpose of the research. For instance, if one is interested in following stratigraphic horizons, then showing all the information irrespective of amplitude fidelity can be the correct approach. In such a case, a continuously adaptive gain, such as automatic gain control (AGC), can be used. Systematic gain based on physical phenomena, such as exponential compensation gain (SEC), attempt to emulate the variation of signal amplitude as it propagates through the ice. As an option, it may be desirable to filter the data with a wide band-pass filter before deconvolution.

Temporal and spatial filtering are often the next processing steps. Note that, in a radargram, the X and Y axis are used to represent space and time, respectively, but the sequences of data corresponding to either constant time or constant space can both be treated as classical time series and filtered either in the time or frequency domains. Filtering can be done either before or after the application of time gain, as far as the effects of the latter are clearly understood, because time gain is a non-linear process. The aim of filtering is to increase the signal-to-noise ratio. Different types of temporal filters may be applied, from bandpass filtering using fast Fourier transform (FFT) to various types of linear and non-linear time-domain convolution filter operators. Similar filtering techniques can be applied in the space domain, in order e.g. to remove the background noise. Most often this takes the form of a high-pass filter or an average trace removal. The latter can be very effective in some situations where transmitter reverberation or time synchronous system artefacts may mask the signal of interest. However, caution should be taken because average trace removal (or other forms of spatial filtering such as subtracting the average) can delete the images of horizontal reflectors. As a rule, high-pass spatial filters retain dipping features and suppress flat lying features, while low-pass filters have the opposite effect.

Deconvolution is the following step. It is aimed at improving temporal resolution by compressing the effective source wavelet contained in the radar trace (the source time function modified by various effects of the medium and the receiver) to a spike (spike deconvolution). Contrary to the case of seismic reflection data processing, deconvolution of radar data does not often yield much benefit. This is partly because the radar pulse is often as short and compressed as can be achieved for the given bandwidth and signal-to-noise conditions, and partly because some of the more standard deconvolution procedures have underlying assumptions required for wavelet estimation, such as minimum phase and stationarity, which are not always appropriate for radar data. Practical implementation issues may further complicate the deconvolution step. As an example, the rapid decrease in radar-signal amplitude implies that deconvolution artefacts may mask weaker deeper events if time gain is not applied before deconvolution. However, if a gain such as AGC is applied before deconvolution then the non-linear nature of time gain may substantially alter the wavelet character, though this does not happen if an exponential gain is employed instead.

Reflections from point sources traversed by common-offset surveys appear in the radargram as downward-open diffraction hyperbolae. Migration is a process that collapses diffractions and maps dipping events on a radar section to their true subsurface locations. This is achieved by associating spatial variations in wave-path length with wave velocity to collapse hyperbolae limbs back to their apex or source. In the same way as deconvolution improves vertical resolution (as discussed in Section 11.3.3), migration is a process that improves horizontal resolution. Migration requires a good knowledge of the velocity structure in the ice. Because of this, migration is usually accompanied by an iterative procedure of radio-wave velocity analysis, aimed at extracting an average velocity or a velocity versus depth functions, which can be done e.g. from CMP data (see Section 11.4) by means of semblance analysis (Sheriff and Geldart 1995).

An additional problem is that, since radar energy is transmitted conically, in cases such as those of valley glaciers reflections from the inclined slopes of the valley walls can be received in addition to those of the point directly underneath the radar, making the radargram difficult to interpret. Three-dimensional radar data processing is the most straightforward solution to this problem.

For visualization purposes, radar traces are plotted as returned signal amplitude on the x axis against time in the y axis. Many such traces are plotted side by side, at a distance proportional to the space between collected traces, to produce a radargram of the wiggle type. The radargram can also be generated in light intensity format, plotting signal intensity as a colour or greyscale shading, such as those shown later in this chapter.

11.6 INTERNAL STRUCTURE AND ICE PROPERTIES

11.6.1 Internal layering

Signals from the internal part of the ice body can have very different characteristics, depending on the physical properties of the medium. They can be related to inhomogeneities in density, temperature, or the amount of impurities. The most prominent signals are internal reflections in the dry snow regimes of glaciers, ice caps and ice sheets. In regions with higher accumulation rates (>100 kg m^{-2} a^{-1}), such as the Greenland ice sheet or alpine glaciers, the layers are often caused by interannual changes in density, which show prominent differences for snow deposited and reworked under colder conditions in winter and spring as compared to that deposited at higher temperatures in summer. These layers therefore represent isochrones on the annual scale. This means that such a layer has the same age everywhere. As GPR maps the variation in depth of these continuous internal layers along the profile, it is possible to deduce information about the accumulation pattern. On temperate glaciers and in percolation zones, it is possible to map the depth of the previous summer surface, and thus to determine the depth of the winter snow cover. It is, however, necessary to measure at the end of the accumulation season, before strong melt at the surface sets in. A number of studies over the last decade have shown that GPR profiling of firn stratigraphy is capable of complementing traditional methods like stakes, snow pits, and cores to map accumulation rates and improve the understanding of spatial accumulation patterns (e.g. Sinisalo et al. 2003). From the radar data, internal layers can be tracked along the profile in the travel time-domain. To derive accumulation rates, it is necessary to know (i) the wave speed to convert travel time to depth, (ii) the density distribution, to calculate the mass between the tracked layer and the surface, or between adjacent layers, (iii) the age of the layer. Once all these properties are known, the accumulation follows at each point along the profile as the quotient of cumulative mass over age difference, as shown in Figure 11.4. To date the internal layers, it is necessary to tie them to firn or ice cores or snow pits, which provide an age-depth profile. This has the advantage that density profiles can also be easily determined. If the layers are tracked over larger distances, it is necessary to consider the variations of density. It might therefore be necessary to measure it at several locations along the profile. Uncertainties in the derived accumulation distribution are based mainly on three components: layer thinning due to ice advection (that is, the influence of ice dynamics), the procedure for depth calibration, and the isochronal accuracy of each horizon. Results indicate that, under homogeneous conditions, as e.g. those present on the Antarctic plateau, uncertainties at a firn depth of 10 m are about 4% of the calculated snow accumulation and decrease to 0.5% at a firn depth of 60 m. In general, conservative uncertainty estimates of surface mass balance derived from GPR are some 5% on the polar plateau, most of which stems from the uncertainty in dating. It is difficult to provide a generally valid estimate for accumulation uncertainties on glaciers, because of the very particular conditions present at different locations. Nevertheless, even without knowledge of the age of the layer and the density profile it is still possible to deduce the qualitative variation of accumulation (Vaughan et al. 2004). The distribution of accumulation is an important parameter for ground-truthing of satellite data, such as those derived from altimeters or gravimetric missions.

A very different type of internal reflector can occur at the boundary between the firn and the ice. At this depth, the density has increased enough by compaction that the pore volume is no longer connected, but individual bubbles form. This process is called pore close-off. In wet snow regimes, where significant melt occurs during summer, it is possible that the melt water drains from the surface down to the firn-ice interface. Depending on the topographic setting of the glacier, it can happen that further percolation of the water through cracks, crevasses, or the formation of moulins is not possible. Therefore considerable amounts of liquid water collect at this impermeable interface. A water table forms (Fountain and Walder 1998). Because of the large differences in the dielectric constant values of ice and liquid water, a very strong radar reflector is observed in radar profiles. In contrast to the isochronous layers, whose age

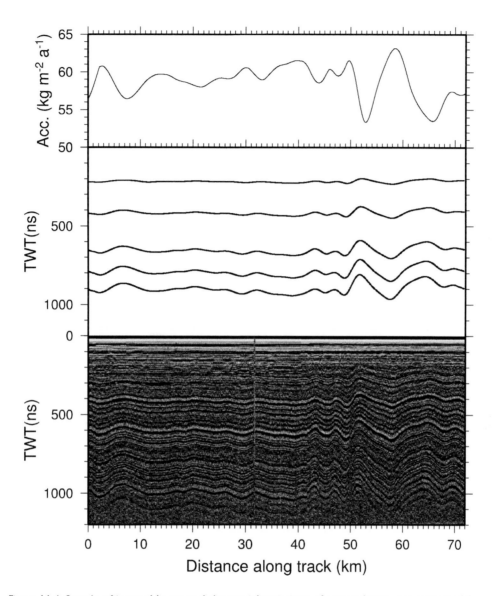

Figure 11.4 Sample of internal layers and the spatial variations of accumulation rate estimated from the uppermost one. TWT = Two-way travel time.

is determined by the time of deposition at the surface of the ice, the reflector from the water table is not an isochrone. It represents a proxy for the post-deposition densification of the firn. Another type of internal reflector is related to the thermal structure within the ice of polythermal glaciers. Although these are most commonly observed in polar or subpolar regions, also in non-maritime mid-latitudes regimes a considerable percentage of glaciers can be considered polythermal (Haeberli and Hoelzle 1995, Suter et al. 2001). Every body of ice contains small amounts of liquid water, e.g. at

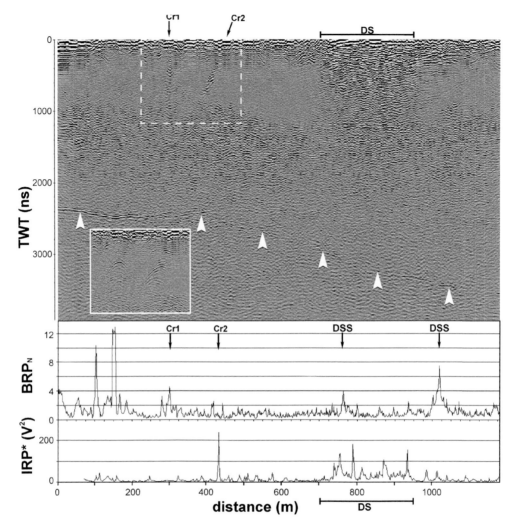

Figure 11.5 Upper panel: Migrated radargram for a transverse radar profile obtained with a 25 MHz GPR in the ablation area of Hansbreen glacier in Svalbard during the melt season of 2003. The cleaner image shown in most of the upper part of the figure represents cold ice, while the lower ice layer, full of diffractions, corresponds to temperate ice. The area rich in diffractions penetrating through the cold ice layer, labeled DS, corresponds to a well-developed drainage system made up of many crevasses and moulins. Cr1 and Cr2 denote crevasse locations. The white arrows show the bedrock reflection. The inset is a non-migrated version of the broken-line square box on top, showing the typical stack of diffraction hyperbolae found at crevasse locations. Lower panels: Spatial changes of BRP_N (bed reflection power normalized) and IRP^* (internal reflection power) along the same profile. The asterisk on IRP is used to denote a modified version of IRP, which does not divide by the number of samples in the time window, so giving a measure of total power reflected by the ice column (excluding bedrock reflection). DSS = drainage system side.

grain boundaries, in veins or water pockets. The fraction of liquid water is very sensitive to temperature. Especially for temperatures close to or at the pressure melting point (PMP), a small increase in temperature leads to a large increase in liquid water. A larger fraction of liquid water results in an increased backscatter observed with radar systems. The possibility of detecting thermal regimes within glaciers by means of radar surveys has been established by several studies (see Pettersson et al. 2003, for a summary). When cold ice (with a temperature below the PMP) overlies temperate ice the boundary between both appears as an internal reflector. This boundary is termed the cold-temperate transition surface (CTS). A sample is shown in the upper panel of Figure 11.5. As the attenuation and backscatter in the cold ice are much lower than in the temperate ice, this reflector should actually be considered a scattering surface, as it does not display continuous reflections and preserved phases. Where exactly the scattering surface observed with radar is located in relation to the CTS depends on the radar system characteristics, especially wavelength and bandwidth (Pettersson et al. 2004).

11.6.2 Density, water content, hydrological aspects

The glacier material (either snow, firn or ice), if considered impurity-free, can be regarded as a mixture of three main constituents, pure ice, air and water, whose volume fractions are denoted by θ, A and W, respectively, the latter being referred to as water content. These volume fractions are related to density and porosity. Water content and density are parameters necessary to characterize the dynamics and thermal regime of glaciers and ice sheets, and to estimate their mass.

The relative amounts and permittivities of pure ice, air and water will determine the permittivity of their mixture, which is, in turn, related to the radio-wave velocity (RWV) in the media under consideration. Therefore, the computation of RWV from e.g. CMP data provides a means for estimating, among others, the water content of temperate ice, which exerts an important influence in the dynamics of glaciers (e.g. Lliboutry and Duval 1985), or the permittivity of cold ice, from which it is possible to estimate its density. There exist 3-component mixture formulae (for glaciology, see e.g. Macheret and Glazovsky 2000). However, their use is not straightforward and it is even unfeasible in most cases. Looyenga (1965) derived a 2-component mixture formula, which has become quite popular in glaciology (e.g. Moore et al. 1999, Murray et al. 2000). However, the assumption that the mixture is made up of only two components implies that it can only be applied under certain conditions. For instance, for water-saturated ice, firn or snow (i.e. no air present in the pores), it relates the permittivity of the mixture to that of ice and water. Or, for dry ice, firn or snow, it links the permittivity of the mixture with that of ice and air. Denoting by ε and f the permittivity and volume fraction, and using the subscripts 1 and 2 to refer to the two constituent materials, and no subscript to refer to the mixture, the Looyenga equation is expressed as

$$\varepsilon^{1/3} = \varepsilon_1^{1/3} + f_2\left(\varepsilon_2^{1/3} - \varepsilon_1^{1/3}\right) \tag{11.9}$$

(for simplicity of notation, and following a common practice in radioglaciology, we have removed the prime in ε even if it denotes the real part of permittivity). In what follows, we will use the subscripts i, w, d, s to refer to pure solid ice, water, dry and soaked material. For dry ice, firn or snow, the above equation becomes

$$\varepsilon_d^{1/3} = 1 + \theta(\varepsilon_i^{1/3} - 1) \tag{11.10}$$

Considering that in dry material the water content is zero, so that $\theta = \rho_d/\rho_i$, and that the relationship between permittivity and velocity is given by Equation 11.2, Equation 11.10 can be rewritten as

$$\rho_d = \frac{\rho_i}{\varepsilon_i^{1/3} - 1} \left[\left(\frac{c_0}{c_d} \right)^{2/3} - 1 \right] \tag{11.11}$$

Because c_0, ρ_i and ε_i are all known quantities, Equation 11.11 allows us to estimate the density of dry ice from the field-measured value of radio-wave velocity in the dry medium c_d.

For water-saturated ice, firn or snow, the Looyenga Equation 11.9 becomes

$$\varepsilon_s^{1/3} = \varepsilon_i^{1/3} + W \left(\varepsilon_w^{1/3} - \varepsilon_i^{1/3} \right) \tag{11.12}$$

from which

$$W = \frac{(c_0/c_s)^{2/3} - \varepsilon_i^{1/3}}{\varepsilon_w^{1/3} - \varepsilon_i^{1/3}} \tag{11.13}$$

This allows us, considering that c_0, ε_i and ε_w are all known quantities, to estimate the water content from the field-measured radio-wave velocity in the soaked medium c_s. Examples of such applications can be found e.g. in Benjumea et al. (2003) and Navarro et al. (2005). There exist alternative formulae for computing the water content of water-saturated glacier media from the permittivities of the constituents and the mixture, such as that of Paren (1970). The Looyenga formula has been pointed out to tend to overestimate the water content of ice. This behaviour has been attributed to the preference of dissolved ions to remain in the liquid water phase of mixtures during freezing (West et al. 2005, 2007), and at least in part to the fact that the non-minimum phase nature of the GPR wavelet (i.e. the fact that the first break of the reflected signal does not correspond to the wavelet peak with highest amplitude) has been ignored in most studies of RWV (Murray et al. 2007). Nevertheless, we have chosen the Looyenga formula for the purpose of illustrating the possibilities, because it can be applied to both dry and water-saturated glacier media. In general, among the variety of methods available for estimating the water content of ice, radar methods give the highest values (e.g. Pettersson et al. 2004). Measurements using radar integrate all water bodies smaller than the radar resolution and will therefore comprise both water contained within the vein network and larger water bodies, such as channels and linked fractures. This could explain the higher water content values shown by radar methods as compared to measurements from ice cores, and points out that radar methods could be measuring at a different scale from that of other methods (Murray et al. 2007). Barrett et al. (2007) have thoroughly analysed the accuracy of RWV determination from CMP data and the corresponding errors in the water content estimates derived from RWV values, showing that these errors can be very large, particularly those referred to velocities for individual layers, unless careful procedures are followed in both data collection and processing.

The power reflection coefficients computed from radar data, such as the bedrock reflection power (BRP) and the internal reflection power (IRP) (Gades et al. 2000), can be used to provide some insight into the hydrological conditions of the glacier and its interface with bedrock. The normalized BRP (BRP_N) provides a measure of the power reflected by the bedrock and corrected for changes in ice thickness, while the IRP is a measure of the power attenuated/scattered in the full ice column from surface to bed. Both are computed directly from the radar traces, as sums of the squared amplitudes within given time windows divided by the number of samples within the time window. For BRP, the time window is a narrow one, centred in the bedrock reflection, while for IRP the time window encompasses most of the ice column from surface to bed. BRP_N is computed as measured BRP divided by calculated BRP, the latter representing the average bed reflection properties as a function of depth. BRP_N and IRP should always be plotted both together, because, as pointed out by Gades et al. (2000), an increase (decrease) in the BRP_N could be just caused by a decrease (increase) in IRP, while a simultaneous increasing (decreasing) trend of both magnitudes might be an indicator of changes in the power transmitted into the ice.

Jania et al. (2005) present some nice examples of the application of power reflection coefficients to understand the hydrological conditions of Hansbreen, a polythermal glacier in Svalbard. We will comment on them referring to Figure 11.5, which shows the radargram for a transverse profile across the ablation area of Hansbreen, and the spatial changes in BRP_N and IRP computed for this profile.

The high values BRP_N at 150–200 m, while IRP remains low, are interpreted as an indicator of a large amount of water at the ice-bed interface, which shows a depression in this area. The IRP plot shows a clear region of high values between 725 and 950 m, approximately (while BRP_N remains nearly constant), that correlates perfectly with the location of a zone with a well-developed drainage system penetrating through the cold ice and manifested as a crevasse and moulin system at the surface. The BRP_N plot shows relative maxima at both sides of the area of the well-developed drainage system. This is interpreted as follows: this area provides an easy way for transfer of water to the temperate ice layer and to bedrock; however, the very high IRP in this area implies a low BRP_N. In contrast, the increased amount of water at the ice-bed interface in this zone is manifested as high BRP_N values at both of its sides, as a result of their relatively low IRP. Such low values of IRP (especially evident at ca. 650 m) at both sides of the crevasse-moulin system correlate well with the location of the thickest cold ice layer, for which a much smaller level of scattering/attenuation is expected than for temperate ice. Cr1 and Cr2 both denote crevasse locations, but have very different signatures in the BRP_N and IRP plots. Cr1 is shown as a local maximum in the BRP_N plot but does not have a counterpart in the IRP plot, while Cr2 appears as a local maximum in the IRP plot, which does not a counterpart in the BRP_N plot. This different pattern could be interpreted as Cr1 having a good link, through the hydraulic system, with the bedrock underneath, so transmitting higher amounts of water to it, resulting in a higher ice-bed interface reflectivity, while Cr2 might not have such a link, so the water remains in the crevasse and does not reach the bed.

11.6.3   Crevasses

A by-product of many radar surveys on glaciers is the detection of crevasses. As these can be filled partly with water, partly with snow and air, they have a considerable reflection coefficient and result in opposing phase signature of the radar return signal. In unmigrated radargrams, crevasses typically appear as hyperbolic diffractions (Glover and Rees 1992), such as those shown in the inset of Figure 11.5. However, if the radargrams are migrated, then the hyperbolae collapse, as happens for the crevasses shown in Figure 11.5 (square box defined by broken line). The geometry of the diffraction in a radargram depends on the horizontal distance of the apex of the crevasse from the profile line, the apex depth, the crevasse width and filling, and its orientation in respect to the radar profile. As the diffraction hyperbola is detectable already several metres to tens of metres before a crevasse is crossed, radar systems are nowadays routinely employed on traverses in regions where crevasses are expected. To this end the radar antennae are mounted at the tip of a boom extending several metres ahead of a vehicle. Whenever a diffraction-like signal appears during real-time monitoring, the vehicle speed can be reduced and the area be investigated in further detail. Under favourable conditions, e.g. if the full hyperbola is recorded, it is also possible to deduce the wave speed from the shape of a hyperbola.

11.6.4   Englacial channels

In regions of strong melt at the surface, the meltwater often penetrates the ice and forms englacial channels. As for crevasses, these channels can be fully filled with water or air, or with a mixture of both. The geometry of such channels is important for studies of glacier hydrology. In favourable cases, it has been possible to detect and map the depth and position of such a channel by GPR. Moreover, detailed analysis of the reflection phase and amplitude also provide the possibility to determine the state of filling of the channel (Stuart et al. 2003).

11.7   BASAL PROPERTIES

Ice thickness is one of the primary characteristics of a body of ice. At the same time, it is also one that is most easily detectable by radar. Consequently, the basal interface has been the primary focus of radar applications for many years. Over time, as radar systems and processing algorithms evolved, other properties were deduced from the ice-bed reflection.

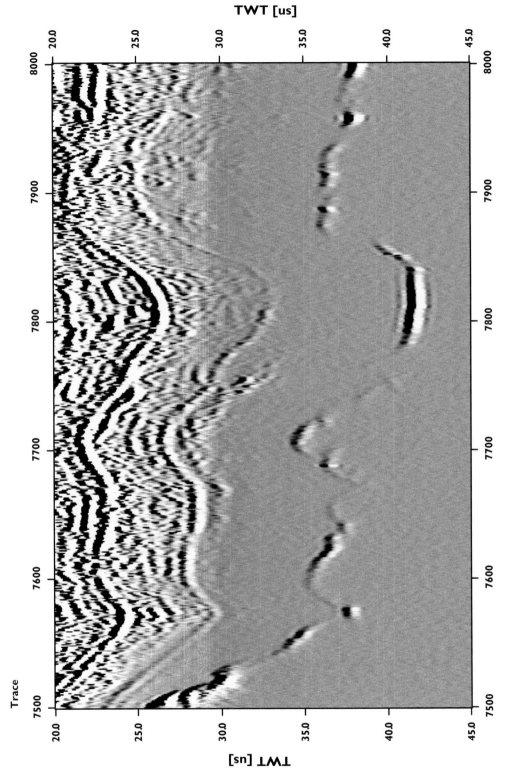

Figure 11.6 Example of radargram showing a subglacial lake, corresponding to the bright nearly flat reflector in the lower part of the figure around trace 7800, also showing a very low reflection at the steep flanks of the lake. Image courtesy of Daniel Steinhage, AWI.

11.7.1   Ice thickness and bedrock topography

As the reflection from the ice-bed interface is usually stronger than any internal reflection, it is very often possible to track it over large portions of the profile. Mapping the depth of this reflector provides the ice thickness, from which the total ice volume of the glacier can be estimated. If accurate data on surface elevation is simultaneously obtained, or available otherwise, it is possible to determine directly the bedrock topography by subtracting the ice thickness from the surface elevation. These subjects will be discussed in detail in Section 11.8.

11.7.2   Conditions at the glacier bed

The basal interface plays an important role in the flow behaviour of a glacier. The ice is frozen to the bedrock if temperatures are below the pressure melting point. In this case the glacier moves slowly by internal deformation only (cf. Chapter 5). At temperatures at the PMP, liquid water is present at the interface and sliding sets in. The amount of sliding depends on the roughness of the bed and the rheological properties of the bed itself. In the presence of sediments, for instance, these can deform as well and contribute to the sliding velocity. In extreme cases, the water layer at the bed can become very thick. If this happens in a localized area away from the sea, one refers to a subglacial lake, which are now commonly observed in Antarctica (an example is shown in Figure 11.6), or other locations such as the Grimsvötn caldera in Iceland. As all these cases involve different physical dielectric properties, they can often be distinguished by means of radar. The results are especially important for studies of the subglacial hydrology, as reflection characteristics show a temporal behaviour, like daily and seasonal cycles. To determine the basal roughness the coherence of the basal reflection over a certain distance is investigated, typically something comparable to the ice thickness. This involves the determination of slopes in bedrock elevation and characterisation of a mean elevation, correlation length of bedrock undulation, and roughness.

To determine the physical properties of the bedrock, the amplitude and phase of the bedrock reflection of a single trace are analysed. However, repeating this analysis for a number of traces provides more reliable results. Only in cases where the permittivity of the bedrock is lower, a phase reversal occurs. This is only the case if a partly air-filled gap is present at the interface, for instance a former drainage channel now void of or only partly filled with water. The analysis of the signal amplitude is based on solving the radar equation introduced in Section 11.2 for the bed reflectivity.

11.8   ESTIMATING ICE VOLUME AND BED TOPOGRAPHY FROM ICE THICKNESS DATA

The primary interest of ice-radar in the context of this book is its use as a tool to determine the ice thickness of glaciers and, consequently, the ice volume (or mass, assuming that density is known) stored by them. It could be thought that radar is also an appropriate tool to determine the volume changes of glaciers, by repeating radar surveys at different times. However, if a non-deformable bedrock is assumed, it is sufficient to perform an accurate combined radar and glacier surface topography campaign, aimed at determining the surface topography, the total ice volume and the bed topography – the latter by subtracting the ice thickness from the surface topography – and compute the volume changes by repeated surface topography measurements, either ground-based (differential GPS, total station) or using remote sensing methods (such as aerial photographs, altimetry or satellite images), depending on the resolution envisaged. A sample of ice thickness change map constructed in such a way is shown in Figure 11.7. Only in cases where a highly deformable bedrock is expected would it be justified– in terms of fieldwork cost – to estimate volume changes by means of repeated radar surveys. One could argue that we should add the cases where notice-able changes in firn compaction are suspected, or isostatic rebound of unknown – but assumed to be large – magnitude is underway. However, these processes involve only minute (firn compaction changes) or null (isostatic rebound) variations in ice thickness, and are only relevant in terms of mass change rather than volume change.

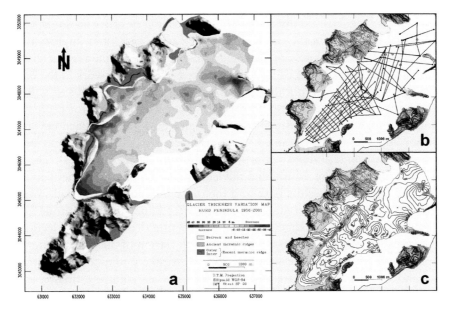

Figure 11.7 a) map of ice thickness changes during 1956–2000 of Johnsons-Hurd glacier, Livingston Island, Antarctica (modified from Molina et al. 2007). It was constructed by subtracting the digital elevation models for both years, retrieved from aerial photographs and satellite images, respectively. Moraines on the proglacial areas are displayed to show earlier Holocene glacier margin positions. The volume change was −0.108 ± 0.048 km³, which represents a −10.0 ± 4.5% change from the 1956 total volume of 1.076 ± 0.055 km³ computed from the radar-measured ice thickness data; b) network of radar profiles, in which the small circles denote points of CMP measurements; c) resulting ice thickness map. (See colour plate section).

11.8.1   Procedure for constructing glacier surface, ice thickness and bed topography maps and for estimating ice volume

The ice thickness is the most immediate product of radar profiling. In order to construct the ice thickness map of a glacier, an appropriate net of radar profiles should be designed, as described in Section 11.4. In practice the main departures from the ideal net design are imposed by the glacier morphology and conditions (location of areas with steep slopes or crevassed areas, melting conditions at the surface, etc.). The radar traces should be properly georeferenced, for instance by simultaneous GPS measurements. After processing the radar data and picking the bedrock reflection, the two-way travel times shown in the radargram are converted to thickness using the radio-wave velocity in ice, either using a standard velocity value appropriate to the glacier under study (depending on whether it is a temperate, polythermal or cold glacier, and the melting conditions) or – much better – RWV determined from in situ CMP measurements, as discussed in Section 11.4. We will discuss later some issues concerning the error in ice thickness. From the ice thickness data, commercial mapping software can be used to construct the surface topography and ice thickness maps. The most usual procedure is to construct the maps from a regular grid interpolated (using kriging or other methods, e.g. Mitas and Mitasova 1999) from the original data. In such a case, it is important to have both surface and ice thickness maps constructed using the same regular grid. There are other alternatives, such as using a triangulated irregular network (TIN; e.g. Wang et al. 2001), which we will not discuss here. The choice of the grid size will have an impact on the surface interpolation error. Usually it will decrease for decreasing grid size (implying higher computational cost), though a grid size is reached beyond which no further improvement is obtained. Such grid size depends on the configuration of the net of radar profiles (in particular, the separation between profiles). The bedrock map is just constructed by subtracting the ice thickness from the glacier surface maps (subtracting the grid values for both and then interpolating the surface). The surface interpolation errors and their statistics should be computed (this is done automatically by usual software packages, though it sometimes requires the selection of the report generation option), because the standard deviation of the surface interpolation error provides a measure of the accuracy of the vertical coordinate of the interpolated surface and will have an impact on the error of the ice volume estimate, as will be discussed later. Most mapping software packages also include an option to calculate the volume from the ice thickness data, using methods such as the trapezoidal rule or Simpson's method (e.g. Burden and Faires 2004, Chapter 4). The latter is preferable, because of its higher accuracy.

11.9   ERROR IN ICE THICKNESS

There are several sources of uncertainty involved in the ice thickness computation, which we discuss below.

11.9.1   Vertical resolution of radar data

The vertical resolution of the radar equipment employed does not constitute a measure of error. However, as it limits the capability of the equipment to distinguish between adjacent radar-detected layers, it can be considered as a lower bound for the error associated to each individual ice thickness measurement. As discussed in Section 11.3.3, the theoretical vertical resolution is about one-quarter of the wavelength, i.e. $\lambda/4$, though in practice this may be worsened to $\lambda/2$, which would be a conservative estimate. Considering a RWV of 165 m μs^{-1} (typical of temperate ice with a moderate water content), this gives e.g. 0.8 m for a 100 MHz radar.

11.9.2   Error in thickness due to error in RWV

The ice thickness H is computed as half the radio-wave velocity c times the two-way travel time τ for the bottom reflection, i.e.

$$H = \frac{1}{2}c\tau \qquad\qquad (11.14)$$

The error in ice thickness will therefore depend on the errors in both RWV and time. Applying error propagation (e.g. Bevington and Robinson 2002) to the above equation, we get

$$e_H = \frac{1}{2}(\tau^2 e_c^2 + c^2 e_\tau^2)^{1/2} \qquad\qquad (11.15)$$

where e is used to denote error estimates. The above equation assumes that the fluctuations in velocity and time are uncorrelated, so that the covariance term in the error propagation equation can de disregarded. We may consider that this holds because the propagation time (obtained from the radargrams) and the velocity (obtained by methods such as CMP) are measured from independent data sets.

Quite often, when RWV is computed from CMP data, the error quoted for RWV is just the root-mean square of the deviations of the fit to the corresponding hyperbola (in reality, to a straight line in the $x^2 - \tau^2$ plane) and is usually very small (typically, <1 m μs^{-1}). Such practice, however, does not provide a realistic measure of error. Barrett et al. (2007) present a complete analysis of the accuracy of RWV determination from CMP data, discussing the influence of the different error sources on the RWV estimate. These include the static errors (errors in travel time), errors in the measured transmitter-receiver offset (which give the highest contribution to error) and those related to the violation of the assumptions inherent to the CMP method. In their model-case study, full-column averaged errors in RWV range from 0.2 to 3.5% for the best-and worst-case scenarios concerning acquisition and processing practices. Errors are much larger for interval (layer) velocities, from 2.5 to 11%, but fortunately these do not play a role in the estimate of error in total thickness.

Regarding the error in travel time, the worst-case scenario in the Barrett et al.'s (2007) analysis considers that it equals one period of the radar wave. A more common scenario would consider half the radar wave period, e.g. 5 ns for a 100 MHz radar.

Let us now quantitatively compare the contributions of the errors in RWV and travel time to the error in ice thickness, as given by Equation 11.15. First note that, while the contribution of the first term depends (through τ) on the thickness of the measurement point, the contribution of the second term is independent of the thickness, and depends only on the frequency of the radar used. Considering a 100 MHz radar, a RWV of 165 m μs^{-1}, a relative error in RWV of 2%, and typical two-way travel times for valley glaciers of $\tau = \{2, 1, 0.5, 0.25\}$ μs (corresponding to ice thickness of $H = \{165, 83, 41, 21\}$ m), the evaluation of equation (11.15) would give errors in thickness of $\{3.33, 1.70, 0.92, 0.58\}$ m, respectively, if both terms of Equation 11.15 are taken into account, and $\{3.30, 1.65, 0.83, 0.41\}$ m, if only the first term is considered. We therefore see that only for thickness below ca. 40 m the difference in error estimates for the cases considering both terms or only the first term becomes relevant (>10%). Moreover, for thickness below ca. 20 m the second term in (11.15) becomes dominant (for $\tau = 0.25$ μs, the two terms within the square root are of the same order). However, for thickness below ca. 40 m the errors in thickness are of the same order as (for $H = 40$ m) or lower than (for $H < 40$ m) the vertical resolution of the radar ($\lambda/2 = 0.83$ m for a 100 MHz radar), so that any further improvement in the error estimate becomes meaningless. Consequently, for estimating the error in ice thickness we could consider in practice a simplified version of Equation 11.15,

$$e_H = e_c \tau / 2 \qquad (11.16)$$

and, using Equation 11.14, the relative error in thickness becomes equal to the relative error in RWV:

$$\frac{e_H}{H} = \frac{e_c}{c} \qquad (11.17)$$

All of the above discussion refers to the estimated error in ice thickness as a function of the error in RWV calculated from a CMP measurement made at a certain location on the glacier. However, our radar profiling will cover most of the glacier surface, and quite noticeable RWV spatial (and also temporal) variations are observed in glaciers, particularly in temperate and polythermal ones (e.g. Jania et al. 2005, Navarro et al. 2009). Therefore, it would be not realistic to take errors corresponding to a given location to make an overall estimate, using Equation 11.15, of the error induced in ice thickness, unless careful practices in fieldwork procedures are followed, as discussed below. The temporal and spatial changes of RWV in temperate and polythermal glaciers are mainly related to variations in water content, so it is of interest to quantify the influence of water content variations on the measured velocity. Barrett et al. (2007) assess how errors in RWV influence the errors in the derived water content. Here we take the reciprocal approach. Benjumea et al. (2003) show a straightforward way to do it. They plot (see Figure 11.8) the curve of RWV versus water content given by Equation 11.13, showing that a change by 1% in the water content fraction (e.g. from 1% to 2%; notice that this represent a relative 100% change in water content for a reference value of 1%) implies a 3% change in RWV, and spatial variations in water content of that order are not rare in temperate and polythermal glaciers (e.g. Navarro

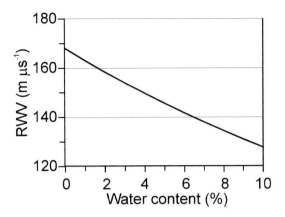

Figure 11.8 Radio-wave velocity (RWV) versus water content, according to Equation 11.13.

et al. 2009). We here refer to horizontal spatial variations of full column-averaged water content; vertical variations can be much higher (e.g. Murray et al. 2000). Temporal variations, related to melting conditions, are even higher than the spatial ones. For instance, changes by 2% in water content volume fraction have been observed in polythermal glaciers (Jania et al. 2005) and would imply up to 6% changes in RWV. Careful procedures in fieldwork design should be followed in order to minimize such possible sources of error in RWV and, by inference, in ice thickness. The ideal setup is to have the radar profiling and CMP measurements both done before the start of the melting season (to minimize the water content and its spatial variations) and as close in time to each other as possible (to minimize the temporal variations). Additionally, it would be convenient to have CMP velocities measured at both accumulation and ablation areas, because full-column (from glacier surface to bed) averaged velocities from CMP measurements in the accumulation area will be biased towards higher values, due to the presence of the firn layer, which has much higher average velocities, of the order of 190 m μs^{-1}. Finally, we recommend to follow careful practices in both data collection and processing referred to RWV calculation from CMP measurements, such as those given by Barrett et al. (2007). The combination of the above practices can lead to quite satisfactory relative errors in ice thickness of the order of 1–2%.

11.9.3 Error in thickness associated with lack of migration

If the glacier bedrock has a noticeable slope, migration becomes a necessary processing step in order to correct the radargrams before picking bedrock reflections to determine the ice thickness. Otherwise, the errors involved in ice thickness determination can be huge e.g. in the areas of steep bed slopes near the walls of a valley glacier. Moran et al. (2000), using three-dimensional Kirchhoff migration and a synthetic-aperture approach, showed improved depth accuracies, in the regions of steepest dip of a valley glacier, up to 36% compared with unmigrated data, and up to 16% compared with standard two-dimensional migration. Though these are perhaps extreme cases, they illustrate the importance of migration in the processing of radar data when steep bed slopes are involved.

11.9.4 Surface interpolation error

All available thickness measurements are taken into account for constructing the ice thickness map by surface interpolation methods. The error in surface interpolation has thereby to be considered. This is usually done through a cross-validation process, which allows the assessment of the relative quality of the grid by computing the gridding errors. This is usually done as follows. Given the known values at N observation locations in the original data set, the gridding errors are calculated by removing the first observation from the data set, and using the remaining data and the specified algorithm to interpolate a value at the first observation location. Using the known observation value at this location, the interpolation error is computed as the difference between the interpolated and the observed values. Then, the first observation is replaced and we proceed with the second observation as we did with the first one, repeating this process up to observation N. This process generates N interpolation errors. Various statistics computed for the errors can then be used as a quantitative, objective measure of quality for the gridding method. The standard deviation of the interpolation error can be taken as a measure of the accuracy of the vertical coordinate of the interpolated surface. Mapping software packages usually include options for estimating the surface interpolation error and associated statistics as described.

11.10 ERROR ESTIMATES FOR ICE VOLUME AND BED TOPOGRAPHY COMPUTATIONS

Because the total volume estimate results from the sum of products, at the cell level, of cell areas (constant for a regular grid) times an average ice thickness for the cell, i.e. $V = \sum_{i=1}^{N} A_i H_i$, the error in volume can be approximated by

$$e_V = \left[\sum_{i=1}^{N} H_i^2 e_{A_i}^2 + A_i^2 \hat{e}_{H_i}^2 \right]^{1/2} \tag{11.18}$$

where, in the light of the discussion in the previous section, we shall take as an estimate for the error in ice thickness for the individual points

$$\hat{e}_{H_i} = \max\{\lambda/2, \, e_{H_i}, \sigma_H\} \tag{11.19}$$

i.e. the maximum among the vertical resolution of the radar data, the error in the ice thickness measurement as given by Equation 11.15, or its simplified version Equation 11.16, and the standard deviation σ_H of the ice thickness interpolation error. The vertical resolution of the radar is included as a lower bound for the error; a less conservative approach would be to set it to $\lambda/4$.

Equation 11.18 can be simplified, particularly in the case of regular rectangular grids. Let us consider such a grid with cell area A_c. Taking into account that the error in area is zero for the portion of glacier covered by inner cells (this also holds for triangulated irregular networks, TINs), and considering $A_c/2$ as the error estimate for the area of each portion of glacier covered by a boundary cell (for a TIN, it would be much lower) as well as an estimate of the area of each of such portions, it follows that

$$e_V = \left[\frac{1}{4}A_c^2 \sum_{i=1}^{NB} H_i^2 + \frac{1}{4}A_c^2 \sum_{i=1}^{NB} \hat{e}_{H_i}^2 + A_c^2 \sum_{i=1}^{NI} \hat{e}_{H_i}^2 \right]^{1/2} \tag{11.20}$$

where NB and NI represent the number of boundary and inner cells, respectively. The first and second terms in (11.20) are usually very small as compared with the third, because: 1) the number of boundary cells is usually much lower than the number of inner cells; 2) the factor $1/4$; 3) H_i in the first term represent the thickness of the boundary cells, which are rather small for most glacier margins (except calving fronts, very steep sidewalls or ice divides; the latter, however, are artificial boundaries which do not contribute to the error in area); 4) the error in thickness for the boundary cells is smaller than that for the inner cells, as implied by Equations 11.15 and 11.16. Consequently, disregarding the first and second terms in Equation 11.20, it simplifies to

$$e_V = A_c \left[\sum_{i=1}^{NI} \hat{e}_{H_i}^2 \right]^{1/2} \tag{11.21}$$

For the case of a TIN, the areas of the cells are different and the corresponding simplified equation would be

$$e_V = \left[\sum_{i=1}^{N} A_i^2 \hat{e}_{H_i}^2 \right]^{1/2} \tag{11.22}$$

Note that in the above equation we are considering the total number of cells N. If a TIN is used, it is more obvious that the term representing the error in area for the portion of glacier covered by boundary cells can be ignored, because the triangles fit the actual boundary much better than the rectangles do. However, the latter also implies that A_i rather than $A_i/2$ should be used as estimates for the areas of the portions of glacier covered by boundary cells.

If volume changes were to be computed by repeated ice thickness measurements at two different times (a procedure which we only recommend for cases where a highly deformable bedrock is expected), the volume change would be determined simply by subtracting the corresponding ice volumes, i.e. $\Delta V = V_2 - V_1$, and the error in volume change should be taken as

$$e_{\Delta V} = \sqrt{e_{V_1}^2 + e_{V_2}^2} \tag{11.23}$$

with the error estimates for V_1 and V_2 determined using Equation 11.18 or its simplified versions.

Similarly, if the error in vertical coordinate of the available surface topography is characterised by a standard deviation σ_S (e.g. the standard deviation of the surface interpolation error), then the error in vertical coordinate of the bed topography constructed by subtracting the ice thickness from the surface topography data will be characterised by a standard error

$$\sigma_b = \sqrt{\bar{e}_H^2 + \sigma_S^2} \tag{11.24}$$

$\bar{e}_H$ being the average of the individual thickness errors e_{H_i}.

REFERENCES

Arcone, S.A., D.E. Lawson and A.J. Delaney (1995). Short-pulse radar wavelet recovery and resolution of dielectric contrasts within englacial and basal ice of Matanuska Glacier, Alaska, U.S.A. *Journal of Glaciology* 41(137), 68–86.

Arcone, S.A. and N.E. Yankielun (2000). 1.4 GHz radar penetration and evidence of drainage structures in temperate ice: Black Rapids Glacier, Alaska, U.S.A. *Journal of Glaciology* 46(154), 477–490.

Arcone, S. (2002). Airborne-radar stratigraphy and electrical structure of temperate firn: Bagley Ice field, Alaska, U.S.A. *Journal of Glaciology* 48(161), 317–334.

Barrett, B.E., T. Murray and R. Clark (2007). Errors in radar CMP velocity estimates due to survey geometry, and their implication for ice water content estimation. *Journal of Environmental and Engineering Geophysics* 12, 101–111.

Benjumea, B., Yu.Ya. Macheret, F.J. Navarro and T. Teixidó (2003). Estimation of water content in a temperate glacier from radar and seismic sounding data. *Annals of Glaciology* 37, 317–324.

Bevington, P.R. and D.K. Robinson (2002). *Data Reduction and Error Analysis for the Physical Sciences*. 3rd ed., McGraw-Hill, Boston, 336 p.

Bogorodsky, V.V., C.R. Bentley and P.E. Gudmandsen (1985). *Radioglaciology*. D. Reidel Publishing Company, Dordrecht, Holland, 272 p.

Burden, J.D. and R.L. Faires (2004). *Numerical Analysis*. 8th ed., Brook-Cole and Thomson Learning, Belmont, 847 p.

Damm, V. (2004). Ice thickness and bedrock map of Matusevisch Glacier drainage basin (Oates Coast). *Terra Antarctica* 11(1), 85–90.

Daniels, D.J. (2004). *Ground Penetrating Radar: Theory and Applications*. 2nd ed., The Institution of Electrical Engineers, London, 726 p.

Dowdeswell, J.A. and S. Evans (2004). Investigations of the form and flow of ice sheets and glaciers using radio-echo sounding. *Reports on Progress in Physics* 67, 1821–1861.

Eisen, O., U. Nixdorf, F. Wilhelms and H. Miller (2002). Electromagnetic wave speed in polar ice: validation of the common-midpoint technique with high resolution dielectric-profiling and γ-density measurements. *Annals of Glaciology* 34, 150–156.

Evans, S. (1967). Progress report on radio echo soundings. *Polar Record* 13(85), 413–420.

Fountain A.G. and J.S. Walder (1998). Water flow Through Temperate Glaciers. *Reviews of Geophysics* 36(3), 299–328.

Gades, A.M., C.F. Raymond, H. Conway and R.W. Jacobel (2000). Bed properties of Siple Dome and adjacent ice streams, West Antarctica, inferred from radio-echo sounding measurements. *Journal of Glaciology* 46 (152), 88–94.

Glover, J.M. and H.V. Rees (1992). Radar investigations of firn structure and crevasses. In: J. Pilon (ed.), *Ground Penetrating Radar*. Geological Survey of Canada Special Paper 90-4, pp. 75–84.

Gogineni, S., T. Chuah, C. Allen, K. Jezek and R.K. Moore (1998). An improved coherent radar depth sounder. *Journal of Glaciology* 44(148), 659–669.

Haeberli, W. and M. Hoelzle (1995). Application of inventory data for estimating characteristics of and regional climate-change effects on mountain glaciers: a pilot study with the European Alps. *Annals of Glaciology* 21, 206–212.

Hawley, R.L., E.M. Morris, R. Cullen, U. Nixdorf, A.P. Shepherd and D.J. Wingham (2006). ASIRAS airborne radar resolves internal annual layers in the dry-snow zone of Greenland. *Geophysycal Research Letters* 33, L04502, doi: 10.1029/2005GL025147.

Helm, V., W. Rack, R. Cullen, P. Nienow, D. Mair, V. Parry and D. J. Wingham (2007). Winter accumulation in the percolation zone of Greenland measured by airborne radar altimeter. *Geophysical Research Letters* 34, L06501, doi:10.1029/2006GL029185.

Jania, J., Yu.Ya. Macheret, F.J. Navarro, A.F. Glazovskiy, E.V. Vasilenko, J. Lapazaran, P. Glowacki, K. Migala, A. Balut and B.A. Piwowar (2005). Temporal changes in the radiophysical properties of a polythermal glacier in Spitsbergen. *Annals of Glaciology* 42, 125–134.

Kovacs, A., A. Gow and R. Morey (1995). The in-situ dielectric constant of polar firn revisited. *Cold Regions Science and Technology* 23, 245–256.

Lliboutry, L. and P. Duval (1985). Various isotropic and anisotropic ices found in glaciers and polar ice caps and their corresponding rheologies. *Annales Geophysicae* 3, 207–224.

Looyenga, H. (1965). Dielectric constants of heterogeneous mixtures. *Physica* 31(3), 401–406.

Macheret, Yu.Ya. and A.B. Zhuravlev (1980). Radilokatsiommoye zondirovanie lednikov Shpitsbergea s vertolyrta [Radio-echo sounding of Spitsbergen glaciers from helicopter]. *Materialy Glyatsiologicheskikh Issledovaniy* [*Data of Glaciological Studies*] 37, 109–131. (in Russian)

Macheret, Yu.Ya., M.Yu. Moskalevsky and Ye.V. Vasilenko (1993). Velocity of radio waves in glaciers as an indicator of their hydrothermal state, structure and regime. *Journal of Glaciology* 39(132), 373–384.

Macheret, Yu.Ya. and A. Glazovsky (2000). Estimation of absolute water content in Spitsbergen glaciers from radar sounding data. *Polar Research* 19(2), 205–216.

Machguth, H., O. Eisen, F. Paul and M. Hoelzle (2006). Strong spatial variability of snow accumulation observed with helicopter-borne GPR on two adjacent alpine glaciers. *Geophysical Research Letters* 33, L13503, doi: 10.1029/2006GL026576.

Marshall, H.-P., G. Koh and R.R. Foster (2005). Estimating alpine snowpack properties using FMCW radar. *Annals of Glaciology* 40, 157–162.

Mitas, L. and H. Mitasova (1999). Spatial Interpolation. In: P. Longley, M.F. Goodchild, D.J. Maguire, D.W. Rhind (eds.), *Geographical Information Systems: Principles, Techniques, Management and Applications*. Vol. 2, pp. 481–492. John Wiley & Sons, New York, 580 p.

Molina, C., F.J. Navarro, J. Calvet, D. García-Sellés and J.J. Lapazaran (2007). Hurd Peninsula glaciers, Livingston Island, Antarctica, as indicators of regional warming: ice volume changes during the period 1956–2000. *Annals of Glaciology* 46, 43–49.

Moore, J.C., A. Pälli, F. Ludwig, H. Blatter, J. Jania, B. Gadek, P. Glowacki, D. Mochnacki and E. Isaksson (1999). High resolution hydrothermal structure of Hansbreen,

Spitsbergen mapped by ground penetrating radar. *Journal of Glaciology* 30(151), 524–532.

Moran, M.L., R.J. Greenfield, S.A. Arcone and A.J. Delaney (2000). Delineation of a complexly dipping temperate glacier bed using short-pulse radar arrays. *Journal of Glaciology* 46(153), 274–286.

Murray, T., G.W. Stuart, M. Fry, N.H. Gamble and M.D. Crabtree (2000). Englacial water distribution in a temperate glacier from surface and borehole radar velocity analysis. *Journal of Glaciology* 46(154), 389–398.

Murray, T., A. Booth and D. Rippin (2007). Limitations of glacier ice-water content estimated using velocity analysis of surface ground-penetrating radar surveys. *Journal of Environmental and Engineering Geophysics* 12, 87–99.

Musil, G.J. and C.S.M. Doake (1987). Imaging subglacial topography by a synthetic aperture radar technique. *Annals of Glaciology* 9, 170–175.

Musil, M., H. Maurer, K. Hollinger and A.G. Green (2006). Internal structure of an alpine rock glacier based on crosshole georadar traveltimes and amplitudes. *Geophysical Prospecting* 54(3), 273–285, doi: 10.1111/j.1365-2478.2006.00534.

Narod, B.B. and G.K.C. Clarke (1994). Miniature high power impulse transmitter for radio-echo sounding. *Journal of Glaciology* 40(134), 190–194.

Navarro, F.J., Yu.Ya. Macheret and B. Benjumea (2005). Application of radar and seismic methods for the investigation of temperate glaciers. *Journal of Applied Geophysics* 57, 193–211.

Navarro, F.J., J. Otero, Yu.Ya. Macheret, E.V. Vasilenko, J.J. Lapazaran, A.P. Ahlstrøm and F. Machío (2009). Radioglaciological studies on Hurd Peninsula glaciers, Livingston Island, Antarctica. *Annals of Glaciology* 50(51), 17–24.

Paren, J.G. (1970). *Dielectric properties of ice*. Ph.D. Thesis, University of Cambridge.

Paren, J.G. (1981). PRC at a dielectric interface. *Journal of Glaciology* 27(95), 203–204.

Pettersson, R., P. Jansson and P. Holmlund (2003). Cold surface layer thinning on Storglaciären, Sweden, observed by repeated ground penetrating radar surveys. *Journal of Geophysical Research* 108(F1), 6004, doi: 10.1029/2003JF000024.

Pettersson, R., P. Jansson and H. Blatter (2004). Spatial variability in water content at the cold-temperate transition surface of the polythermal Storglaciären, Sweden. *Journal of Geophysical Research* 109(F2), 2009, doi: 10.1029/2003JF000110.

Plewes, L. and B. Hubbard (2001). A review of the use of radio-echo sounding in glaciology. *Progress in Physical Geography* 25(2), 203–236.

Raju, G., W. Xin and R.K. Moore (1990). Design, development, field operations and preliminary results of the coherent Antarctic radar depth sounder (CARDS) of the University of Kansas, U.S.A. *Journal of Glaciology* 36(123), 247–258.

Robin, G. de Q., S. Evans and J.T. Bailey (1969). Interpretation of Radio Echo Sounding in Polar Ice Sheets. *Philosophical Transactions of the Royal Society of London, Series A, Mathematical and Physical Sciences* 265(1166), 437–505.

Safaeinili, A., J. Plaut, J. Holt, R. Phillips, Y. Gim, R. Orosei, D. Biccari, R. Seu and G. Picardi (2008). Recent results from MARSIS and SHARAD radar sounders. *Geophysical Research Abstracts* 10, EGU2008-A-04516.

Sheriff, R.E. and L.P. Geldart (1995). *Exploration Seismology*. Cambridge University Press, Cambridge, 592 p.

Sinisalo, A., A. Grinsted, J.C. Moore, E. Kärkäs and R. Petterson (2003). Snow accumulation studies in Antarctica with ground penetrating radar using 50, 100 and 800 MHz antenna frequencies. *Annals of Glaciology* 37, 194–198.

Stuart, G., T. Murray, N. Gamble, K. Hayes and A. Hodson (2003). Characterization of englacial channels by ground-penetrating radar: An example from Austre Brøggerbreen, Svalbard. *Journal of Geophysical Research* 108(B11), 2525, doi: 10.1029/2003JB002435.

Suter, S., M. Laternser, W. Haeberli, M. Hoelzle and R. Frauenfelder (2001). Cold firn and ice of high altitude glaciers in the Alps: Measurements and distribution modelling. *Journal of Glaciology* 47(156), 85–96.

Sverrisson, M., A. Jóhanesson and H. Björnsson (1980). Radio-Echo equipment for depth sounding of temperate glaciers. *Journal of Glaciology* 25(93), 477–486.

Thorning, L. and E. Hansen (1987). Electromagnetic reflection survey 1986 at the Inland Ice margin of Pakitsoq basin, central Greenland. *Rapport Grønlands Geologiske Undersøgelse* 135, 87–95.

Ulaby, F.T., R.K. Moore and A.K. Fung (1982). *Microwave Remote Sensing: Active and Passive*. Vol. 2. Radar Remote Sensing and Surface Scattering and Emission Theory. Addison-Wesley, Reading, 609 p.

Vasilenko, E.V., V.A. Sokolov, Yu. Ya. Macheret, A.F. Glazovsky, M.L. Cuadrado and F.J. Navarro (2002). A digital recording system for radioglaciological studies. *Bulletin of the Royal Society of New Zealand* 35, 611–618.

Vasilenko E.V., F. Machío, F.J. Navarro and R. Rodríguez-Cielos (2008). VIRL7: A new radar system for glaciological applications. *Abstracts of the International Symposium on Radioglaciology and its Applications*, Madrid, Spain, June 9–13, 2008.

Vaughan, D.G., P.S. Anderson, J.C. King, G.W. Mann, S.D. Mobbs and R.S. Ladkin (2004). Imaging of firn isochrones across an Antarctic ice rise and implications for patterns of snow accumulation rate. *Journal of Glaciology* 50(170), 413–418.

Waite, A.H. and S.J. Schmidt (1961). Gross errors in height indication of pulsed radar altimeters operating over thick ice or snow. *Institute of Radio Engineers International Convention Record* 5, 38–53.

Walford, M.E.R., P.C. Holdorf and R.G. Oakberg (1977). Phase sensitive radio-echo sounding at the Devon Island ice cap, Canada. *Journal of Glaciology* 18(79), 217–229.

Wang, K., C.-P. Lo, G.A. Brook and H.R. Arabnia (2001). Comparison of existing triangulation methods for regularly and irregularly spaced height fields. *International Journal of Geographical Information Science* 15(8), 743–762.

Watts, R.D. and A.W. England (1976). Radio-echo sounding of temperate glaciers: ice properties and sounder design criteria. *Journal of Glaciology* 17(75), 39–48.

Watts, R.D. and D.L. Wright (1981). Systems for measuring thickness of temperate and polar ice from the ground or from the air. *Journal of Glaciology* 27(97), 459–469.

Welch, B.C., W.T. Pfeffer, J.T. Harper and N.F. Humphrey (1998). Mapping subglacial surfaces below temperate valley glaciers using 3-dimensional radio-echo sounding techniques. *Journal of Glaciology* 44(146), 164–170.

West, J., D.M. Rippin, A.L. Endres and T. Murray (2005). Dielectric properties of ice-water systems: laboratory characterization and modeling. *EOS, Transactions, American Geophysical Union, Joint Assembly Supplement* 86(18), abstract NS43A-06.

West, L.J., D. Rippin and T. Murray (2007). TDR probes for measuring dielectric permittivity of glacial ice cores: design and performance. *Journal of Environmental and Engineering Geophysics* 12, 37–45.

Wu, T.T. and R.W.P. King (1965). The cylindrical antenna with non-reflecting resistive loading. *IEEE Transactions on Antennas and Propagation* 13, 369–373 (May 1965; correction p. 998, Nov. 1965).

Yilmaz, O. (1987). *Seismic Data Processing*. Investigations in Geophysics 2, Society of Exploration Geophysicists, Tulsa, 526 p.

Yilmaz, O. (2001). *Seismic Data Analysis: Processing, Inversion and Interpretation of Seismic Data* (Vols. 1 & 2). Investigations in Geophysics 10, Society of Exploration Geophysicists, Tulsa, 2027 p.

Chapter 12

Detection and visualization of glacier area changes

Frank Paul
Department of Geography, University of Zurich, Switzerland

Johan Hendriks
Department of Geosciences and Geography, University of Helsinki, Finland

12.1 INTRODUCTION

The area of a glacier is a basic property in a glacier inventory that is frequently used for upscaling of only locally available information to large ensembles of glaciers (e.g. size dependent calculations of overall area change or sea level rise). The land surface area covered by glaciers and ice caps is also an important boundary condition for climate models that calculate the energy fluxes according to the surface type. As the globally available data sets of glacier covered area are either not complete or very rough, new initiatives like Global Land Ice Measurements from Space (GLIMS) have started to compile a global glacier inventory (location, size, digital outlines) from satellite data. Glacier area change (percent per year) could be compared on a global scale to quantify regional climate change effects. However, glacier-specific changes could only be determined when the same entities are compared. This is quite troublesome when only the point information from the former World Glacier Inventory (WGI) is available. Glacier area changes as obtained from satellite data over large regions do also help to assess the representativeness of the sparser sample of field-based length change and mass balance measurements or to identify what else is going on (Haeberli et al. 2007). For example, it is possible to observe whether the change is restricted to the glacier terminus or if a glacier with a stagnant tongue is really in a good health. In the European Alps, such downwasting (i.e. stationary thinning) and disintegrating glaciers with little change at the terminus could be observed widely using spaceborne sensors (Paul et al. 2007). In the sections below, various ways of qualitative and quantitative area change assessment and visualisation are described. Thereby, Landsat is in general used as a synonym for Landsat-type satellites that have similar spatial resolution and spectral bands.

12.2 SIMPLE IMAGE OVERLAY

A very efficient tool for rapid change detection analysis are animated image time series from false colour composites which make use of the special spectral characteristics of ice and snow (Paul et al. 2003 and 2007). For example, the combination of TM bands 5 (SWIR wavelength), 4 (NIR) and 3 (red) as red, green and blue, respectively, in a false colour composite yield very detail rich images as the grey levels in these three bands have little correlation. They have thus also widely been used to display image quick-looks from TM/ETM+ data in the worldwide web. Image time series from Landsat TM/ETM+ require only a relative image matching with a few ground control points (GCPs), as the orbit of Landsat has been stable for more than 20 years. The images should be acquired around the same time of the year in order to reduce illumination changes in the shadowed zones. Moreover, the size of the image frames is restricted to the screen size, i.e. about 1200 pixels or 40 km at a resolution of 30 m (Paul et al. 2003). Colour balance is in general not a serious problem and public domain image display tools can be used for the animation. Such animations clearly show changes in glacier length, emerging new rock outcrops, the formation of pro-glacial lakes and the overall glacier shrinkage, thus pointing to the regions of rapid environmental change (Paul et al. 2007).

For a more quantitative assessment of glacier area change, glacier maps can be created (see Chapter 8) and the number of pixels for specific glaciers times the area covered by a pixel (spatial resolution) provides the corresponding glacier area. Area changes are then calculated from a second data set acquired at a different point in time (Paul 2002a). However, the analysis for several glaciers and the consistency of the derived areas is strongly facilitated when techniques of a Geographic Information System (GIS) are applied. Visualisations of the glacier area change from Landsat can be generated by

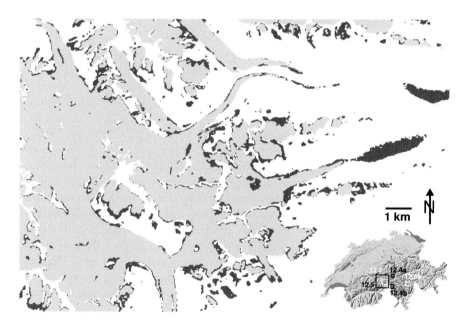

Figure 12.1 Change detection of glacier areas in the Oberaar glacier using simple image overlay of classified glacier maps without orthorectification or any correction for misclassification of debris cover and lakes. Glacier classification from 1998 is depicted in light grey, the larger 1985 extent in dark grey (the two large isolated dark grey areas to the right are lakes). The region shown in this figure is somewhat larger (17 km by 11 km) than Figure 12.3. The inset shows the approximate location of the regions depicted in Figures 12.3 to 12.6.

digital overlay of the relatively matched glacier maps (Paul 2002b, Paul et al. 2002). In Figure 12.1 such an overlay is presented for the Oberaar Glacier in Switzerland and its surroundings (cf. Figure 8.4) as mapped in 1985 and 1998 without any correction for debris cover or lakes and snow, but with a median filter applied to both glacier maps. This simple method generally fails when comparisons with different data sources (i.e. other satellite sensors, geocoded vector data sets, DEM fusion) are performed and orthorectification is required.

12.3   ORTHORECTIFICATION OF SATELLITE IMAGES

In high-mountain topography an accurate digital elevation model (DEM) is needed to compensate for the effects of panoramic distortion (Figure 12.2a) and creating orthoimages (i.e. each image pixel as seen from a nadir position) from the satellite data (Schowengerdt 1997). In Figure 12.2b the shift in pixel position as a function of terrain height and distance from the image centre for Landsat TM is given. For example, a mountain peak at 2500 m elevation with a distance of 45 km to the image centre is shifted by about 150 m from its real position corresponding to five pixels. Moreover, appropriate topographic maps or ground-based surveys with GPS receivers are required for collection of GCPs which should have x, y, and z (elevation) coordinates in a metric map coordinate system. Both data sets (DEMs, maps) are often not easily available with a sufficient spatial resolution/accuracy in glacierized regions outside of Europe.

The Shuttle Radar Topography Mission (SRTM) 3 arc second (about 90 m) resolution DEM, that can be down-loaded for free from a NASA ftp-server, provides DEM information between 61,0°N and 57,4°S (Farr et al. 2007). In glacierized regions outside these latitudes, DEMs can also be created from stereo sensors like ASTER along-track (Paul and Kääb 2005), SPOT across-track or interferometric SAR data (Joughin et al. 1996, Chapter 9). While the DEM created from ASTER data may exhibit large errors at north facing slopes as well as in regions of cast shadow and uniform snow cover (Kääb 2002, Gonçalves and Oliveira 2004, Eckert et al. 2005), the SRTM DEM suffers from foreshortening and layover effects due to the sensor geometry and characteristics of radar data, both causing data voids in rugged topography (Hall et al. 2005). Partly, the DEMs derived from optical sensors can be used to fill the data voids of the SRTM DEM or vice versa (Kääb 2005). Despite the somewhat higher spatial resolution obtained from an ASTER DEM (e.g. 30 m), the accuracy of the elevation values are similar to the SRTM3 DEM, at least in rugged topography (Hirano et al. 2002, Kääb 2002, Toutin 2002, Eckert et al. 2005). Both the ASTER and the SRTM DEM have already been applied successfully for calculation of glacier elevation changes by comparison with previous DEMs (e.g. Larsen et al. 2007, Paul and Haeberli 2008, Rignot et al. 2003, Surazakov and Aizen 2006). In addition to DEM generation from spaceborne remote sensing data, airborne remote sensing data, such as laser scanner data (Geist and Stötter 2007), airborne SAR data (Høgda et al. 2007) and aerial photography as presented in Chapters 7, 9 and 10 in this book can be applied for high resolution (e.g. 1 m spacing) DEM generation in small regions.

A further critical point is the use of a common map projection and datum (ellipsoid) in all data sets, as only this ensures proper image overlay. In order to allow the

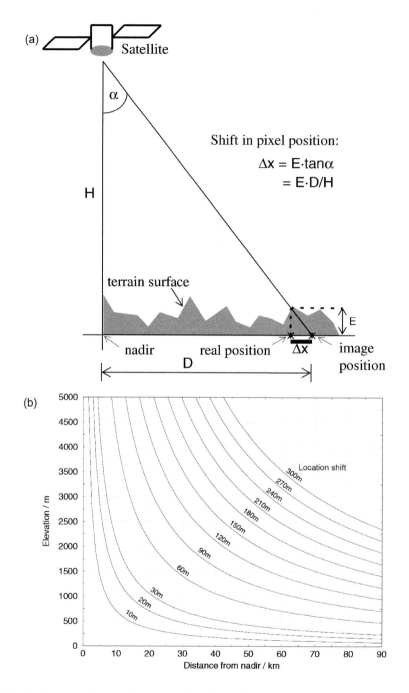

Figure 12.2 a) schematic diagram illustrating the pixel shift due to terrain height and the oblique viewing geometry (panoramic distortion) from a satellite sensor; b) illustration of the pixel shift as a combined effect of terrain height and distance from the scene centre for Landsat TM.

calculation of glacier areas, images are generally reprojected to a metric system like UTM with the WGS84 or other ellipsoid (e.g. Paul and Kääb 2005). For example, in the OMEGA project all the image material and DEMs produced were in UTM with the WGS84 ellipsoid. In smaller countries often national metric systems are used as they are provided on the topographic maps that are used for GCP collection. If properly set, transformations between different projections is without problems and can be performed by common GIS or other software. The entire DEM generation and orthorectification issue for satellite data is integrated in various commercial image processing software packages and can be applied by unexperienced users as well (Gonçalves and Oliveira 2004). However, one has less control on the stereo matching process during DEM generation and additional post-processing is required for reduction of gross errors (Kääb 2002, 2005).

12.4   GIS-BASED CALCULATIONS

Once glacier maps represented as raster data have been generated from the multispectral satellite data (Figure 12.3a), the regions covered by glacier should be converted to glacier outlines represented as vector data for further analysis (Figure 12.3b). This so called raster-vector conversion is straight forward and is included in most GIS and digital image processing software. In general, the latter can be used for creating the glacier map and exporting it to a geocoded image file (e.g. GeoTIFF) and the GIS software imports this image to a raster format and then converts it to polygon outlines (Paul 2007, Paul et al. 2002). Additionally, glacier boundaries delineating catchment areas can be defined in a separate vector layer (Figure 12.3c), when possible according to a former inventory. This allows the automated extraction of individual glaciers from the classified map (Figure 12.3d) and a consistent analysis through time (Paul 2002b, Paul et al. 2002, Heiskanen et al. 2003). This separate vector layer can also be used to remove misclassification (e.g. snow patches or detachment from a lake) and assign a joint identification code (ID) to several glaciers at the same time (e.g. from the parent glacier in a former inventory).

Another technique is used for data storage in the GLIMS database at National Snow and Ice Data Center (NSIDC), where the glacier shape is stored as a group of individual segments which keep their function (e.g. terminus, ice divide, lateral boundary) in a metadata table (Raup et al. 2007a). However, this storage format requires some additional digitizing of individual glacier segments or assignment of their function. The implementation of an automatic conversion from the polygon to the segments format is an ongoing effort (Raup et al. 2007b).

Once the geocoded glacier outlines are available in a digital format as a vector layer, a huge number of further calculations can be performed, in particular in combination with a DEM (Kääb et al. 2002, Paul et al. 2002). This includes the calculation of 3D glacier parameters (e.g. slope, aspect, lowest and highest glacier elevation) on a glacier specific basis that can be used for further modelling (Haeberli and Hoelzle 1995), the creation of glacier inventory data (Paul 2007), or assessment of glacier hypsographies (area-elevation distribution) for assessment of future climate change impacts (Paul et al. 2004; Zemp et al. 2006), to name not all but a few. If several glacier outlines of specific years are available, they can also be used for the more

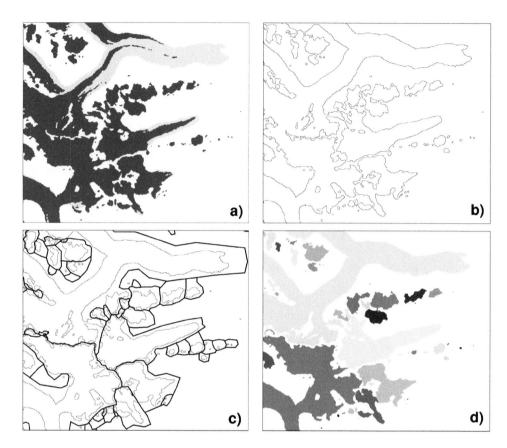

Figure 12.3 Processing steps of the classified glacier map for the Oberaar glacier (sub-section of Figure 12.1). a) glacier map (dark grey) with manually added debris cover (light grey) in a raster format (cf. Figure 8.4); b) glacier outlines in vector format after raster-vector conversion with the GIS; c) glacier catchments (thick black lines) from manual delineation for extraction and identification of individual glaciers; d) individual glaciers colour coded and converted back to the raster format for zonal (glacier specific) calculations with various input grids (DEM, slope, aspect, etc.).

qualitative visualization of glacier change (Figure 12.4) and the communication of the changes to the public (Figure 12.5).

The calculation of glacier specific area changes and other parameters can be performed by external programs that may also include output format changes for later display in scatter plots, bar graphs or other diagrams (Paul et al. 2002, Paul 2007). It can be noted that the entire processing chain (glacier mapping, area calculation, change assessment, visualization) can be performed with non-commercial software when orthorectified satellite data are available. However, more comfortable processing environments can be implemented in several commercial software products (Hendriks and Pellikka 2007, Paul 2007).

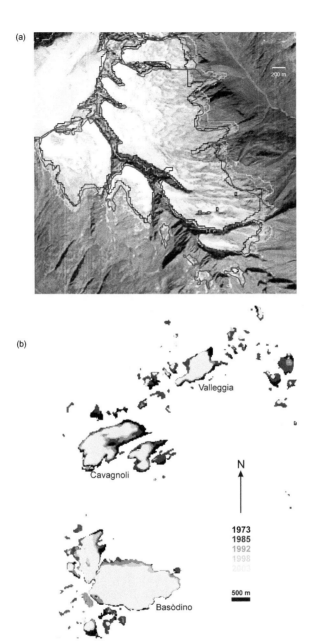

Figure 12.4 a) digital overlay of glacier outlines from 1973 (dark grey), 1985 (white), 1992 (light grey) and 1998 (black) in the Göschener Alp Region, Switzerland. The 1973 outlines are from the digitized glacier inventory, all others are obtained from Landsat TM; b) glacier areas for five individual years on a pixel-by-pixel basis coded by different shades of grey with advancing Basòdino, disintegrating Cavagnoli and stable Valleggia glaciers. Several snow patches were present in the 1985 image (dark grey). (See colour plate section).

Figure 12.5 Synthetic oblique perspective view of the region to the west of the Grimsel Pass with a fused satellite image (Landsat TM and IRS-1C) draped over a DEM and glacier outlines from 1850 (white), 1973 (light grey) and 1998 (black). DEM25 reproduced with permission from swisstopo (BA091556). (See colour plate section).

12.5 VISUALISATION OF GLACIER CHANGE

There are abundant possibilities to visualize temporal glacier changes. They can be sorted according to several aspects like animated or static, 2D or 3D, nadir or oblique, greyscale or colour, thematic or realistic rendering, and all combinations thereof (Kääb et al. 2003, Paul et al. 2003, Lobben 2003). In respect to the comparable fast glacier changes, the appropriate method mainly depends on the scale and time frame depicted for the images, as well as on the forum of the dissemination. For this contribution we have selected three examples in Switzerland (see Figure 12.1 for location): (1) Outlines from a small group of glaciers in the Göschener Alp Region in a 2D nadir view with a satellite image in the background (Figure 12.4a), (2) glacier changes on a pixel-by-pixel basis for five distinct years in the region around Basòdino Glacier (Figure 12.4b), and (3) a synthetic oblique perspective view of the region to the west of the Grimsel Pass, with glacier outlines from 1850, 1973 and 1998 draped over a fused true colour satellite image and over a DEM (Figure 12.5). The selected examples illustrate the possibility of tracking glacier changes with Landsat TM over very short periods of time (a few years), but also the need for proper orthorectification of the satellite data when they are combined with digital geocoded data from other sources such as a DEM, digitized outlines, and different sensors.

While in Figure 12.4a the intermittent advance period of Alpine glaciers until about 1985 is nicely illustrated without obscuring any of the other positions, the direct overlay of the classified glacier maps as in Figure 12.4b (e.g. for Basòdino Glacier) obscures some of the former glacier positions and change assessment has to be done with the original data. However, the disintegration caused by retreat, shrinkage and splitting of Cavagnoli Glacier can be followed very clearly. The perspective view (Figure 12.5) has a two-fold message: On the one hand such a pseudo-realistic rendition is visually more attractive than their 2D counterparts, and on the other hand some terrain is obscured and the entire range of changes is not visible. Moreover, a certain distance to the mountain range is required to yield an attractive effect, implying that only large changes (i.e. over a longer period of time) can be shown. As such, this latter kind of images better addresses the general public, while illustrations in Figure 12.4 are more interesting for a scientific community.

12.6   RECENT GLACIER CHANGES IN THE ALPS

The focus of this section is on observations that have been made from satellite data in the Swiss Alps. The comparison of the former glacier inventories with the recently compiled ones for Austria (Lambrecht and Kuhn 2007) and Italy (Citterio et al. 2007) confirm the observations discussed below. An overview of glacier area changes on a global scale is given by Barry (2006).

During the past 25 years glaciers in the Alps experienced a dramatic decline. This has been documented from direct mass balance observations (Zemp et al. 2005) as well as from the new Swiss glacier inventory, where area changes from 1985 to 1998/99 are compared to the 1850–1973 period (Paul 2007, Paul et al. 2004). The related analysis reveals 18% glacier area loss in Switzerland (−22% when scaled to the entire Alps) between 1985 and 1998/99 (i.e. −1.4% per year), which is about seven times faster than for the 1850–1973 period. Small glaciers less than one square kilometer in size contribute about 44% to the total loss although they cover only 18% of the total area in 1973, i.e. their relative decrease in area is much higher (−39%) than the mean value. Moreover, a large scatter in the relative area changes was observed towards smaller glaciers (Figure 12.6a), indicating that the behaviour of individual glaciers smaller than about 5 km^2 might not reveal climate change impacts, as only the total sample of a larger region provides a realistic picture. This is also obvious from the changes observed in the Rheinwald Region in Switzerland (Figure 12.6b). Most often glaciers with little change are side-by-side to glaciers with a dramatic area loss. These observations encourage the application of satellite data for assessment of glacier change, as they cover entire mountain ranges at the same point in time and the data processing from thresholded ratio images is comparably straight forward as well as robust (see Chapter 8) and thus widely applicable.

Analysis of glacier area changes in the Alps revealed further, that downwasting and disintegration instead of a dynamic retreat was a dominant response of the glaciers to the extraordinary warm conditions of the past 20 years. Several adverse effects

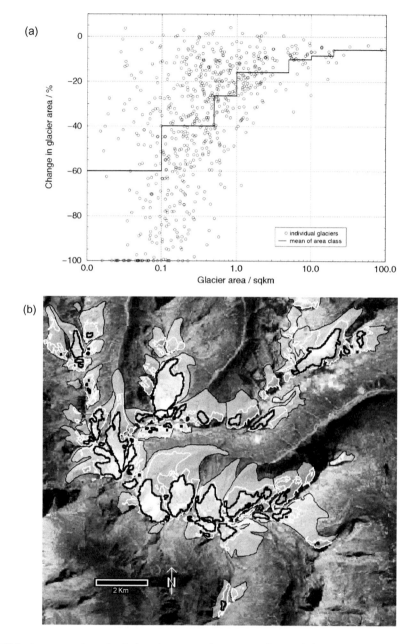

Figure 12.6 a) scatter plot of relative change in glacier area from 1973 to 1998 vs. glacier size in 1973 for a sample of 713 Swiss glaciers. The increasing scatter towards smaller glaciers is obvious. The mean change for distinct size classes increase towards smaller glaciers (black line); b) glacier outlines in the Rheinwald Region for 1850 (black, thin), 1973 (white) and 1999 (black, thick) on a Landsat TM band 3 from 1999. The high local variability of the area changes is remarkable.

had been initiated that are related to positive feedbacks (e.g. lake formation, tongue separation, albedo lowering, loss of firn reservoirs) which will cause further glacier decline in the near future, independent of climate development. The rapid area changes observed throughout the Alps are also causing new challenges for the international coordinated glacier monitoring strategies (Haeberli et al. 2007, Paul et al. 2007). Apart from losing glaciers with comparably long direct mass balance series (e.g. Caresèr in Italy), the annual measurements of length changes are getting more difficult and might not be related to glacier health any more. As such, the current rapid glacier wastage is not only related to problems for tourism, hydrology and potential new hazards, but also to the future continuity of the long data sets obtained by global glacier monitoring.

REFERENCES

Barry, R. (2006). The status of research on glaciers and global glacier recession: a review. *Progress in Physical Geography* 30, 285–306.

Citterio, M., G. Diolaiuti, C. Smiraglia, C. D'Agata, T. Carnielli, G. Stella and G.B. Siletto (2007). The fluctuations of Italian glaciers during the last century: a contribution to knowledge about alpine glacier changes. *Geografiska Annaler, Series A* 89(3), 167–184.

Eckert, S., T. Kellenberger and K. Itten (2005). Accuracy assessment of automatically derived digital elevation models from ASTER data in mountainous terrain. *International Journal of Remote Sensing* 26(9), 1943–1957.

Falorni, G., V. Teles, E.R. Vivoni, R.L. Bras and K.S. Amaratunga (2005). Analysis and characterization of the vertical accuracy of digital elevation models from the Shuttle Radar Topography Mission. *Journal of Geophysical Research* 110, F02005.

Farr, T.G., and 17 others (2007). The Shuttle Radar Topography Mission *Reviews of Geophysics*. 45, RG2004, DOI: 10.1029/2005RG000183.

Geist, T. and J. Stötter (2007). Documentation of glacier surface elevation change with multi-temporal airborne laser scanner data – case study: Hintereisferner and Kesselwandferner, Tyrol, Austria. *Zeitschrift für Gletscherkunde und Glazialgeologie* 41, 77–106.

Gonçalves, J.A. and A.M. Oliveira (2004). Accuracy analysis of DEMs derived from ASTER imagery. *International Archives of Photogrammetry, Remote Sensing and Spatial Information Sciences* 34(B3), 5 p.

Haeberli, W. and M. Hoelzle (1995). Application of inventory data for estimating characteristics of and regional climate-change effects on mountain glaciers: a pilot study with the European Alps. *Annals of Glaciology* 21, 206–212.

Haeberli, W., M. Hoelzle, F. Paul and M. Zemp (2007). Integrated monitoring of mountain glaciers as key indicators of global climate change: the example of the European Alps. *Annals of Glaciology* 46, 150–160.

Hall, O., G. Falorni and R.L. Bras (2005). Characterization and quantification of data voids in the Shuttle Radar Topography Mission data. *IEEE Geoscience and Remote Sensing Letters* 2(2), 177–181.

Heiskanen, J., K. Kajuutti and P. Pellikka, 2003. Mapping glacier changes, snowline altitude and AAR using Landsat data in Svartisen, Northern Norway. *Geophysical Research Abstracts* 5, EAE03-J-10328.

Hendriks, J.P.M. and P. Pellikka (2007). Semi-automatic glacier delineation from Landsat imagery over Hintereisferner in the Austrian Alps. *Zeitschrift für Gletscherkunde und Glazialgeologie* 38(2), 139–154.

Hirano, A., R. Welch and H. Lang (2002). Mapping from ASTER stereo image data: DEM validation and accuracy assessment. *ISPRS Journal of Photogrammetry and Remote Sensing* 57, 356–370.

Høgda, K.A., T. Geist, M. Jackson, H. Elvehøy, J. Stötter and I. Lauknes (2007). Comparison of digital elevation models from airborne SAR technology and airborne laser scanner technology at Engabreen, Svartisen, Norway. *Zeitschrift für Gletscherkunde und Glazialgeologie* 41, 205–226.

Joughin, I., R. Kwok and M. Fahnestock (1996). Estimation of ice sheet motion using satellite radar interferometry: Method and error analysis with application to the Humboldt glacier, Greenland. *Journal of Glaciology* 42(142), 564–575.

Kääb, A. (2002). Monitoring high-mountain terrain deformation from air- and spaceborne optical data: Examples using digital aerial imagery and ASTER data. *ISPRS Journal of Photogrammetry and Remote Sensing* 57, 39–52.

Kääb, A., F. Paul, M. Maisch, M. Hoelzle and W. Haeberli (2002). The new remote-sensing-derived Swiss glacier inventory: II. First Results. *Annals of Glaciology* 34, 362–366.

Kääb, A., Y. Isakowski, F. Paul and A. Neumann (2003). Glaziale und periglaziale Prozesse: Von der statischen zur dynamischen Visualisierung. *Kartographische Nachrichten* 5, 206–212.

Kääb, A. (2005). Combination of SRTM3 and repeat ASTER data for deriving alpine glacier flow velocities in the Bhutan Himalaya. *Remote Sensing of Environment* 94(4), 463–474.

Lambrecht, A. and M. Kuhn (2007). Glacier changes in the Austrian Alps during the last three decades, derived from the new Austrian glacier inventory. *Annals of Glaciology* 46, 177–184.

Larsen, C.F., R.J. Motyka, A.A. Arendt, K.A. Echelmeyer and P.E. Geissler (2007). Glacier changes in southeast Alaska and northwest British Columbia and contribution to sea level rise. *Journal of Geophysical Research* 112, F01007.

Lobben, A. (2003). Classification and application of cartographic animation. *The Professional Geographer* 55(3), 318–328.

Paul, F., A. Kääb, M. Maisch, T.W. Kellenberger and W. Haeberli (2002). The new remote sensing derived Swiss glacier inventory: I. Methods. *Annals of Glaciology* 34, 355–361.

Paul, F. (2002a). Changes in glacier area in Tyrol, Austria, between 1969 and 1992 derived from Landsat 5 TM and Austrian Glacier Inventory data. *International Journal of Remote Sensing* 23(4), 787–799.

Paul, F. (2002b). Combined technologies allow rapid analysis of glacier changes. *EOS, Transactions, American Geophysical Union* 83(23), 253, 260, 261.

Paul, F., A. Kääb, M. Maisch, T.W. Kellenberger and W. Haeberli (2003). Das neue Schweizer Gletscherinventar: Anwendungen in der Gebirgskartographie. *Kartographische Nachrichten* (5), 212–217.

Paul, F., A. Kääb, M. Maisch, T.W. Kellenberger, and W. Haeberli (2004). Rapid disintegration of Alpine glaciers observed with satellite data. *Geophysical Research Letters* 31, L21402.

Paul, F. and A. Kääb (2005). Perspectives on the production of a glacier inventory from multispectral satellite data in the Canadian Arctic: Cumberland Peninsula, Baffin Island. *Annals of Glaciology* 42, 59–66.

Paul, F. (2007). *The new Swiss glacier inventory 2000 – Application of remote sensing and GIS.* Schriftenreihe Physische Geographie, Universität Zürich 52, 210 p.

Paul, F., A. Kääb and W. Haeberli (2007). Recent glacier changes in the Alps observed from satellite: Consequences for future monitoring strategies. *Global and Planetary Change* 56, 111–122.

Paul, F. and W. Haeberli (2008). Spatial variability of glacier elevation changes in the Swiss Alps obtained from two digital elevation models. *Geophysical Research Letters* 35, L21502.

Raup, B.H., A. Kääb, J.S. Kargel, M.P. Bishop, G. Hamilton, E. Lee, F. Paul, F. Rau, D. Soltesz, S.J.S. Khalsa, M. Beedle and C. Helm (2007a). Remote sensing and GIS technology in the

Global Land Ice Measurements from Space (GLIMS) Project. *Computers and Geosciences* 33, 104–125.

Raup, B., A. Racoviteanu, S.J.S. Khalsa, C. Helm, R. Armstrong and Y. Arnaud (2007b). The GLIMS geospatial glacier database: A new tool for studying glacier change. *Global and Planetary Change* 56, 101–110.

Rignot, E., A. Rivera and G. Casassa (2003). Contribution of the Patagonia Icefields of South America to sea level rise. *Science* 302, 434–437.

Schowengerdt, R.A. (1997). *Remote Sensing – Models and Methods for Image Processing.* Academic Press, New York, 522 p.

Surazakov, A.B. and V.B. Aizen (2006). Estimating volume change of mountain glaciers using SRTM and map-based topographic data. *IEEE Transactions on Geoscience and Remote Sensing* 44(10), 2991–2995.

Toutin, T. (2002). Three-dimensional topographic mapping with ASTER stereo data in rugged topography. *IEEE Transactions on Geosciences and Remote Sensing* 40(10), 2241–2247.

Zemp, M., R. Frauenfelder, W. Haeberli and M. Hoelzle (2005). Worldwide glacier mass balance measurements: general trends and first results of the extraordinary year 2003 in Central Europe. *Data of Glaciological Studies* [Materialy glyatsiologicheskih issledovanii], Moscow, Russia, 99, 3–12.

Zemp, M., W. Haeberli, M. Hoelzle and F. Paul (2006). Alpine glaciers to disappear within decades? *Geophysical Research Letters* 33, L13504.

Chapter 13

Detection of distortions in digital elevation models: simultaneous data acquisition at Hintereisferner glacier

Olli Jokinen
Department of Surveying, Helsinki University of Technology, Finland

13.1 INTRODUCTION

Various methods and data exist that can be used for producing digital elevation models (DEMs) of glacial areas. For comparison of different methods, it is advantageous that the data are acquired as near to simultaneously as possible. Such a simultaneous data acquisition was successfully carried out on August 12, 2003, at Hintereisferner, a valley glacier in the Central Eastern Alps, Austria, as a part of the OMEGA project funded by the European Commission during 2001–2004. The synchronous acquisition included very high resolution optical images by the Ikonos satellite, aerial photographs, airborne digital camera images, and airborne laser scanner data. For the georeferencing and verification of the spaceborne and airborne data, ground truth data were simultaneously acquired including terrestrial photographs in a small test area on the glacier, ground control points (GCPs) on bedrock surrounding the glacier, and checkpoints on ice along the glacier snout.

The high number of data sets acquired simultaneously makes the data unique and opens up new possibilities for comparison of different methods in a difficult mountainous environment. New information on the accuracy of the methods and especially of DEMs produced from Ikonos very high resolution satellite imagery will be obtained in changeable glacial areas where reliable reference data are difficult to obtain. Analyzing differences between multiple DEMs leads to the detection of shape distortions, blunders, and other inaccuracies in the DEMs. A surface matching algorithm will be considered for registration of the DEMs accurately into the same coordinate system and thus for correction of differences in georeferencing of the DEMs which may be considerable due to difficulties in establishing a reliable and well distributed network of ground control points covering the whole area.

This chapter is organized as follows. Related work dealing with comparison and accuracy of two or more DEMs is reviewed in Section 13.2. The DEMs produced from the data acquired simultaneously at Hintereisferner are described in Section 13.3. Section 13.4 presents methods for correction of differences in georeferencing of DEMs and detection of possible distortions in DEMs. The results obtained with the DEMs of Hintereisferner are presented in Section 13.5 and conclusions summarized in Section 13.6.

13.2 RELATED WORK

We have found no related work where more than two DEMs produced from simultaneously acquired data are compared in areas where considerable changes may occur in a fairly short time. In Jokinen et al. (2007), spaceborne Ikonos, airborne interferometric synthetic aperture radar (InSAR), airborne laser scanner, and terrestrial photography DEMs, GPS profiles and GCPs are compared at Engabreen, an outlet glacier of Svartisen ice caps, Norway. The data were acquired within one month during which changes occurred on the glacier. Nevertheless, qualitative differences between the DEMs such as gross errors in the Ikonos DEM, artificial edges in the InSAR DEM, and different georeferencing of the laser scanner DEM were detected. The elevations of the DEM derived from laser scanner data were the most accurate, followed by those of the DEM produced from Ikonos images after differences in georeferencing had been corrected.

In Baltsavias et al. (2001), a laser scanner DEM and three aerial photography DEMs produced by different digital photogrammetric systems were evaluated against a reference DEM generated manually in an analytical plotter at Unteraargletcher in Bernese Alps, Switzerland. There was a time difference of 1–2 weeks between the acquisitions of the aerial photographs and the laser scanning, during which the glacier may have flowed approximately 0.8–1.6 metres. For the DEMs from aerial photography, an elevation accuracy of 0.6–0.9 m is reported for glacial areas while for cliffs and breaklines, the accuracy is lower. Large differences in elevation have also been observed in the overlapping areas between neighbouring models in the DEMs from aerial photography while blunders occur particularly in areas of rough terrain. The elevations of the laser scanner DEM are found to have lower accuracy than the elevations of the DEMs from aerial photography due to differences in orientation and due to blunders in laser scanner data in debris-covered areas, cliff regions, and areas of low elevation where the maximum range for distance measuring was exceeded. In the upper part of the glacier, laser scanning performed as well as photogrammetry.

Favey (2001) shows further that the change in elevation during one year calculated independently from two laser scanner DEMs and two DEMs from aerial photography over Unteraargletcher is similar for both techniques and can be determined with an accuracy of 0.5–0.7 m. Systematic effects related to the laser scan pattern have also been detected in a difference image between the DEMs from aerial photography and laser scanner data after a heuristically determined systematic offset of 0.6 m between the elevations of the data sets had been corrected.

The accuracy and geomorphologic quality of aerial photography and laser scanner DEMs are compared further in other applications, not dealing with glaciers, by Baltsavias (1999a) and Kraus and Pfeifer (1998). The elevation accuracy of laser scanner data is affected by uncertainties in the GPS/INS positioning (location and attitude) of the sensor, and errors in distance measurement (Baltsavias 1999a). The elevation accuracy decreases with increasing terrain slope and roughness because of uncertainty in the horizontal coordinates (Kraus and Pfeifer 1998). There may also be height discrepancies in overlapping areas of neighbouring laser scanner strips. Different factors affecting the accuracy of 3-D coordinates measured by laser scanning are discussed by Baltsavias (1999b). Laser scanner DEMs are usually processed to give a smooth surface, while breaklines and natural features are preserved in DEMs manually measured from aerial photographs (Baltsavias 1999a). The accuracy of elevations measured manually from aerial stereo images depends mainly on the flying altitude above the ground, camera focal length, surface slope, and accuracy of image measurements (Kraus 1987).

Various high resolution optical satellite stereo images including Ikonos II, Spot 5, Eros A, and Quickbird are tested for DEM production and compared to a reference laser scanner DEM by Toutin (2004). An accuracy of 1.5 m is reported for the elevations of the Ikonos DEM on bare surfaces including soils and lakes.

Eisenbeiss et al. (2004) investigate the accuracy of Ikonos for point positioning and digital surface model (DSM[1]) generation using a block of images over a difficult terrain including snow, long shadows, and occlusions due to mountains, in Thun, Switzerland. The image quality was not the best as artefacts were visible especially in homogeneous areas in the Ikonos images. Image enhancement improved the quality of images especially in areas covered by snow. An elevation accuracy of 1–5 m was achieved when compared to a reference laser scanner DSM, which had an accuracy of 0.5–1 m in open areas and 1.5 m in vegetation areas. The elevation accuracy of the Ikonos DSM was highest in open-textured areas and lower in urban and vegetated areas where the laser scanner and Ikonos DSMs measured heights of different objects. A bias of 1 m between the laser scanner and Ikonos DSMs was reported. There was a time difference of 3–4 years between laser scanning and Ikonos acquisition so that the snow-cover was different in the mountains and the comparison inadequate there.

Baltsavias et al. (2006) detected stripes in the difference image between the same Ikonos and laser scanner DSMs of Thun. It was found that the stripes were due to problems in the interior orientation of the Ikonos images as there seemed to be unmodelled relative shifts between the three CCD arrays, which form the whole Ikonos image. Poon et al. (2005) reported a similar RMS difference of 2–5 m for an Ikonos DSM as compared to a reference laser scanner DSM in urban and rural areas in Hobart, Australia. The accuracy varies according to the land cover and surface slope.

Small format digital imagery provides a low-cost alternative to aerial photographs but processing of digital imagery requires more work. The number of images is large and orientation of them using conventional methods would require establishing a large number of GCPs which may be impossible or impractical in difficult terrain such as alpine glaciers or other areas undergoing rapid changes.

Mills et al. (2003) produced a wire-frame DEM from data collected by a kinematic GPS and then applied surface matching to solve the absolute orientation of a digital camera DEM with respect to the wire-frame DEM. An elevation accuracy of about 0.5 m has been obtained with Kodak DCS660 digital images at a scale of 1:22,000 in a study area for coastal erosion. Before absolute orientation, errors up to 60 m appeared towards the borders of the image block.

Mostafa and Schwarz (2001) have developed an airborne system consisting of two Kodak DCS420c digital cameras, a nadir and an oblique one, and a GPS/INS for direct georeferencing of the imagery. Elevation accuracies of 0.7 m using direct

---

1 A DSM represents the "top" surface including also vegetation, buildings, bridges, and other above ground structures, which are excluded in a DEM.

georeferencing and of 0.3 m using GPS/INS-aided block triangulation with images at a scale of 1:12,000 was achieved in an urban test area. Mason et al. (1997) reported an elevation accuracy of 0.6 m with stereo pairs of Kodak DCS460 digital images at a scale of 1:18,500 in an urban area.

Only a few experiments have been reported for airborne digital imagery of alpine glaciers. Gleitsmann and Kappas (2004) investigated DEM production from oblique medium format Rollei6008 metric digital images taken from a light aircraft flying below clouds at Wolverine glacier in Alaska. The approach is suitable for monitoring relatively small targets in detail despite adverse weather conditions.

13.3   SIMULTANEOUS DATA ACQUISITION

All the remote sensing data considered in this chapter were acquired on the same day, August 12, 2003, at Hintereisferner in the Central Eastern Alps, Austria (Pellikka 2007). The unique data included Ikonos satellite images, aerial photographs, airborne digital camera images, airborne laser scanner data, terrestrial images, and GPS measurements on the glacier. A set of ground control points on bedrock surrounding the glacier was also measured with a differential GPS during a field campaign from August 10 to 13, 2003. Different partners of the OMEGA project produced the DEMs of Hintereisferner as summarized in Table 13.1.

Table 13.1 Summary of DEMs and ground truth data. The format includes the grid spacing for raster DEMs and the average length of sides of triangles for TIN models.

	Data	Method/ software	Format	Georeferencing	Coordinate system
I_M	Ikonos	manual	8.8 m TIN	GCPs	M28
I_L	Ikonos	automatic/ LPS	2 m raster	GCPs	M28
I_R	Ikonos	automatic/ RSG	1 m raster	GCPs	M28
I_T	Ikonos	manual	10.8 m TIN	GCPs	M28
AP	aerial photographs	semi- automatic	10 m raster	GCPs	M28 and UTM32/WGS84
DC	digital camera	automatic/ EnsoMOSAIC	10 m raster	GCPs	UTM32/WGS84
LS	laser scanner	automatic/ SCOP++	5 m raster	GPS and football field	UTM32/WGS84
LS_T	laser scanner	original data	1.2 m TIN	GPS and football field	UTM32/WGS84
TP	terrestrial photographs	Manual	0.9 m TIN	GCPs	UTM32/WGS84
G_R	GPS bedrock	–	14 points	–	M28 and/or UTM32/WGS84
G_G	GPS glacier	–	9 points	–	M28

Two partners produced separate DEMs from the same pair of Ikonos images. WM-data Novo produced an Ikonos DEM using an automatic image matching technique of Leica Photogrammetry Suite 8.7 beta version (LPS 8.7) software and manual editing in ESRI ArcView. Two triangulated irregular network (TIN) DEMs were generated by manual stereoscopic interpretation for the glacial area of Hintereisferner and for a terrestrial photography (Chapter 6) site located on the northwestern side of the ablation area of Hintereisferner. The Institute of Digital Image Processing at Joanneum Research (JR) produced another Ikonos DEM for the whole area using their own RSG software package (Joanneum Research 1997).

Aerial photographs of scale 1:15,000 were acquired with a RMK TOP 15 aerial camera having a focal length of 153 mm. The images were digitized with a scanning resolution of 14 μm and a DEM of 10 m grid spacing was produced using a semi-automatic approach by the Department of Photogrammetry and Remote Sensing at the Technische Universität München (TUM). An orthophotograph was generated and borders of glacial areas were extracted by detecting ridge and valley lines and using texture filters. The borders of some smaller glaciers were extracted manually and a few tiny patches were detected by grey-level thresholding of the red wavelength band of the orthophotograph.

The airborne laser scanner data acquisition was performed by TopScan GmbH with an Optech ALTM 2050 laser scanning system. A preliminary laser scanner DEM of 10 m and a refined, more carefully generated one of 5 m grid spacing were produced using SCOP++ software with the implemented linear prediction algorithm (Pfeifer et al. 2001) by the Institute of Geography at the University of Innsbruck (Chapter 10). The original laser scanner points were also available at the terrestrial photography site and triangulated into a TIN DEM.

The airborne digital camera images were acquired simultaneously with the laser scanning by TopScan GmbH with a 4000 × 4000 pixel digital metric camera. An image mosaic was generated and a digital camera DEM was produced using EnsoMOSAIC software of Stora Enso Ltd (EnsoMOSAIC 2006) by the Department of Geography at the University of Helsinki (Parviainen 2006).

During the field campaign, concentric terrestrial image sequences were captured from different camera positions at the terrestrial photography site. A set of control points was measured with a tacheometer for the orientation of the image sequences. A TIN DEM was measured manually from a stereo pair of panoramic images produced from concentric images by the Institute of Photogrammetry and Remote Sensing at Helsinki University of Technology.

There existed 14 GCPs in the neighbourhood of Hintereisferner (Kuhn 1980, Schneider 1975, personal communication with H. Lechner, AVT ZT GmbH, and K. Eder, TUM). These were specified in zone M28 of the reference system of the Military Geographic Institute in Austria based on the Gauss-Krüger projection of the Bessel 1841 ellipsoid with the central meridian at 10 degrees 20 minutes East (www.geocities.com/CapeCanaveral/1224/prj/at/at.html). Two of the GCPs were also given in the WGS84 system and seven additional ones were measured with a differential GPS during the field campaign. The map projection used with the WGS84 ellipsoid was zone 32 of the Universal Transverse Mercator projection having the central meridian at 9 degrees east (Pearson 1990).

The Ikonos, aerial photography, digital camera, and terrestrial photography DEMs were georeferenced independently by each producer in one or both of the reference coordinate systems using the established GCPs. The initial positioning of the laser scanner data was based on GPS measurements on board which were differentially corrected using data from two permanent GPS receiving stations located at Patscherkofel and Krahberg in Austria. The georeferencing of the laser scanner DEM was refined using a control area surveyed with a tacheometer by the Austrian Mapping Agency, Bundesamt für Eich- und Vermessungswesen (BEV) (Chapter 10). The tacheometer measurements were linked to a known control point existing in the neighbourhood. The coordinates of the control point were given in the M28 system and transformed to the WGS84 system by BEV. For the evaluation of the DEM accuracy, nine checkpoints were measured by JR with a differential GPS on the glacier.

13.4 METHODS

Systematic differences in elevation between two DEMs produced from simultaneously acquired data may result from differences in georeferencing of the DEMs, from internal errors in either or both of the DEMs, or from the method of comparison itself such as resampling of the other DEM when the difference is evaluated. In the following, methods are considered for correction of differences in georeferencing of the DEMs in Section 13.4.1 and detection of distortions in the elevations of the DEMs in Section 13.4.2. Resampling errors will be quantified later in Section 14.3.1 below.

13.4.1 Correction of differences in georeferencing

Georeferencing means the process of referencing an image to a geographic location defined by coordinates in given coordinate system. For multiple spaceborne, aerial, or terrestrial optical images, the georeferencing is usually solved indirectly by triangulation using ground control points with known geographic coordinates, their corresponding image coordinates measured from the images, an appropriate sensor model which relates the image coordinates to object coordinates, and tie points measured between overlapping images. A bundle block adjustment yields the exterior orientations of the images and object coordinates of the tie points in a reference coordinate system where a DEM is further generated via stereoscopic measurements. Direct georeferencing of airborne data such as laser scanner data, digital camera images, or aerial photographs is based on GPS/IMU measurements on board and usually refined using additional ground truth data.

When DEMs are georeferenced using different methods or different ground truth data, the results are also more or less different depending on the quality of GCPs and their image coordinates, correctness of the sensor model, accuracy of GPS/IMU measurements and other ground truth data. Possibilities for establishing an adequate and dense network of GCPs in the mountains are limited and difficult flying conditions may decrease the accuracy of direct georeferencing. In the case study, two reference coordinate systems were used and transferring the data from one system to another introduces a major error source for comparison of DEMs georeferenced in different systems.

Possible differences in the original georeferencing of the DEMs are compensated by registering the DEMs accurately into the same coordinate system using an algorithm based on surface matching as introduced by Jokinen (1998, 2000). The algorithm estimates the parameters of rigid body transformations (three rotations and three translations) between the coordinate systems of different DEMs by minimizing simultaneously the mean of the squares of weighted differences in elevation between multiple DEMs in the overlapping areas. The weighting includes adaptive thresholds for the direction of the surface normal and the distance between corresponding points. The corresponding points are established on the underlying grids of the DEMs according to the current values of the unknown rotation and translation parameters updated during the iteration by the Levenberg-Marquardt algorithm. The adaptive thresholds become tighter as the iteration proceeds according to the idea proposed by Zhang (1994).

The matching algorithm provides a high accuracy, large pull-in range, and short computing time if the data can be represented as single valued parametric surfaces (Jokinen 2000) such as DEMs on regular grids. Both planimetric coordinates and elevations are subject to measuring errors in the original laser scanner data and photogrammetric measurements. When a DEM is produced on a regular grid, only the elevations are corrupted with measuring errors while the grid points are accurate. For photogrammetrically measured TIN DEMs, the planimetric ground coordinates are usually more accurate than the elevations (Baltsavias 1999a) and the matching algorithm will show good performance for TIN DEMs of Hintereisferner, too. Gross errors in the correspondences are eliminated using adaptive thresholds. Systematic distortions in the DEMs affect the performance and result of surface matching. The matching should be ideally performed on areas free of distortions.

13.4.2 Detection of distortions

The focus is on detecting small-scale distortions in the shape of the elevation surface. This requires that the DEMs are accurately in the same coordinate system which is achieved by surface matching as described above. The detection is based on analyzing pair-wise difference images between the elevations of three or more DEMs covering the same area and acquired at the same time. We are looking for ridge edges, areas of positive or negative values where the differences in elevation exceed error bounds to be discussed in Section 14.3.1, and trends existing in all difference images involving a particular DEM but missing in all other difference images between other DEMs. This suggests the existence of a distortion or another systematic error and shows which DEM is distorted or erroneous. The detection of artificial edges is ensured checking visually the DEM in question in the area of detection. It is essential that the number of DEMs is at least three as the difference between two DEMs does not show which of them contains the error assuming that neither of them is a reference DEM known to be correct. A simultaneous data acquisition is also essential for detection of distortions in glacial and snow-covered areas.

In Chapter 14, the error bounds for the difference in elevation are used for change detection and accuracy estimation. Since changes are now excluded, possible reasons for areal differences exceeding the error bounds, calculated according to expected noise levels in the elevations, include the following. There may be systematic errors in the elevations of one or both of the DEMs, the noise levels in the elevations may be larger than expected, or the surface matching algorithm may have converged to an incorrect solution.

Previous knowledge of weaknesses in different methods of producing DEMs helps the detection of distortions. Shape distortions may appear in overlapping areas between adjacent laser strips where height discrepancies are usually adjusted during DEM production. Similarly, the overlaps between adjacent stereo models may be subject to distortions in DEMs produced from multiple aerial photographs or digital camera images. Deformation of the image block with increasing distortion towards the scene borders is also a typical problem. The reasons for distortions vary. The quality of the original data, weather conditions during data acquisition, appropriateness of the sensor model and its parameters, properties of the target area such as the amount of texture, surface reflectivity and roughness all influence the quality of the DEM produced. If the shape of the DEM surface is distorted, distances and directions between points may be incorrect and surface curvature erroneous. The distortions may be so large that the DEM is useless in practice.

13.5   RESULTS

In this section, the results obtained with the DEMs of Hintereisferner are presented. The elevation accuracy is first evaluated against ground truth data in Section 13.5.1. Problems related to having two reference coordinate systems are discussed and the method of surface matching for dealing with the differences in georeferencing is illustrated in Section 13.5.2. Mean and RMS differences in elevation between the DEMs are then calculated according to original georeferencing and after surface matching and analyzed in ice- and snow-covered test areas in Section 13.5.3. Finally, distortions detected in each of the DEMs are described and illustrated in Section 13.5.4.

13.5.1   Accuracy against ground truth data

The results of accuracy estimation are given in Table 13.2. The abbreviations of the DEMs and ground truth data were given in Table 13.1. The positions of the GCPs, checkpoints, and terrestrial photography site are illustrated on the DEM produced from aerial photographs in Figure 13.1.

13.5.1.1   Accuracy against GCPs

The sample means and standard deviations of differences in elevation and RMS differences in elevation between the DEMs and GCPs shown in Table 13.2 indicate how well the image orientations could be solved for the DEMs produced from Ikonos images, aerial photographs, and digital camera images while for the laser scanner DEM, the GCPs act as checkpoints as they were not used in the georeferencing. Since the measured image coordinates of the GCPs and the corresponding positions in the ground coordinates of the DEMs are not known to us and cannot be recovered afterwards from the DEMs without having the original image measurements and calibration data, only the differences in elevation are evaluated while no positional differences can be investigated.

Table 13.2 Sample mean and standard deviation of differences in elevation, RMS differences in elevation, and bounds for the interpolation error (BIE) on the average between the DEMs and ground truth in metres according to the original georeferencing and after matching. The abbreviations are given in Table 13.1.

DEM-ground truth	Original (m)				After matching (m)			
	Mean	Standard deviation	RMS	BIE	Mean	Standard deviation	RMS	BIE
$I_M - G_G$	−1.4	1.2	1.8	0.2	0.0	0.3	0.3	0.6
$I_M - G_R$	0.4	3.9	3.4	1.7	0.0	3.9	3.3	1.7
$I_L - G_G$	−1.4	1.1	1.7	0.1	0.0	0.4	0.4	0.1
$I_L - G_R$	−4.9	10.3	11.0	0.2	0.0	10.3	9.9	0.2
$I_R - G_G$	−2.9	1.2	3.2	0.03	0.0	0.5	0.4	0.1
$I_R - G_R$	−2.9	2.7	3.9	0.3	0.0	2.7	2.6	0.3
$I_T - LS_T$	−0.1	1.2	1.2	0.9	−0.2	1.1	1.1	0.6
$AP - G_G$	−0.5	1.4	1.4	0.03	0.0	0.1	0.1	0.02
$AP - G_R$	−6.6	6.8	9.2	1.7	0.0	6.8	6.4	1.7
$DC - G_G$	30.3	8.5	31.3	5.8	−2.3	2.3	3.1	5.1
$DC - G_R$	5.7	15.7	16.0	7.3	0.0	15.7	14.9	7.3
$LS - G_G$	−3.2	1.6	3.5	0.1	0.0	0.2	0.2	0.2
$LS - G_R$	−2.3	0.7	2.4	0.4	0.0	0.7	0.6	0.4
$LS - LS_T$	−0.5	0.3	0.6	0.8	0.0	0.3	0.3	0.7
$LS_T - TP$	−0.7	0.5	0.9	0.1	0.02	0.19	0.19	0.13

According to the original georeferencing, the elevations of the DEM manually generated from the Ikonos images are closest to the GCPs on the average. However, this DEM covers only the GCPs which are close to the glacier, while the other DEMs cover also GCPs farther away where the differences are larger than close to the target area. The elevations of the Ikonos LPS, Ikonos RSG, aerial photography, and laser scanner DEMs are several meters below the GCPs on the average. The bias is clearest for the elevations of the laser scanner DEM as the standard deviation of differences is low when compared to the mean difference. The mean differences exceed the bounds for the interpolation error for bilinearly interpolated DEMs while for TIN DEMs, the bounds are more tolerant as there are some large triangles where interpolation errors may be considerable.

The differences in elevation were adjusted by matching the GCPs with the DEMs using the registration algorithm described above without weighting for surface normal and only a shift in elevation as an unknown. The RMS differences are reduced for all the DEMs after matching and the laser scanner DEM has the lowest RMS difference. The RMS differences after matching are mainly affected by random measuring errors in the elevations, possible residual systematic errors, and interpolation errors.

13.5.1.2   Accuracy against checkpoints

The elevation accuracy of the DEMs was evaluated at nine checkpoints measured on the glacier at the time of the simultaneous data acquisition on August 12, 2003. Figure 13.2a shows the differences in elevation between the aerial photography DEM

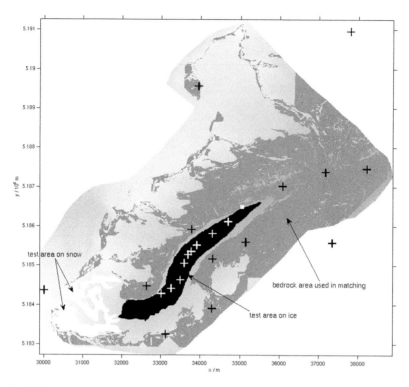

Figure 13.1 GCPs on the bedrock are shown as black crosses, checkpoints on the glacier as white crosses, and the terrestrial photography site as a white square. The test areas on ice and snow are illustrated by black and white, respectively. The dark grey areas are on bedrock. The coordinate system is M28.

and checkpoints with 95% confidence intervals for the differences. The confidence intervals include a bound for the interpolation error plus 1.96 times the standard deviation of the difference obtained through error propagation from a supposed noise level of 0.5 m for the elevations of the aerial photography DEM and an estimated standard deviation of 0.01–0.02 m for the elevations of the checkpoints. A systematic error appears as the differences increase in absolute value up the glacier. A similar trend appears also in the residual for the laser scanner DEM, which was georeferenced using a separate control area, and for all the Ikonos DEMs georeferenced using the GCPs. It is concluded that the checkpoints on the glacier are not compatible with the other data.

Since the checkpoints were not located at marked positions or close to natural features that could have been identified in the original images or DEMs, the checkpoints were registered with the aerial photography DEM using the matching algorithm with all the translation and rotation parameters as unknowns. The systematic error is reduced after matching in Figure 13.2b where the checkpoints have moved southeast with respect to the aerial photography DEM. The solution obtained with the aerial photography DEM was given as an initial estimate for matching with the other DEMs as this resulted in the reduction of the systematic error for all the DEMs while no good

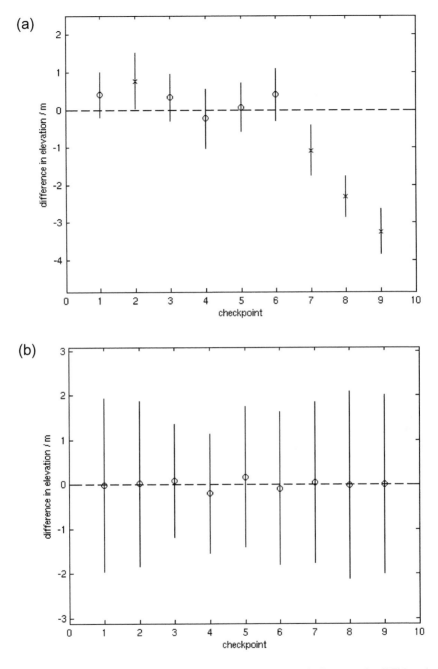

Figure 13.2 Systematic differences in elevation between the aerial photography DEM and check-points appearing according to the original georeferencing (a) are reduced after correcting the positions of the checkpoints by matching (b). The vertical lines illustrate the confidence intervals. Differences marked by crosses differ from zero and those marked by circles are about zero within the error bounds.

solutions were obtained when started from the original orientation of the data sets. For the Ikonos manual and LPS DEMs, the translations in the ground coordinates and the rotation around the vertical axis were kept fixed during registration. The results in Table 13.2 show that the aerial photography DEM outperforms all the other DEMs at the checkpoints.

13.5.1.3 Accuracy against terrestrial photography DEM

The DEM from the terrestrial photography provided the surface topography for a test area of 226 square metres in front of a longitudinal hillock. The point density of the laser scanner data was high enough for an adequate comparison (127 points in the overlapping area) while the Ikonos DEM was too sparse (only three points within this area). The RMS difference of 0.19 m achieved for the TIN laser scanner and terrestrial photography DEMs after correcting differences in georeferencing is comparable to the accuracy reported previously for the laser scanner data. More on this subject appears in Kajuutti et al. (2007).

13.5.2 Uncertainties due to different reference coordinate systems

The accuracy estimation described above was partly hampered by uncertainty in the values of the transformation parameters between the two reference coordinate systems used at Hintereisferner. In order to avoid this uncertainty, each DEM was compared to GCPs given in the same reference coordinate system as the DEM. The DEMs produced from Ikonos images and aerial photographs were thus compared to GCPs given in the M28 system while the digital camera and laser scanner DEMs were compared to GCPs in the WGS84/UTM32 system. The terrestrial photography DEM was in the WGS84/UTM32 system but the checkpoints were given in the M28 system and they were transformed to the WGS84/UTM32 system for evaluation of the accuracy of the digital camera and laser scanner DEMs. The transformation involved an inverse map projection of Gauss-Krüger to the Bessel ellipsoid, change of datum from the Bessel to the WGS84 ellipsoid, and a map projection to the UTM32 system. The heights of the geoid with respect to the Bessel and WGS84 ellipsoids were obtained from BEV, as a $3' \times 5'$ grid which was bilinearly interpolated for intermediate areas. The Bursa-Wolf model (Lankinen et al. 1992) was applied to the change of the reference ellipsoid. The values of the parameters of the Helmert transformation were estimated using the correspondences between GCPs given in the M28 and WGS84/UTM32 systems. However, it was not known which height of the geoid had been used for the GCPs given in the M28 system and measured earlier so the height given by the BEV was used. When different values for the parameters of the Helmert transformation given in (www.geocities.com/CapeCanaveral/1224/dat/dat.html; Schmidt 2003) were tested, the average position of the checkpoints varied from 1.6 to 11.0 m and the average elevation varied from 0.3 to 2.9 m. The values calculated using the locally established GCPs gave a difference of 2.0 m in position (x, y) and 0.5 m in elevation (z) with respect to the values regarded most accurate of those given on the www-page mentioned above. The values based on the local GCPs were applied in Tables 13.2–13.5.

Another drawback for the accuracy evaluation was that the number of GCPs and checkpoints was small and the terrestrial photography site covered only a small

Table 13.3 Average differences in elevation between the DEMs according to the original georeferencing and after surface matching when the matching is performed in the whole area and restricted to the bedrock area only. The differences are shown for the whole area and for three smaller areas covered by ice, snow, and bedrock, respectively. The abbreviations are given in Table 13.1.

DEMs	Original All	Matched on all (m)				Matched on rock (m)		
		All	Ice	Snow	Rock	Ice	Snow	Rock
$I_M - I_R$	2.2	0.25	−0.001	0.38	0.98	−0.17	0.56	0.82
$I_M - AP$	0.52	0.53	−0.03	−0.03	0.37	−0.25	0.28	0.18
$I_M - DC$	−20.2	1.3	−0.24	1.9	1.3	−3.1	−0.03	1.5
$I_M - LS$	2.0	0.09	0.02	−0.25	−0.25	−0.03	−0.24	−0.11
$I_L - I_R$	8.8	6.7	−0.23	0.70	−0.27	0.33	1.6	0.14
$I_L - AP$	−0.08	0.19	−0.26	0.75	−0.26	0.16	1.7	−0.16
$I_L - DC$	−20.3	7.4	−9.2	6.0	9.4	0.15	36.6	1.3
$I_L - LS$	0.45	−0.78	0.11	−0.007	−0.80	0.41	0.42	−0.62
$I_R - AP$	−2.5	−0.03	0.24	1.22	−0.14	0.58	1.7	0.02
$I_R - DC$	−25.3	6.3	−8.3	7.0	7.4	1.3	22.0	11.0
$I_R - LS$	−1.6	−0.75	0.26	0.29	−0.57	0.40	0.47	−0.49
AP − DC	−19.4	7.2	−8.0	5.5	10.9	2.1	35.6	10.8
AP − LS	−0.09	−0.80	0.30	−1.0	−0.61	0.04	−1.6	−0.68
DC − LS	24.3	−6.9	8.7	−6.4	−9.8	−2.3	−38.0	−10.6

Table 13.4 RMS differences in elevation between the DEMs according to the original georeferencing and after surface matching when the matching is performed in the whole area and restricted to the bedrock area only. The differences are shown for the whole area and for three smaller areas covered by ice, snow, and bedrock, respectively. The abbreviations are given in Table 13.1.

DEMs	Original All	Matched on all (m)				Matched on rock (m)		
		All	Ice	Snow	Rock	Ice	Snow	Rock
I_M, I_R	8.7	8.5	0.4	13.1	19.6	0.5	13.0	19.6
I_M, AP	11.0	10.9	0.5	3.9	7.8	0.6	3.9	7.8
I_M, DC	25.2	10.0	4.6	5.3	18.7	5.7	5.7	19.1
I_M, LS	9.0	8.8	0.5	3.9	6.4	0.5	3.9	6.4
I_L, I_R	62.8	62.2	0.6	14.0	17.2	0.7	14.2	17.3
I_L, AP	21.2	21.1	0.8	3.9	14.2	0.8	4.1	14.2
I_L, DC	29.3	22.2	11.1	10.1	25.1	11.5	37.3	20.4
I_L, LS	20.6	20.5	0.6	3.8	14.5	0.7	3.8	14.4
I_R, AP	10.0	9.7	0.7	1.6	10.6	0.9	2.0	10.6
I_R, DC	31.2	20.5	10.5	7.9	22.3	7.4	22.2	23.4
I_R, LS	10.9	10.5	0.6	1.0	11.3	0.7	1.1	11.3
AP, DC	29.2	19.7	10.0	7.9	21.2	7.8	36.2	20.2
AP, LS	9.7	9.7	0.6	1.5	5.0	0.5	2.0	5.0
DC, LS	28.8	18.9	10.6	8.6	20.4	8.2	38.7	19.9

area. Based on the results in Table 13.2, none of the DEMs can be regarded as a reference DEM, which would be known to have an elevation accuracy an order of magnitude higher than the others within the whole area. Differences in elevation can, however, be studied by comparing the DEMs to each other. The first observation was

Table 13.5 Average bounds for the interpolation error for the whole area and for three smaller areas covered by ice, snow, and bedrock, respectively. The abbreviations are given in Table 13.1.

	Bound for the interpolation error (m)			
DEMs	All	Ice	Snow	Rock
$I_M - I_R$	0.17	0.11	0.13	0.41
$I_M - AP$	0.15	0.03	0.08	0.47
$I_M - DC$	5.57	0.78	4.31	6.87
$I_M - LS$	0.23	0.15	0.21	0.33
$I_L - I_R$	0.74	0.18	0.24	0.59
$I_L - AP$	0.25	0.03	0.08	0.28
$I_L - DC$	0.25	0.16	0.26	0.23
$I_L - LS$	0.65	0.17	0.35	0.74
$I_R - AP$	0.36	0.12	0.18	0.41
$I_R - DC$	0.29	0.12	0.14	0.41
$I_R - LS$	0.37	0.12	0.16	0.41
AP − DC	0.16	0.03	0.07	0.22
AP − LS	0.26	0.15	0.20	0.26
DC − LS	0.24	0.15	0.24	0.24

that systematic differences of various magnitudes exist between the elevations due to differences in the original georeferencing of the DEMs. For DEMs georeferenced in different reference coordinate systems, the differences depend on the values of the parameters of the Helmert transformation applied. Figure 13.3a shows the differences between the elevations of the laser scanner and Ikonos RSG DEMs according to the original georeferencing. There are large positive differences of several metres on the southeastern side and large negative differences on the northwestern side of the valley although the DEMs should represent the same surface. The difference image suggests that there is a horizontal shift between the DEMs. In Figure 13.3b, the DEMs have been registered using the surface matching algorithm and the differences in elevation are now smaller and more evenly distributed although there appear some distortions to be discussed later in Section 13.5.4.

13.5.3 Mean and RMS differences in elevation between DEMs

The mean and RMS differences in elevation between various DEMs are presented in Tables 13.3 and 13.4, respectively. The differences were calculated according to the original georeferencing and after surface matching within the whole overlapping area and restricting the matching into a bedrock area shown in Figure 13.1. The differences were evaluated for the whole and bedrock areas and for two test areas covered by ice and snow, respectively. The ice and snow areas were extracted manually from the aerial orthophotograph and they are also illustrated in Figure 13.1. The bedrock areas were obtained as a complement of areas inside the borders of glacial areas extracted from the aerial orthophotograph as described above in Section 13.3. Points where the surface slope exceeded 45 degrees were excluded from the mountainous bedrock areas

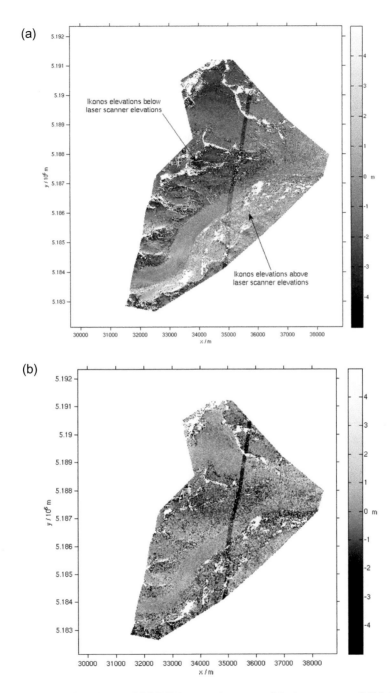

Figure 13.3 Elevations of the Ikonos RSG DEM minus elevations of the laser scanner DEM a) according to the original georeferencing and b) after surface matching. Differences in georeferencing have been corrected and distortions can be analyzed. The coordinate system is M28. (See colour plate section).

so that the matching was performed only in areas where the data were supposed to be accurate.

The results show that the mean differences are somewhat larger on ice and snow while the RMS differences are almost the same when the matching is restricted to the bedrock area as compared to matching on every surface type. The RMS differences between the Ikonos, aerial photography, and laser scanner DEMs vary from 0.4 to 0.9 m in the test area covered by ice. The area is well textured by dirty ice and rather smooth as opposed to rough bedrock areas where the differences are largest in most cases. The RMS differences on snow vary from 1.0 to 4.1 m between the Ikonos manual, Ikonos LPS, aerial photography, and laser scanner DEMs. The RMS differences between the Ikonos RSG DEM and the other two Ikonos DEMs are 13.0 to 14.2 m on snow but only 1.0 to 2.0 m when the Ikonos RSG DEM is compared to the aerial photography and laser scanner DEMs. The Ikonos RSG DEM covers only a small part of the snow test area and this may affect the results. None of the DEMs shows a superior performance with respect to the others as concerns the statistics calculated but the lowest RMS differences are obtained between the Ikonos manual, aerial photography, and laser scanner DEMs. The digital camera DEM is distinguished as the differences against GCPs, checkpoints, and other DEMs are clearly larger than for the other DEMs. This is caused by a systematic distortion in the digital camera DEM, which will be discussed in more detail in Section 13.5.4. Due to memory limitations, the dense Ikonos RSG and LPS DEMs were subsampled onto grid spacings of 2 m and 6 m, respectively, and these sparser DEMs were used in Tables 13.3–13.5 and also when the Ikonos RSG DEM was compared to the GCPs in Table 13.2.

A comparison between the laser scanner DEM on a regular grid of 5 m spacing and the TIN DEM generated from the laser scanner data at the terrestrial photography site shows that the elevations of the former are 0.5 m below the elevations of the latter on the average within an area of 6.6 hectares covered by the TIN laser scanner DEM. The RMS difference in elevation is 0.3 m after surface matching as reported in Table 13.2. The RMS difference in elevation includes an error due to interpolation of the TIN laser scanner DEM.

Table 13.5 shows the bounds for the interpolation error on the average for the whole area and the smaller areas covered by ice, snow, and bedrock, respectively. The bounds are smaller than the mean and RMS differences in elevation in most cases. The true interpolation error at a point can also be much lower in magnitude than the worst case bound based on the maximum of the Laplacian. Moreover, positive and negative interpolation errors at different points partly cancel each other out in the mean difference but not in the RMS difference.

The standard deviation of the mean difference in elevation is typically of the order of millimetres if calculated according to supposed noise levels in the elevations while the contribution propagated from the uncertainty of registration is of the order of 10^{-5} m. The interpolation and random errors thus do not explain all the observed differences. This suggests that there may be systematic errors in the elevations or in the registration as the noise levels in the elevations are hardly so high that their influence on the averaged quantities would explain the differences. One possible reason is an error introduced when a DEM on a regular grid is generated from original data points. This is especially the case for the Ikonos LPS DEM in the

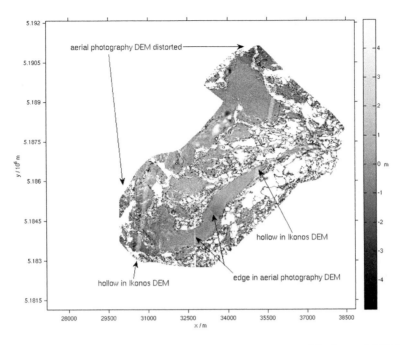

Figure 13.4 Elevations of the aerial photography DEM minus elevations of the Ikonos LPS DEM. There are two longitudinal hollows in the Ikonos LPS DEM. The artificial edges and systematic shape distortion appear in the aerial photography DEM. The coordinate system is M28. (See colour plate section).

mountain areas where sparse measurements were interpolated into a dense grid of 2 m spacing. We are unfortunately not able to estimate the magnitude of this error without knowing the original measurements of the Ikonos data. Any errors introduced during production of DEMs on regular grids are included in the elevation errors reported in Tables 13.2–13.4. For TIN DEMs, there is no such error but the bound for the interpolation error is much larger.

13.5.4 Distortions

The differences in elevation between various DEMs after surface matching are shown in Figures 13.3b–13.7 where differences larger than five (Figures 13.3b–13.6) or fifty metres (Figure 13.7) in absolute value have been omitted in order to observe small-scale differences better. The following distortions or other errors were detected from the difference images or the DEMs themselves.

13.5.4.1 Distortions in DEMs produced from Ikonos images

The Ikonos manual and Ikonos LPS DEMs contain two longitudinal hollows, one of which appears also in the Ikonos RSG DEM while the other one is outside its area of coverage. The hollows are oriented roughly from south to north across the whole DEMs and appear as approximately vertical stripes in all difference images

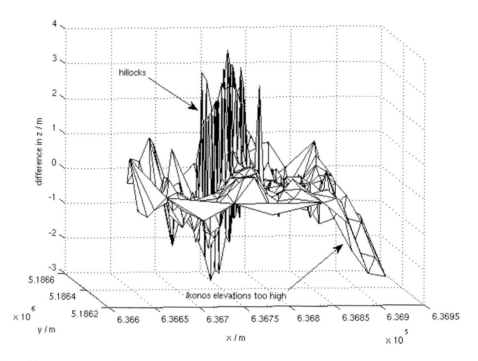

Figure 13.5 Elevations of the TIN laser scanner DEM minus elevations of the Ikonos manual DEM at the site where terrestrial photographs were acquired (Chapter 6). Large differences appear by the hillocks and the elevations of the Ikonos DEM are too high near the eastern border of the area. The coordinate system is WGS84/UTM32.

involving one of the Ikonos DEMs such as in Figures 13.3b and 13.4, but not in any other difference images between other DEMs nor in the difference image between two Ikonos DEMs where the hollows in both of the DEMs cancel each other out. The stripes are not straight lines in the ground coordinates but curving according to the terrain profile, as can be seen in Figure 13.3b. The orientation of the stripes is about 10 degrees from the north, which is close to the flight direction of the Ikonos satellite having a sun synchronous orbit of 98.1 degrees inclination. The depths of the hollows are about 1.6 ± 0.4 m, which is in accordance with the height of jumps observed in the Ikonos DEM studied by Baltsavias et al. (2006). The hollows are evidently caused by unmodelled relative shifts between the three CCD arrays forming the Ikonos images, as proposed by Baltsavias et al. (2006).

The Ikonos LPS and RSG DEMs contain a major peak near the tip of Hintereisferner but at different locations. Based on the field knowledge (Pellikka 2009) there are no topographical peaks in this area nor in the other DEMs so the peak is obviously erroneous. There are also a few smaller peaks in both of these Ikonos DEMs and some minor peaks here and there in the Ikonos RSG DEM, which look like noise. The Ikonos LPS DEM differs noticeably from the Ikonos RSG and other DEMs on the slopes of the Hintereisferner valley. The differences exceed error bounds in these areas and elsewhere on steep slopes. There is no clear pattern in the differences, which would be an indication of inaccurate registration of the DEMs. It is more likely that

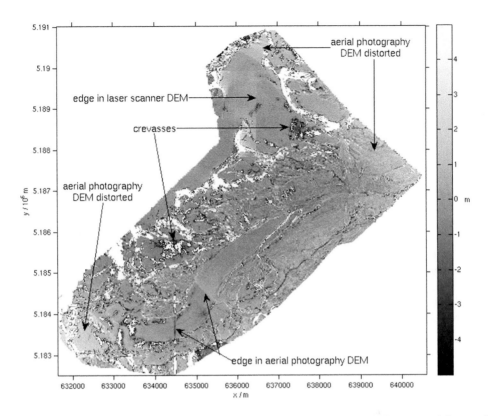

Figure 13.6 Elevations of the laser scanner DEM of 10 m grid spacing minus elevations of the aerial photography DEM. The artificial edges and differences by crevasses have been corrected in the refined laser scanner DEM of 5 m grid spacing. There appear artificial edges and a shape distortion in the aerial photography DEM. The coordinate system is WGS84/UTM32. (See colour plate section).

the large differences and peaks are due to problems in the automatic stereo matching (e.g. caused by occlusion in rough terrain) or interpolation of sparse measurements into a dense grid during DEM production.

When the Ikonos RSG DEM is compared to the aerial photography and laser scanner DEMs, it is found that there are large areas on the bedrock where the elevations of the Ikonos RSG DEM are somewhat too low or high as illustrated in Figure 13.3b. If the surface matching is restricted to the bedrock areas, the difference images change only little and the areas of positive and negative differences on the bedrock remain the same while the differences are small on the glacier. It seems that the elevations on the glacier and mountains are not consistent in the Ikonos RSG DEM. However, the differences are mostly within error bounds calculated according to expected noise levels of 1.0 m, 0.5 m, and 0.32 m for the elevations of the Ikonos, aerial photography, and laser scanner DEMs, respectively. It may thus be that the areas of positive and negative differences are only due to interpolation and random errors.

When the Ikonos DEM, which was manually generated to cover the terrestrial photography site, is compared to the laser scanner data, elevation differences of several metres appear by the hillocks before and after surface matching as shown in Figure 13.5. The hillocks are located at slightly different places in these data sets. The differences in elevation exceed error bounds by the hillocks and near the eastern border of the area where the differences increase in absolute value. The same increasing trend appears when the Ikonos DEM is compared to the aerial photography DEM in this area but not in the difference between the aerial photography DEM and laser scanner data. The elevations of the Ikonos DEM are thus too high there.

13.5.4.2 Distortions in DEM produced from aerial photographs

There appear artificial edges across Hintereisferner in difference images between the aerial photography and Ikonos LPS DEMs in Figure 13.4 and between the aerial photography and laser scanner DEMs in Figure 13.6. These edges are missing from difference images between other DEMs not involving the aerial photography one. The detection of rising elevations of about two metres in height and an edge of about one meter in height was verified by investigating the aerial photography DEM around the area in question. The rising elevations and artificial edges are located along borderlines between adjacent stereo models, which were combined, when the aerial photography DEM for the whole area was produced. This reveals that differences between adjacent models existed and adjustment was needed to produce a continuous surface in the overlapping area.

In Figure 13.4, there are also somewhat larger negative differences near the northeastern and southwestern borders of the area and slightly positive values in the middle of the area. The same appears in Figure 13.6 except that the differences are negative in the middle and positive near the borders. Such a trend does not exist between any of the Ikonos LPS or RSG DEMs and the laser scanner DEM. This suggests that the aerial photography DEM is slightly distorted in the direction from southwest to northeast so that the elevations are too high in the middle and too low near the northeastern and southwestern borders. A similar trend was detected when the aerial photography DEM was compared to the GCPs. The distortion is within ± 2 m in the elevations of the aerial photography DEM. The aerial photography DEM contains also some areas in the northwestern mountainous part of the DEM where the elevations differ by more than 100 m from the elevations of the other DEMs.

13.5.4.3 Distortions in laser scanner DEM

An artificial edge was found in the laser scanner DEM of 10 m grid spacing by comparing the laser scanner, aerial photography, and Ikonos DEMs at Kesselwandferner glacier next to Hintereisferner. The edge is visible in difference images involving the low resolution laser scanner DEM such as in Figure 13.6 while it is missing in the difference image between the aerial photography and Ikonos LPS DEMs in Figure 13.4 and in difference images involving the refined laser scanner DEM of 5 m grid spacing such as in Figure 13.3b. The edge was verified by inspecting the laser scanner DEM in this area. It is believed that the step edge is located between adjacent laser scans and possibly caused by inaccurate orientation of a single scan. The magnitude of the edge is about one metre in elevation (z).

The elevations of the laser scanner DEM of 10 m grid spacing are about 3 metres below the elevations of the Ikonos RSG and aerial photography DEMs on the average in areas containing crevasses near the tip of Kesselwandferner and on the mountain northwest from Hintereisferner as pointed out in Figure 13.6. The elevations in the crevassed areas have been corrected in the refined laser scanner DEM of 5 m grid spacing as there is no observable difference to the Ikonos RSG DEM in these areas in Figure 13.3b. In fact, no distortions have been found in the refined laser scanner DEM nor in the laser scanner data at the terrestrial photography site.

13.5.4.4 Distortions in DEM produced from digital camera images

Figure 13.7 shows the difference between the laser scanner and digital camera DEMs. There are large negative differences in the middle and large positive differences near the borders. Similar differences appear if the digital camera DEM is compared to the aerial photography or any of the Ikonos DEMs. The elevations of the digital camera DEM are thus systematically distorted and the shape of the distortion looks like an ellipsoid with decreasing values towards the borders of the DEM. This kind of distortion suggests that the interior and relative orientations of the images were not correct after block adjustment. Due to lack of a dense set of ground control points, the exterior orientations of the images with respect to an external coordinate system could not be solved separately but tie points were mainly used. It turned out that lens

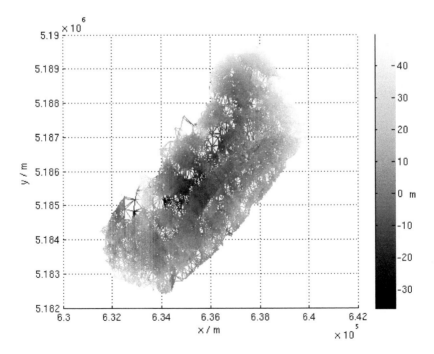

Figure 13.7 Elevations of the laser scanner DEM minus elevations of the digital camera DEM. The digital camera DEM is distorted as the elevations decrease towards borders. The coordinate system is WGS84/UTM32. (See colour plate section).

distortions had not been corrected properly so that the coordinates of the tie points in adjacent images were most probably erroneous, which resulted in an incorrect solution for the image orientations. The digital camera DEM was also matched to the other DEMs locally within areas of a few square kilometres in size in the middle of the original DEM so that the overall shape distortion towards the borders of the DEM would not have dominated the difference images after registration. However, the RMS differences were still several metres and the artificial edges in the aerial photography and laser scanner DEMs were not visible in the difference images between these DEMs and the digital camera DEM.

13.6   CONCLUSIONS

In this chapter, surface matching techniques were applied to correct differences in georeferencing of DEMs on alpine glaciers where a dense and reliable network of ground control points was hard to establish. A method for detecting distortions in DEMs was devised based on analyzing differences in elevation between simultaneously acquired and accurately aligned three or more DEMs with the help of error bounds derived for the differences.

A unique set of DEMs and ground truth data acquired simultaneously at Hintereisferner glacier was investigated and a variety of distortions detected. Rising elevations and an artificial edge were found from the aerial photography DEM in areas where adjacent stereo models had been combined into a larger elevation model for the whole area. The elevations of the aerial photography DEM were slightly distorted towards the southwestern and northeastern borders of the area covered by the DEM. For the digital camera DEM, a similar distortion towards all the borders and across the whole area was striking. An artificial edge, presumably due to misalignment of a strip, and large differences by crevasses were detected in the preliminary laser scanner DEM investigated while these problems had been corrected in the refined laser scanner DEM produced more carefully. Two longitudinal hollows were found in the Ikonos DEMs, which were most probably caused by problems in the interior orientation of the original Ikonos images. The Ikonos LPS DEM differed significantly from the other DEMs on the slopes of the glacier valley and in the mountains, a minor inconsistency (which was mostly within error bounds) between elevations on the glacier and in the mountains was found in the Ikonos RSG DEM, and gross errors were detected in both of these Ikonos DEMs. The positions of hillocks were somewhat different in the Ikonos DEM and laser scanner data at the terrestrial photography site. The checkpoints on the glacier were found to be mislocated.

The RMS differences between the elevations of most of the DEMs were from 0.4 to 0.9 m in a smooth well textured test area covered by ice while larger differences appeared in another test area covered by snow and in the surrounding mountains of rough terrain. The systematic errors detected in the DEMs obviously explain part of the observed differences in elevation exceeding the bounds for the interpolation error and standard deviations of differences in elevation propagated from supposed noise levels in the elevations and estimated uncertainty of registration. It is a future challenge to develop methods for correcting the distortions and other errors detected and thus to improve the accuracy of current techniques for DEM production on alpine glaciers.

REFERENCES

Baltsavias, E. (1999a). A comparison between photogrammetry and laser scannning. *ISPRS Journal of Photogrammetry and Remote Sensing* 54(2–3), 83–94.

Baltsavias, E. (1999b). Airborne laser scanning: basic relations and formulas. *ISPRS Journal of Photogrammetry and Remote Sensing* 54(2–3), 199–214.

Baltsavias, E., E. Favey, A. Bauder, H. Bösch and M. Pateraki (2001). Digital surface modelling by airborne laser scanning and digital photogrammetry for glacier monitoring. *The Photogrammetric Record* 17(98), 243–273.

Baltsavias, E., Z. Li and H. Eisenbeiss (2006). DSM generation and interior orientation determination of IKONOS images using a testfield in Switzerland. *Photogrammetrie, Fernerkundung, Geoinformation* 1, 41–54.

Eisenbeiss, H., E. Baltsavias, M. Pateraki and Z. Li (2004). Potential of IKONOS and Quickbird imagery for accurate 3D point positioning, orthoimage and DSM generation. *International Archives of Photogrammetry, Remote Sensing and Spatial Information Sciences* 35(B7), 1250–1256.

EnsoMOSAIC software, http://www.storaenso.com/CDAvgn/main/0„1_-1362-2381-,00.html (last visited 28 March, 2006).

Favey, E. (2001). *Investigation and Improvement of Airborne Laser Scanning Technique for Monitoring Surface Elevation Changes of Glaciers*. Dissertation No. 14045, ETH Zürich.

Gleitsmann, L. and M. Kappas (2004). Digital multi-image photogrammetry combined with oblique aerial photography enables glacier monitoring survey flights below clouds in Alaska. *Proceedings of IGARSS'04, Anchorage, Alaska, U.S.A.*, September 20–24, 2004, Vol. 2, 1148–1151.

Joanneum Research. RSG in Erdas Imagine, Field Guide, Version 1.0, 1997.

Jokinen, O. (1998). Area-based matching for simultaneous registration of multiple 3-D profile maps. *Computer Vision and Image Understanding* 71(3), 431–447.

Jokinen, O. (2000). *Matching and Modeling of Multiple 3-D Disparity and Profile Maps*. Doctoral thesis, Acta Polytechnica Scandinavica, Mathematics and Computing Series 104, 117 p.

Jokinen, O., T. Geist, K.-A. Høgda, M. Jackson, K. Kajuutti, T. Pitkänen and V. Roivas (2007). Comparison of digital elevation models of Engabreen glacier. *Zeitschrift für Gletscherkunde und Glazialgeologie* 41, 185–204.

Kajuutti, K., O. Jokinen, T. Geist and T. Pitkänen (2007). Terrestrial photography for verification of airborne laser scanner data on Hintereisferner in Austria. *Nordic Journal of Surveying and Real Estate Research* 4(2), 24–39.

Kraus, K. (1987). *Photogrammetrie, Band 2, Verfeinerte Methoden und Anwendungen.* 2nd ed., Dümmler, Bonn, 504 p.

Kraus, K. and N. Pfeifer (1998). Determination of terrain models in wooded areas with airborne laser scanner data. *ISPRS Journal of Photogrammetry and Remote Sensing* 53(4), 193–203.

Kuhn, M. (1980). Begleitworte zur Karte des Hintereisferners 1979, 1:10.000. *Zeitschrift für Gletscherkunde und Glazialgeologie* 16(1), 117–124.

Lankinen, U., M. Martikainen and J. Santala (1992). Maastomittauksen laskentakaavoja. *Reports of Institute of Geodesy and Cartography, Helsinki University of Technology*, 21, 99–103.

Mason, S., H. Rüther and J. Smit (1997). Investigation of the Kodak DCS460 digital camera for small-area mapping. *ISPRS Journal of Photogrammetry and Remote Sensing* 52(5), 202–214.

Mills, J., S. Buckley and H. Mitchell (2003). Synergistic fusion of GPS and photogrammetrically generated elevation models. *Photogrammetric Engineering and Remote Sensing* 69(4), 341–349.

Mostafa, M. and K.-P. Schwarz (2001). Digital image georeferencing from a multiple camera system by GPS/INS. *ISPRS Journal of Photogrammetry and Remote Sensing* 56(1), 1–12.

Parviainen, P. (2006). *Detection of glacier facies and parameters using aerial false colour digital camera data—case study Hintereisferner*, Austria. M.Sc. thesis, Department of Geography, University of Helsinki, 74 p. + appendices.

Pearson, F. (1990). *Map Projections: Theory and Applications*. Boca Raton, CRC Press, 372 p.

Pellikka, P. (2007). Monitoring glacier changes within the OMEGA project. *Zeitschrift für Gletscherkunde und Glazialgeologie* 41, 3–5.

Pellikka, P. (2009). Personal communication, 11.5.2009.

Pfeifer, N., P. Stadler and C. Briese (2001). Derivation of digital terrain models in the SCOP++ environment. *Proceedings of the OEEPE Workshop on Airborne Laserscanning and Interferometric SAR for Digital Terrain Models*, Stockholm, Sweden, March 1–3, 2001, 13 p.

Poon, J., C. Fraser, Z. Chunsun, Z. Li and A. Gruen (2005). Quality assessment of digital surface models generated from IKONOS imagery. *The Photogrammetric Record* 20(110), 162–171.

Schmidt, R. (2003). Untersuchung verschiedener digitaler Geländemodelle hinsichtlich ihrer Eignung für die dynamische Lawinensimulation mit dem dreiphasigen Simulationsprogramm SAMOS. Diploma thesis. Institute of Geography, University of Innsbruck.

Schneider, H. (1975). Die Karte des Kesselwandferners 1971 und die Grundlagen der Vermessung. *Zeitschrift für Gletscherkunde und Glazialgeologie* 11(2), 229–244.

Toutin, T. (2004). Comparison of stereo-extracted DTM from different high-resolution sensors: SPOT-5, EROS-A, IKONOS-II, and QuickBird. *IEEE Transactions on Geoscience and Remote Sensing* 42(10), 2121–2129.

Zhang, Z. (1994). Iterative point matching for registration of free-form curves and surfaces. *International Journal of Computer Vision* 13(2), 119–152.

Chapter 14

Accuracy aspects in topographical change detection of glacier surface

Olli Jokinen
Department of Surveying, Helsinki University of Technology, Finland

Thomas Geist
FFG – Austrian Research Promotion Agency, Vienna, Austria
Institute of Geography, University of Innsbruck, Austria

14.1 INTRODUCTION

A digital elevation model (DEM) may be defined as any numeric or digital representation of terrain elevation given as a function of geographic location (Raaflaub and Collins 2006). It is usually given on a regular or irregular grid of ground locations while interpolation yields elevations between the grid points. A DEM is thus a discrete approximation of the true surface at the time of data acquisition.

In this chapter, methods are considered for detecting and quantifying topographical changes occurring between two instants by analyzing differences between respective DEMs. The focus is on distinguishing between true changes occurred in the area and fictitious changes related to uncertainties in the measurements and the method of computing the difference between the surfaces. Error bounds are derived for changes in elevation and for volume lying between the surfaces. There exist related works on change detection while the main contribution of this chapter is to provide accuracy estimates for the volumetric case, which are currently missing.

The chapter is organized as follows. Previous research on topographical change detection on glaciers is reviewed in Section 14.2. Section 14.3 presents methods for estimating the accuracy of change in elevation and volume between two surfaces. The methods are explored with DEMs of two mountain glaciers in Section 14.4, and conclusions are presented in Section 14.5.

14.2 PREVIOUS RESEARCH

Topographic change detection has been much studied in alpine glaciers for both surface elevation (Lundstrom et al. 1993, Etzelmüller et al. 1993, Etzelmüller and Sollid 1997, Kääb 2002) and volume (Kite and Reid 1977, Brecher and Thompson 1993, Vignon et al. 2003). However, only a few papers exist in which uncertainties of the change are considered.

Reynolds and Young (1997) quantify various uncertainties related to estimation of change in volume and elevation of Athabasca glacier from a series of digitized maps. The uncertainties include spatially variable errors in the elevations of the original maps, errors in fitting surfaces to digitized contours, and registration errors between raster DEMs. It is found that the errors are often greater than the actual changes in surface elevation especially if the interval between data acquisitions is short. Albrecht et al. (2000) compare changes in the volume of Storglaciaren of Sweden computed according to a time-dependent ice flow model with changes measured. Uncertainties in the model and measurements are discussed. Magnússon et al. (2005) estimate changes in the volume of Vatnajökull ice cap in Iceland with error bounds.

Etzelmüller (2000) compares various surface descriptors such as altitude, slope, curvature, roughness factor, semivariogram, elevation-relief ratio, and altitude skewness for change detection at Finsterwalderbreen in Svalbard. Statistical measures for change detection include analyzing correlation coefficients between the DEMs in local neighbourhoods and investigating areal changes in classification of DEMs into homogeneous regions. It is pointed out that averaging over larger areas yields more reliable results than differences at individual points for noisy DEMs.

For reliable change detection, it is essential that the DEMs are accurately in the same coordinate system, which can be achieved using robust surface matching techniques. Pilgrim (1996) presents a robust weighted least square method, which can simultaneously match two DEMs and detect gross errors and regional differences between the DEMs. Similar adaptive weights based on the distribution of residuals are proposed for rejecting incompatible matches during surface matching in Zhang (1994). Data snooping is considered for detection of gross errors while matching similar regions in Karras and Petsa (1993). Buckley and Mitchell (2004) apply surface matching techniques to detect and overcome systematic differences between a denser airborne laser scanner DEM and sparser photogrammetric data before integrating them into a fused surface model with optimized point distribution.

14.3   METHODS FOR DETECTING AND MEASURING CHANGES

This section presents methods to detect and measure changes in elevation and volume between two twice continuously differentiable surfaces approximated by discrete DEMs. The focus is on deriving error bounds and confidence limits for the changes detected.

14.3.1   Change in elevation

It is assumed that there are no cavities below the surface of the glacier so that the surface can be represented as a function $z = z(x, y, t)$ defined on $(x, y) \in \Omega \subset \mathbf{R}^2, t \geq 0$, where Ω covers the extent of the glacier within the period under examination. In a digital elevation model, we have a discrete sample of elevations z_i observed on a set of points $(x_i, y_i), i = 1, \ldots, N$, which may constitute a regular grid in the form of an image or an irregular grid in the form of a triangulated set of points. The grid points divide Ω into a set of rectangular or triangular patches $\Omega_k, k = 1, \ldots, K$, respectively.

Computing the differences in elevation between two DEMs observed at different times $t = t_1$ and $t = t_2 > t_1$ involves transforming the DEMs into the same georeference and resampling one DEM to the grid points of the other DEM. DEMs produced at different times using different techniques and different ground truth are usually only approximately in the same coordinate system. This leads to undesirable differences between the DEMs, which can be eliminated in most cases by surface matching as was discussed in Section 13.4.1 above. Let $z_i'(x_i', y_i')$ denote the elevation at a grid point transformed from one system to another by a six-parameter rigid body transformation. The resampling of the elevations of the other DEM into the transformed grid points (x_i', y_i') is carried out within the overlapping area of the DEMs using bilinear interpolation if the DEM to be resampled is given in an image format or computed linearly according to the planar patches if the DEM is a triangulated irregular network (TIN) model. In order to reduce an error due to interpolation, it is usually advantageous to resample the DEM with the higher point density. Let it be the DEM at $t = t_2$. The observed differences in elevation are then given within the overlapping area as $\hat{d}(x_i', y_i') = \tilde{z}_i(x_i', y_i', t_2) - z_i'(x_i', y_i', t_1)$ where $\tilde{z}_i$ denotes the resampled elevation.

Error bounds for the difference in elevation are discussed by Jokinen et al. (2007) concerning bilinear interpolation and uncertainty of registration. Consider now the case where the DEM to be interpolated consists of triangular patches. Denote the partition of the domain of the TIN DEM as $\Lambda_p, p = 1, \ldots, P$. Let the vertices of a triangle Λ_p be located at r_0, r_1, r_2 and let $r_i' = (x_i', y_i') \in \Lambda_p$. Two pieces of straight lines are defined as $L_{12} : \{r | r = r_1 + s(r_2 - r_1), 0 \leq s \leq 1\}$ and $L_{03} : \{r | r = r_0 + u(r_i' - r_0), 0 \leq u \leq u_p\}$ where $u_p \geq 1$ is solved from equation $u_p(r_i' - r_0) + s_p(r_1 - r_2) = r_1 - r_0$ and thus determines the intersection of lines L_{12} and L_{03}. An error bound for linear interpolation (Mäkelä et al. 1984, p. 92) is applied first to interpolation along line L_{12} at point where $s = s_p$ and then along line L_{03} at point where $u = 1$, i.e., $r = r_i'$. It yields

$$\left| \tilde{z}_i(r_i', t_2) - z(r_i', t_2) \right| \leq \tfrac{1}{2} s_p(1 - s_p)|r_1 - r_2|^2 \max_{r \in L_{12}} \left| \frac{\mathrm{d}^2 z(r(s))}{\mathrm{d}s^2} \right|$$

$$+ \tfrac{1}{2}(u_p - 1)|r_i' - r_0|^2 \max_{r \in L_{03}} \left| \frac{\mathrm{d}^2 z(r(u))}{\mathrm{d}u^2} \right| \tag{14.1}$$

where the second derivatives of the elevation curves are approximated as follows. A polynomial surface of third order in x and y is fitted using least squares techniques to the elevation data in a neighbourhood of Λ_p consisting of triangles sharing a common vertex with Λ_p. The polynomial is centred at the mean of the vertices of Λ_p. A second order polynomial is used instead if the condition number of the coefficient matrix of the normal equations for the third order case is large or the rank of the coefficient matrix is less than nine indicating that there are too many parameters to be estimated from the data. The polynomial surface is intersected with vertical planes defined by lines L_{12} and L_{03}. These yield polynomial curves of third or second order, the second derivatives of which are of first or zero order. The maximum of the absolute value of each second derivative is thus obtained at one or both of the end points of the piece of line in question. Equation 14.1 is evaluated for three different alternatives of choosing vertex r_0 and the maximum of the three resulting error bounds is used. For bilinear interpolation within a rectangular patch, the error bound for linear interpolation is first applied in the direction of one ground coordinate and then in the direction of the other ground coordinate. The second derivatives are approximated using a discrete operator and the maximum of the absolute value of this operator at the vertices of the grid cell is applied.

If the elevations of the DEMs were contaminated only by normally distributed random measuring errors with zero means and known standard deviations, then the error in the difference in elevation at point $(x'_i, y'_i) \in \Omega'_k$ would be, with 95% probability, bounded by

$$|d(x'_i, y'_i) - \hat{d}(x'_i, y'_i)| \leq S(x'_i, y'_i) + 1.96 \ \mathrm{Std}(d(x'_i, y'_i)) \equiv H(x'_i, y'_i) \qquad (14.2)$$

where S is the bound for the interpolation error and the standard deviation of the difference in elevation is obtained through error propagation from the standard deviations of the elevations and the covariance matrix of the registration parameters given by the inverse of the Hessian of the weighted least squares merit function minimized in surface matching. On the other hand, if the difference in elevation exceeds the error bound at some point, then it may be that the noise levels in the elevations are larger than expected at this point, the surface matching algorithm has not converged to a correct solution, one or both of the elevations contain systematic errors, or a change has really occurred. Change detection is thus based on analyzing points or areas where the differences in elevation between two DEMs exceed the error bounds. If the other causes for differences in elevation can be ruled out or the magnitude of them estimated, then the error bounds for the difference give an estimate of the accuracy of change in elevation.

A drawback is that the noise levels in the elevations are not well known on alpine glaciers and moreover, they vary within the scene coverage depending on the method of data acquisition, measuring geometry, and surface properties such as roughness, slope, and texture. There exist previous studies on this subject as referenced in Section 13.2 above and some new knowledge was also presented in Section 13.5.1. For change detection purposes, it may be expected that the noise levels are approximately the same as reported in previous studies. There may be single points corrupted with unexpectedly large noise but if there are large areas where the differences in elevation exceed the error bounds derived from the noise levels expected, then it suggests changes have occurred in these areas. For differences averaged over large areas, the standard deviation of the mean decreases as the inverse square root of the number of points while the interpolation error is bounded by the mean of the bounds for the interpolation error at individual points. Consequently, the error bound for the mean difference in elevation over large areas is dominated by the interpolation error while the effect of noise diminishes. The accuracy of change can thus be estimated for large areas while the estimates may not be appropriate at single points.

14.3.2 Change in volume

Recall that it is assumed the terrain can be represented as a surface $z = z(x, y, t)$ on $(x, y) \in \Omega \subset \mathbf{R}^2, t \geq 0$. The volume between two surfaces observed at $t = t_1$ and $t = t_2 > t_1$ is given as an integral

$$\Delta V = \iint\limits_{\Omega} (z(x, y, t_2) - z(x, y, t_1)) dx\, dy. \tag{14.3}$$

If the bottom of the glacier remains stable between t_1 and t_2, then this integral gives also the change in the volume of the glacier.

In the case of digital elevation models, the integral in Equation 14.3 is approximated as

$$\Delta \hat{V} = \sum_{k=1}^{K} \bar{d}_k A_k \tag{14.4}$$

where $\bar{d}_k$ is the average of differences $\hat{d}(x_i', y_i') = \tilde{z}_i(x_i', y_i', t_2) - z_i'(x_i', y_i', t_1)$ at the vertices of Ω_k' and A_k is the area of Ω_k'.

There are two things that affect the accuracy of $\Delta \hat{V}$ in Equation 14.4. These include firstly a truncation error when the integral is replaced by a finite sum and secondly, errors in the differences in elevation due to measurement errors in the elevations, uncertainty of the registration, and interpolation of the other DEM. The truncation error depends on the shape of the elevation surface and density of measurements. Consider first the case where $\Omega_k', k = 1, \ldots, K$, is a regular grid of rectangles. If the twice continuously differentiable function $d(x, y)$ were known, the truncation error E_T could be estimated as

$$|E_T(\Delta V)| \leq \frac{h^4}{12} \sum_{k=1}^{K} \max_{(x, y) \in \Omega_k'} \left| \frac{\partial^2}{\partial x^2} d(x, y) + \frac{\partial^2}{\partial y^2} d(x, y) \right| \tag{14.5}$$

where h is the grid spacing assumed to be equal in x and y. Equation 14.5 can be derived by extending the error estimate of a trapezoidal rule (Davis and Rabinowitz 1967, pp. 110–111) from the one-dimensional case to the two-dimensional one as has been done for Simpson's rule in Burden and Faires (1993, p. 213). Since the data are only known on a discrete grid, the Laplacian of d is approximated once again by the conventional Laplacian operator of $\hat{d}$ and the maximum of the absolute value of this operator at the vertices of Ω_k' is used in Equation 14.5.

In the case $\Omega_k', k = 1, \ldots, K$, is an irregular set of triangles, the second partial derivatives of d are approximated as derivatives of local polynomial surface patches of third or second order as described above in Section 14.3.1 except that the fittings are now carried out in a local coordinate system for each triangle Ω_k' where the x-axis is along the longest side of Ω_k' and the vertices of Ω_k' are located at $(0, 0), (b_k, 0), (a_k, c_k)$ with a_k, b_k, c_k positive for $k = 1, \ldots, K$. Let $M_{k,xx}, M_{k,yy}, M_{k,xy}$ be the maxima of the absolute values of the second partial derivatives with respect to x and y and the mixed partial derivative, respectively, of three local surface patches centred and evaluated at the three vertices of Ω'_k. Expand $d(x,y)$ into a Taylor series at vertex $(0,0)$. The quadrature $Q_k = \bar{d}_k A_k$ integrates the first and zero order terms of the series exactly and the truncation error can be estimated after some calculation as

$$|E_T(\Delta V)| \leq \sum_{k=1}^{K} \left(M_{k,xx} |B_k| + M_{k,yy} |C_k| + M_{k,xy} |D_k| \right) \tag{14.6}$$

where

$$B_k = \iint\limits_{\Omega'_k} \frac{1}{2} x^2 \, dx \, dy - Q_k \left(\frac{1}{2} x^2 \right)$$

$$= \frac{1}{24} \left(3a_k^3 c_k + \frac{c_k}{b_k - a_k} (3a_k^4 + b_k^4 - 4a_k^3 b_k) \right) - \frac{1}{12} (a_k^2 b_k c_k + b_k^3 c_k)$$

$$C_k = \iint\limits_{\Omega'_k} \frac{1}{2} y^2 \, dx \, dy - Q_k \left(\frac{1}{2} y^2 \right) = -\frac{1}{24} b_k c_k^3$$

$$D_k = \iint\limits_{\Omega'_k} xy \, dx \, dy - Q_k \, (xy)$$

$$= \frac{1}{24} \left(3a_k^2 c_k^2 + \left(\frac{c_k}{b_k - a_k} \right)^2 (-3a_k^4 + b_k^4 + 8a_k^3 b_k - 6a_k^2 b_k^2) \right) - \frac{1}{6} a_k b_k c_k^2$$

The effect of interpolation errors on the change in volume is estimated as

$$|E_I(\Delta V)| \leq \sum_{k=1}^{K} G_k A_k \tag{14.7}$$

where G_k is the mean of bounds for interpolation errors in differences in elevation at the vertices of Ω'_k calculated according to Equation 14.1 in case $\Lambda_p, p = 1, \ldots, P$, is a set of triangles and correspondingly for a set of rectangles. Any systematic errors left in the elevations propagate also into the change in volume and can be estimated similarly using first order error propagation.

The precision of the change in volume is characterized by $\text{Var}(\Delta V)$ calculated through error propagation from the standard deviations of the original elevations of the DEMs assumed independent and from the covariance matrix of the registration parameters. The registration is solved by surface matching using the elevations but correlations between the registration parameters and elevations are ignored. The mean differences in elevation at neighbouring triangles or rectangles appearing in Equation 14.4 are correlated as they share a common vertex. Consequently, the derivatives needed in the error propagation are calculated up to the elevations and registration parameters assumed independent.

Let U_T and U_I be the bounds for E_T and E_I, respectively, calculated according to Equations 14.5–14.7. If it is assumed normally distributed noise in the elevations of the DEMs, then the error made in the estimation of change in volume is, with 95% probability, bounded by

$$|\Delta V - \Delta \hat{V}| \leq U_T + U_I + 1.96 \, \mathrm{Std}(\Delta V) \equiv U \tag{14.8}$$

14.4 CASE STUDIES

This section reports the changes in elevation and volume observed between an aerial photography and a laser scanner DEM at Svartisheibreen, an outlet glacier of West Svartisen ice cap in Northern Norway, and in a sequence of ten laser scanner DEMs at Hintereisferner, a valley glacier in the Central Eastern Alps in Ötztal, Austria. The test results were computed using Matlab software in a Sun Fire 25K server of CSC, the Finnish IT Center for Science.

14.4.1 Aerial photography and laser scanner DEMs over Svartisheibreen

A TIN DEM of Svartisheibreen (Figure 3.1) was produced using manual stereoscopic interpretation by Erno Puupponen of Finnish Consulting Engineers – Sito Oy from aerial photographs at a scale of 1:15000 acquired on August 25, 2001, by Fotonor AS, Norway. The points were classified into rock, soil, snow, ice, and boundaries between them by the stereo operator. The length of sides of triangles in the aerial photography DEM was 13 m on the average. The aerial triangulation was based on ground control points existing in the area and points derived from a topographic map.

The laser scanning over Svartisheibreen was performed by TopScan GmbH, Germany, with an Optech ALTM 1225 laser scanning system on September 24, 2001. The average distance between neighbouring raw data points was 1.5 m. A laser scanner DEM on a regular grid of 1 m spacing was produced using SCOP++ software with the implemented linear prediction algorithm (Pfeifer et al. 2001) by the Institute of Geography at the University of Innsbruck, Austria. The initial positioning of the laser scanner data was based on GPS/INS measurements on board and it was differentially corrected using data from a permanent receiving station in Bodø and a temporary station set up in Holandsfjord. The georeferencing was refined using a flat reference area (a football field) surveyed with a total station. The standard deviation between laser scanner measurements and a plane fitted to the reference measurements in the calibration field was 0.09 m (Geist et al. 2006).

The aerial photography and laser scanner DEMs were first compared in stable bedrock areas. Systematic differences in elevation were observed as the elevations of the laser scanner DEM were above the elevations of the aerial photography DEM in the southern bedrock areas and the other way round in the northern and western bedrock areas. The difference in orientation corresponded to a height difference of approximately 2 m at a distance of 3 km. The DEMs were registered to the same coordinate system using the surface matching algorithm where the matching was restricted to areas classified as rock or soil. Changes occurring in the ice- and snow-covered areas were then analyzed using the error bounds derived for the case where the domain was an irregular set of triangles of the aerial photography DEM and bilinear interpolation was used in the resampling of the laser scanner DEM.

The differences in elevation between the laser scanner and aerial photography DEMs are illustrated on a contour map of the aerial photography DEM in Figure 14.1. The points where the differences in elevation $\hat{d}$ are larger in absolute value than the error bounds H calculated using supposed noise levels of 0.5 m and 0.09 m for the elevations of the aerial photography and laser scanner DEMs, respectively, are illustrated in Figure 14.1a while Figure 14.1b shows the points where the differences in elevation are within the error bounds. The bedrock areas surrounding the glacier are clearly visible as areas of no change.

The volume lying in between the true surfaces approximated in the discrete aerial photography and laser scanner DEMs was estimated for the whole glacier area and then separately for areas covered by ice and snow. The change in volume and contributions of different error sources are presented in Table 14.1. The truncation error dominates the error bounds while the effects of interpolation errors and measuring noise are smaller. The relative error measures are about the same for the ice- and snow-covered areas.

The elevation range of the ice- and snow-covered areas was divided into intervals of 20 m. Triangles in the aerial photography DEM were classified into the elevation intervals according to the elevation of the centre point of the triangle. The estimated change in volume $\Delta \hat{V}$ within triangles of each class is shown with error bounds U in Figure 14.2a. The change in volume is proportional to the area of the class, which is largest for the elevation interval around 1000 m a.s.l. Figure 14.2b shows the average change in elevation $\hat{d}$ with error bounds H for points in elevation intervals of 20 m. The lowering has been quite even in elevation intervals above 800 m a.s.l. while the biggest changes have occurred below 760 m a.s.l. where a tongue-shaped steep area of ice projects out of the main stream and flows down to the brink of a 29 m deep canyon. A block of ice may have cleft from the tip and dropped into the canyon, which would be one possible explanation for the increase in elevation in the lowest 20 m interval in the bottom of the canyon and decrease in elevation in the second lowest 20 m interval on the brink of the canyon.

According to field measurements at some stakes, approximately 1.25–1.35 m ice had melted away around 775–820 m a.s.l. and approximately 1.15 m ice had melted away at 900 m a.s.l. between the data acquisitions. The average changes calculated from differences between the DEMs are 1.27 m and 1.76 m, respectively. The first one corresponds well to the field measurement carried out by Hallgeir Elvehøy and

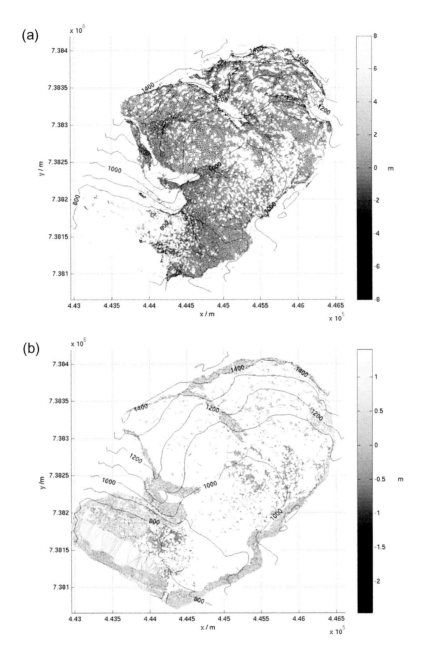

Figure 14.1 Differences in elevation between the laser scanner and aerial photography DEMs at points where the changes a) exceed the error bounds b) are within the error bounds. Differences in elevation larger than 8 m in absolute value occurring at few points have not been plotted in Figure 14.1a to better distinguish spatial variations of smaller differences. White areas contain no data. The contour lines are according to the aerial photography DEM. Svartisheibreen, Norway, August 25–September 24, 2001. (See colour plate section).

Table 14.1 An estimate for the volume in between the surfaces approximated in the aerial photography and laser scanner DEMs, bounds for truncation (U_T) and interpolation (U_I) errors, standard deviation of the change in volume, 95% error bound, and $U/\Delta \hat{V}$ for the whole area and areas covered by ice and snow.

Region	Area/ km^2	$\Delta \hat{V}/$ $10^6 m^3$	$U_T/$ $10^6 m^3$	$U_I/$ $10^6 m^3$	$Std(\Delta V)/$ $10^6 m^3$	$U/$ $10^6 m^3$	$U/\Delta \hat{V}$
Ice and snow	5.4	−8.1	2.1	0.2	0.05	2.4	−0.30
Ice	2.0	−2.8	0.7	0.1	0.02	0.8	−0.30
Snow	3.4	−5.3	1.4	0.1	0.04	1.6	−0.30

Miriam Jackson from the Norwegian Water Resources and Energy Directorate (NVE), but the second one differs considerably. However, the exact positions of the stakes were not known and the standard deviation of differences in elevation between the DEMs within an elevation interval of 890–910 m a.s.l. is 0.93 m so that the difference with respect to the field measurement may be just due to spatial variation of the melting process or some error in the DEMs.

14.4.2   Sequence of laser scanner DEMs over Hintereisferner

A sequence of ten laser scanner DEMs on regular grids of 5 m spacing produced from raw data acquired by TopScan GmbH at Hintereisferner during ten flight campaigns between October 11, 2001, and September 26, 2003, was studied. DEM production was described in detail in Chapter 10 above and in Geist and Stötter (2007). The average distance between adjacent data points in the raw data varied from 0.8 to 1.4 m in different campaigns. A study area for change detection was extracted manually from an aerial orthophotograph produced by the Institute of Photogrammetry and Chartography at Technische Universität München, Germany, from aerial photographs acquired over Hintereisferner on the same day, August 12, 2003, as one of the laser scannings was carried out (see Chapter 13). The study area covered the glacial area of Hintereisferner, snow areas in the upper part of the glacier, and some surrounding bedrock and debris-covered ice-free areas near the tip of the glacier in order to guarantee that glacial areas in all laser scanner DEMs acquired earlier were also included. The study area is illustrated in Figure 14.3.

Two ice-free areas shown in Figure 14.3 were also extracted to estimate the standard deviation of differences in elevation between the laser scanner DEMs and precision of elevations in a single laser scanner DEM. These areas were located on the slopes of the valley around the tip of the glacier where the topography was rather similar to the study area. Differences in elevation were evaluated pair-wise between three laser scanner DEMs produced from data acquired in September 2002, August 2003, and September 2003 when the valley slopes were free from snow. All the three

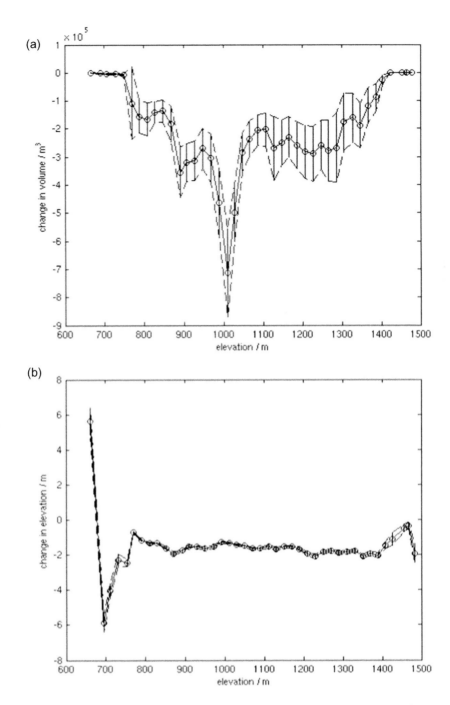

Figure 14.2 Change in a) volume b) elevation (circles and solid line) with error bounds (dashed lines) for elevation classes at 20 m intervals. Svartisheibreen, Norway, August 25–September 24, 2001.

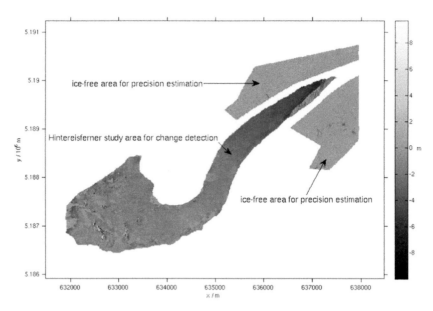

Figure 14.3 Changes in elevation at least occurred in Hintereisferner between August 19, 2002, and August 12, 2003, according to the error bounds and differences in elevation between the laser scanner DEMs in the ice-free test areas for precision estimation. Changes in elevation larger than 10 m in absolute value occurring at few points in the study area have not been plotted in the image to better distinguish spatial variations of smaller differences. (See colour plate section).

DEMs were given in exactly the same ground coordinates so that no interpolation errors appeared in the comparison. The mean differences in elevation varied from 0.05 to 0.12 m in absolute value and the sample standard deviations of differences in elevation varied from 0.40 to 0.49 m between the three DEMs in the test areas. Differences between the DEMs of the same year were smaller than between the DEMs of different years. The average noise level in the elevations of a single laser scanner DEM was estimated to be 0.32 m, given as the sample standard deviation of all the differences in elevation divided by the square root of two. Since the sample size of observed differences in elevation in the test areas was large (352542 observations), the *t* distribution adequate for sample statistics was approximated by the normal distribution and the value of 1.96 was applied in the 95% error bounds in Equations 14.2 and 14.8. No systematic differences in elevation, which could have been due to differences in georeferencing, were detected. No corrections to the original georeferencing provided by TopScan GmbH were thus needed and the uncertainties in the registration parameters were not considered in the estimation of change in elevation and volume in the following examples. In this case, the domain consisted of regular rectangles and bilinear interpolation was used in the resampling of the other DEM.

The smallest change in elevation that has at least occurred according to the error bounds is given by function f defined as $f = (|\hat{d}| - H)\mathrm{sgn}(\hat{d})$ if $|\hat{d}| > H$ and $f = 0$

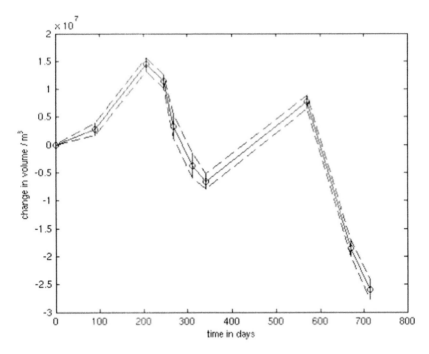

Figure 14.4 Change in volume (circles and solid line) with error bounds (dashed lines) in Hintereisferner during a period of two years.

Table 14.2 Volume between the first laser scanner DEM (October 10, 2001) and the other DEMs, bounds for truncation (U_T) and interpolation (U_I) errors, standard deviation of the change in volume, 95% error bound, and $U/\Delta\hat{V}$.

Date	$\Delta\hat{V}/$ $10^6 m^3$	$U_T/$ $10^6 m^3$	$U_I/$ $10^6 m^3$	$Std(\Delta V)/$ $10^6 m^3$	$U/$ $10^6 m^3$	$U/\Delta\hat{V}$
9 January, 2002	2.8	1.2	0.0	0.005	1.2	0.42
7 May, 2002	14.4	1.1	0.0	0.005	1.2	0.08
15 June, 2002	11.4	1.2	0.0	0.005	1.2	0.11
8 July, 2002	3.4	1.0	1.1	0.005	2.2	0.64
19 August, 2002	−3.7	1.0	1.2	0.005	2.2	−0.60
18 September, 2002	−6.6	1.4	0.0	0.005	1.4	−0.21
4 May, 2003	7.8	1.2	0.0	0.005	1.2	0.15
12 August, 2003	−18.6	1.4	0.0	0.005	1.4	−0.08
26 September, 2003	−25.9	1.8	0.0	0.005	1.8	−0.07

if $|\hat{d}| \leq H$. Figure 14.3 illustrates the least change in elevation between the DEMs of August 19, 2002, and August 12, 2003, in the study area. The figure verifies the loss in elevation, which increases towards the tip of the glacier where the elevations have decreased at least about 5.5–6.5 m and at most about 7.1–8.1 m during a year. The change in elevation varies spatially on the glacier near the tip while $H \approx 0.8$ m in this area.

Figure 14.4 shows the estimated change in volume $\Delta \hat{V}$ with 95% error bounds U for each DEM with respect to the first one. The seasonal variation is clearly discernible. The error bounds are tight enough to confirm the observed change in volume. The contributions of different error sources to the confidence limits are given in Table 14.2. The bounds for truncation and interpolation errors are of the same order of magnitude for those DEMs which were not given in the exactly same ground coordinates. The noise in the elevations is of minor importance. The 95% error bounds vary from 7 to 64% of the changes in volume in absolute value.

14.5   CONCLUSIONS

The main contribution of this chapter has been to derive estimates for the accuracy of change in elevation and volume lying in between two glacial surfaces represented as DEMs. A novel bound was introduced for the truncation error when the domain of the volume integral was a set of triangles. Bounds were also given for the truncation error in the case of a regular set of rectangles and for errors occurring in bilinear interpolation of a raster DEM and in linear interpolation of a TIN DEM. The precision of the change in elevation and volume was estimated by error propagation from either supposed or estimated noise level in the elevations and from estimated uncertainty in the registration of the DEMs after systematic differences in georeferencing had been corrected by a surface matching algorithm.

The test results at Svartisheibreen and Hintereisferner showed that the error bounds were tight enough for reliable detection of even short-term changes occurring during a period of one month when DEMs generated from airborne laser scanner data and aerial photographs were used. The error bound for change in volume included the truncation error, which has not been addressed, to our knowledge, in any previous research dealing with estimation of the change in volume of a glacier.

REFERENCES

Albrecht, O., P. Jansson and H. Blatter (2000). Modelling glacier response to measured mass-balance forcing. *Annals of Glaciology* 31, 91–96.

Brecher, H.H. and L.G. Thompson (1993). Measurement of the retreat of Qori Kalis glacier in the tropical Andes of Peru by terrestrial photogrammetry. *Photogrammetric Engineering and Remote Sensing* 59(6), 1017–1022.

Buckley, S. and H. Mitchell (2004). Integration, validation and point spacing optimization of digital elevation models. *The Photogrammetric Record* 19(108), 277–295.

Burden, R.L. and J.D. Faires (1993). *Numerical Analysis*. 5th ed., PWS Publishing Company, Boston, 768 p.

Davis, P.J. and P. Rabinowitz (1967). *Numerical Integration*. Blaisdell Publishing Company, Waltham, 230 p.

Etzelmüller, B., G. Vatne, R.S. Odegard and J.L. Sollid (1993). Mass balance and changes of surface slope, crevasse and flow pattern of Erikbreen, northern Spitsbergen: an application of a geographical information system (GIS). *Polar Research* 12(2), 131–146.

Etzelmüller, B. and J.L. Sollid (1997). Glacier geomorphometry – an approach for analyzing long-term glacier surface changes using grid-based digital elevation models. *Annals of Glaciology* 24, 135–141.

Etzelmüller, B. (2000). On the quantification of surface changes using grid-based digital elevation models (DEMs). *Transactions in GIS* 4(2), 129–143.

Geist, T., H. Elvehøy, M. Jackson and J. Stötter (2006). Investigations on intra-annual elevation changes using multi-temporal airborne laser scanning data: case study Engabreen, Norway. *Annals of Glaciology* 42, 195–201.

Geist, T. and J. Stötter (2007). Documentation of glacier surface elevation change with multi-temporal airborne laser scanner data – case study: Hintereisferner and Kesselwandferner, Tyrol, Austria. *Zeitschrift für Gletscherkunde und Glazialgeologie* 41, 77–106.

Jokinen, O., T. Geist, K.-A. Høgda, M. Jackson, K. Kajuutti, T. Pitkänen and V. Roivas (2007). Comparison of digital elevation models of Engabreen glacier. *Zeitschrift für Gletscherkunde und Glazialgeologie* 41, 185–204.

Kääb, A. (2002). Monitoring high-mountain terrain deformation from repeated air- and space-borne optical data: examples using digital aerial imagery and ASTER data. *ISPRS Journal of Photogrammetry and Remote Sensing* 57(1–2), 39–52.

Karras, G.E. and E. Petsa (1993). DEM matching and detection of deformation in close-range photogrammetry without control. *Photogrammetric Engineering and Remote Sensing* 59(9), 1419–1425.

Kite, G.W. and I.A. Reid (1977). Volumetric change of the Athabasca Glacier over the last 100 years. *Journal of Hydrology* 32(3/4), 279–294.

Lundstrom, S.C., A.E. McCafferty and J.A. Coe (1993). Photogrammetric analysis of 1984–89 surface altitude change of the partially debris-covered Eliot Glacier, Mount Hood, Oregon, U.S.A. *Annals of Glaciology* 17, 167–170.

Magnússon, E., H. Björnsson, J. Dall and F. Pálsson (2005). Volume changes of Vatnajökull ice cap, Iceland, due to surface mass balance, ice flow, and subglacial melting at geothermal areas. *Geophysical Research Letters* 32(5), L05504.

Mäkelä, M., O. Nevanlinna and J. Virkkunen (1984). *Numeerinen matematiikka*. 2nd ed., Gaudeamus, Mänttä, 331 p.

Pfeifer, N., P. Stadler and C. Briese (2001). Derivation of digital terrain models in the SCOP++ environment. *Proceedings of the OEEPE Workshop on Airborne Laserscanning and Interferometric SAR for Digital Elevation Models*, Stockholm, Sweden, March 1–3, 2001, 13 p.

Pilgrim, L. (1996). Robust estimation applied to surface matching. *ISPRS Journal of Photogrammetry and Remote Sensing* 51(5), 243–257.

Raaflaub, L. and M. Collins (2006). The effect of error in gridded digital elevation models on the estimation of topographic parameters. *Environmental Modelling & Software* 21(5), 710–732.

Reynolds, J.R. and G.J. Young (1997). Changes in areal extent, elevation and volume of Athabasca Glacier, Alberta, Canada, as estimated from a series of maps produced between 1919 and 1979. *Annals of Glaciology* 24, 60–65.

Vignon, F., Y. Arnaud and G. Kaser (2003). Quantification of glacier volume change using topographic and ASTER DEMs: a case study in the Cordillera Blanca. *Proceedings of the IGARSS'03*, Toulouse, France, July 21–25, 2003, Vol. 4, 2605–2607.

Zhang, Z. (1994). Iterative point matching for registration of free-form curves and surfaces. *International Journal of Computer Vision* 13(2), 119–152.

Chapter 15

The role of remote sensing in worldwide glacier monitoring

Andreas Kääb
Department of Geosciences, University of Oslo, Norway

15.1 INTRODUCTION

Presently, worldwide climate-related long-term glacier monitoring supports four major goals. (1) Long-term monitoring of glaciers is a base for understanding of processes that are related to interactions between glaciers and climate. (2) Due to the climate sensitivity of glaciers, their monitoring supports the detection of climate and environmental change that is expressed through glaciers and their changes. Glacier changes reflect for example the regional variability of climate change, or allow for detecting climate change in remote areas where no meteorological measurements are available (Figure 15.1). (3) Results from glacier monitoring aid the validation of models such as global and regional climate models (GCMs and RCMs), impact models, sensitivity studies, scenarios, interpolations and extrapolations in space and time, etc. Finally, (4) glacier monitoring supports the assessment of impacts that result from climate-change induced glacier changes. The most important global-scale impact is global sea level change. The most important regional- and local-scale impacts include change of water resources and glacier runoff, natural hazards, landscape change, effects on tourism, glacier-permafrost interactions and sediment budget.

Other goals of glacier monitoring, which are not necessarily related to climate and climate change, include understanding of other glacial processes, water resources management, hazard assessment and monitoring and engineering and construction tasks.

15.2 THE GLOBAL HIERARCHICAL OBSERVING STRATEGY

Within the framework of the global climate-related terrestrial observing systems (Global Climate Observing System, GCOS; Global Terrestrial Observing System, GTOS), a Global Hierarchical Observing Strategy (GHOST) was developed to be used for all terrestrial variables in order to provide a scientifically sound framework for global terrestrial observations (WMO 1997). Out of 14 key terrestrial variables in global climate-related observations, three cover the cryosphere: frozen ground, snow cover, and glaciers and ice caps. Through a system of five tiers, the regional to global

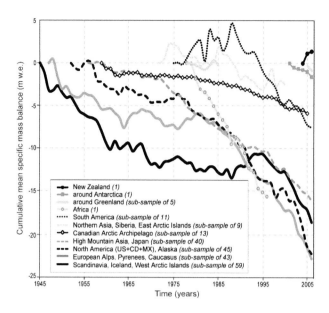

Figure 15.1 Cumulative mean specific mass balance for a number of glacier regions. Data collected by WGMS (Zemp et al. 2008).

representativeness in space and time of the observation records should be assured. Concerning area and mass of glaciers and ice caps these five tiers are described below (Haeberli et al. 2000, Haeberli et al. 2002). Even if GHOST is designed for global-scale observation networks, parts of its strategy are also useful for smaller networks such as national or regional ones, and for better understanding the role of remote sensing in glacier monitoring, compared to field methods.

Tier 1 is *Transects along environmental gradients*, e.g. from continental to maritime environments, employing multiple components of the observation system (see tiers 2–5). The primary emphasis is on spatial diversity at large, continental-type scales or on elevation belts of high-mountain areas. Special attention is given to long-term measurements. For glaciers, both field methods and remote sensing, or combinations of them, are suited to provide transects of changes in glacier extent and mass.

Tier 2 is *Extensive and process-oriented glacier mass-balance and flow studies* within major climatic zones for calibrating numerical models. Such detailed observations form the basis for the parameterization of numerical energy-balance, mass-balance and flow models, which then enable improved process understanding, sensitivity experiments and extrapolation to areas with less comprehensive measurements. Site locations should represent a broad range of climatic zones such as tropical, subtropical, monsoon-type, mid-latitude maritime to continental, sub-polar, or polar. Typically, a combination of a number of glaciological field measurements and terrestrial, airborne or spaceborne remote sensing support such process studies.

Tier 3 is studying *Regional-scale glacier volume/mass changes* within major mountain systems. For this purpose, observations with a limited number of strategically selected index stakes on the glaciers, with seasonal or annual time resolution, should be combined with precise topographic mapping of volume change of entire glaciers at decadal-scale intervals. Related worldwide measurements are collected and re-distributed by the World Glacier Monitoring Service (WGMS) (WGMS 2005) (Figures 15.1 and 15.2). Mass balance measurements at seasonal or annual resolution on selected glaciers are clearly the domain of traditional glaciological field measurements, which can hardly be replaced by remote sensing methods. Measuring decadal-scale volume changes of these mass-balance glaciers, however, is within the domain of remote sensing, so far in particular of airborne remote sensing.

Tier 4 is *Long-term observations of glacier length changes* concentrating typically on a minimum of about 10 sites within each major mountain range. This tier aims at assessing the representativeness of mass-balance and volume change measurements (tier 3). Sites should be selected to represent different climate characteristics, glacier sizes and dynamic responses. Related worldwide measurements are collected and re-distributed by the WGMS (WGMS (2005)). Glacier length changes have so far been mainly collected using in-situ measurements. High resolution airborne and spaceborne remote sensing is, however, well suited to complement, support, or in a number of cases even replace field observations of glacier length changes.

Tier 5 is *Global coverage by glacier inventories* repeated at decadal-scale time intervals and analyses of existing and newly available inventory data. Continuous upgrading of preliminary inventories and repetition of detailed inventories should enable global coverage and permit validation of climate models. Glacier inventorying on large spatial scales is only possible using airborne or spaceborne remote sensing. Tier 5 is thus clearly within the domain of remote sensing. The Global Land Ice Measurements from Space project (GLIMS) is a global initiative that aims to fulfil tier 5 of GHOST.

This integrated and nested multi-level strategy aims at combining in-situ observations with remotely sensed data, detailed process understanding with global coverage,

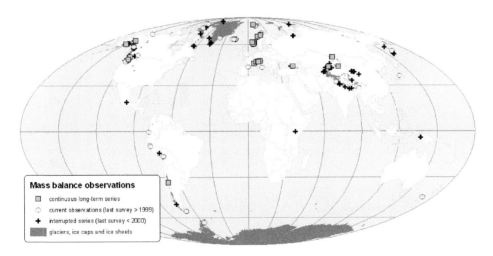

Figure 15.2 Worldwide network of mass balance measurements collected by the WGMS (Zemp et al. 2008).

and traditional measurements with new technologies (Haeberli 1998, Williams and Hall 1998, Haeberli et al. 2000, Haeberli et al. 2002).

15.3   THE ROLE OF REMOTE SENSING

Remote sensing plays an important role in achieving the above tiers of GHOST. Remote sensing technologies applied to glacier monitoring have the potential to facilitate measurements that are traditionally performed using field techniques. Glacier length changes, for example, are often measured on the ground, usually at one or a few points. Airborne or high resolution spaceborne data allow for measuring glacier length changes along entire glacier tongues, and in that way complement or even replace ground-based measurements.

Remote sensing technologies enable the observation of glaciological variables that were difficult or impossible to be measured previously. Glacier area changes, for example, can hardly be measured using ground methods, at least not for large glaciers or for a large number of glaciers (Figure 15.3, Chapter 8). Another example is ice velocity measurements in heavily crevassed glacier sections that are not accessible by direct measurements in the field (Figure 15.4).

Remotely sensed data provide a spatial distribution and representativeness of observations that are otherwise impossible to achieve. Glacier thickness changes and ice velocities, for example, can be measured on the ground for selected locations but hardly over entire glaciers. Thus, remote sensing results can be used to interpolate point data or check the representativeness of scattered ground measurements (Figures 15.4 and 15.5).

Global-scale glacier monitoring requires continuous long-term time series. Economic, political and security problems in a number of regions with glaciers make it in general or over time periods difficult to maintain ground-based glaciological observations. Remote sensing, in particularly satellite-based methods, enables the

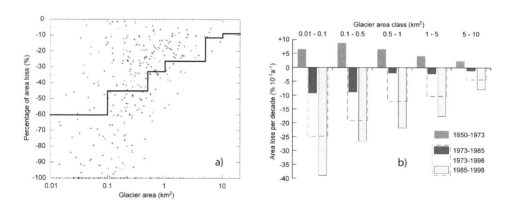

Figure 15.3 Glacier area changes 1850–1973, 1973–1985, 1973–1998 and 1985–1998 in the Valais, Swiss Alps, derived from repeat Landsat imagery. a) area loss of individual glaciers. The stepped line represents the average area loss per size class; b) average decadal area loss per size class (after Kääb et al. 2002, Paul et al. 2002).

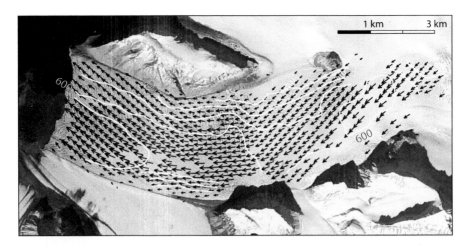

Figure 15.4 Surface velocity field for a section of Kronebreen, Svalbard, derived from ASTER imagery of June 26 and August 6, 2001. Isolines indicate glacier velocity in metres per year. Underlying ASTER image of August 6, 2001 (after Kääb et al. 2006).

globally uniform measurement of a number of glaciological variables, independent of local political limitations. On the other hand, however, the availability and costs of satellite data for scientific purposes depends on the economic and political situation of a small number of countries that develop and maintain suitable satellite missions.

Remote sensing technologies have enabled a major step forward towards a uniform global glacier monitoring. However, understanding glacial processes or detecting climate change requires a combination of remote sensing measurements, in-situ measurements and models. Both the design of monitoring systems and the interpretation of monitoring results should therefore be embedded in an integrated strategy of observations from field, air and space, together with modelling, so that all these components complement and support each other. GHOST adapts such principles for a global monitoring strategy.

Remote sensing technologies are able to substantially support glaciological process studies in tier 2 of GHOST, regional-scale observation of glacier volume/mass changes in tier 3, measurement of glacier length changes in tier 4, and glacier inventorying in tier 5 through providing ice flow components, elevation changes on glaciers, and changes in glacier area.

Ice flow components can be remotely sensed using cross-correlation between repeat terrestrial, airborne or spaceborne optical images (Figure 15.4), radar interferometry applying differential interferometric synthetic aperture radar or DInSAR (mainly ground-based and spaceborne), cross-correlation of repeat SAR images applying speckle tracking, and cross-correlation of repeat high resolution digital elevation models (DEMs) such as from terrestrial and airborne laser scanning (see Chapter 10).

Elevation changes on glacier surfaces and related volume changes can be obtained by differencing repeat DEMs from terrestrial, aerial and satellite

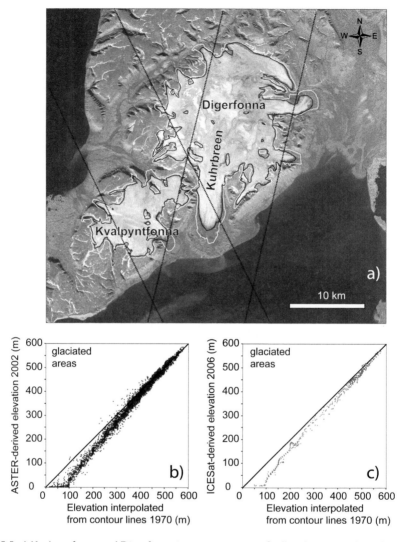

Figure 15.5 a) Kvalpyntfonna and Digerfonna ice caps in eastern Svalbard presented in ASTER satellite image from 2002; b) elevation changes between topographic map of 1970 and DEM derived from the ASTER stereo images; c) elevation changes between topographic map of 1970 and ICESat GLAS LiDAR DEM from 2006. The lines in ASTER image indicate the ICESat ground tracks used (after Kääb 2008). (See colour plate section).

stereophotogrammetry (see Chapters 6 and 7) (Figure 15.5), repeat terrestrial and airborne laser profiling and scanning (Chapter 10), ground-based, air- and space-borne DInSAR (Chapter 9), with the DEMs from the Shuttle Radar Topography Mission (SRTM) forming a key benchmark dataset, and satellite laser (LiDAR) or radar altimetry (see Chapter 10).

Glacier outlines or terminus positions, and their variations over time, can be derived from repeat terrestrial, air- and spaceborne optical imagery (see Chapter 6, 7, 8 and 12). The application of (semi) automatic classification methods to multispectral satellite imagery for the first time in history enables the compilation of an accurate global inventory of land ice areas, which is the major goal of the GLIMS project. Glacier outlines and their changes can also be derived from terrestrial and airborne laser scanning (Knoll 2009, Chapter 10), and terrestrial, air- and spaceborne SAR images (Chapter 9).

Further important glaciological variables are firn and snow lines and snow zones on glaciers, and their changes in space and time. They can efficiently be mapped from optical satellite data and from spaceborne SAR data (Chapter 9).

The technical and scientific challenge for remote-sensing based, worldwide glacier monitoring consists in the meaningful fusion of the above-listed methods and data towards an operational, robust and sustainable monitoring system that integrates also field observations.

15.4 GLOBAL LAND ICE MEASUREMENTS FROM SPACE PROJECT AND OTHER PROJECTS

A number of projects support worldwide, or at least large-scale, glacier mapping and monitoring from space. The most developed of these projects is the Global Land Ice Measurements from Space project (GLIMS; see details below). A more recent initiative is the GlobGlacier project by the European Space Agency (ESA). Started in late 2007 it aims to support and complement WGMS and GLIMS glacier monitoring activities using remote sensing techniques to provide glacier outlines, snow lines, glacier topography, glacier thickness changes and glacier velocities for a substantial number of glaciers worldwide (Paul et al. 2009). Other activities of large-scale glacier monitoring are closely related to satellite missions. The NASA ICESat mission, for example, provides repeat surface elevations, which allows for accurate calculation of thickness changes for a large number of global locations with ice cover, either by comparing repeat ICESat data, or ICESat data and other elevation information (Figure 5) (Zwally et al. 2002). Another example is the IPY SPIRIT (International Polar Year) initiative that provides DEMs over polar ice masses based on the SPOT5 satellite (Korona et al. 2008). Finally a number of institutions run large-scale satellite-based glacier monitoring activities, for example monitoring of the dynamics of the Greenland outlet glaciers by the NASA JPL (Rignot and Kanagaratnam 2006). The OMEGA, funded by the FP5 of the EC, focused on testing various remote sensing data and methods from optical to radar and from terrestrial to spaceborne data for constructing of digital elevation models and delineating glacier outlines and zones (Pellikka and Kajuutti 2005). The project compiled an extensive dataset of test glaciers Hintereisferner in Austria and Svartisen in Norway and studied the glacier changes in detail within those areas.

The international GLIMS project is a global consortium of universities and research institutes, whose purpose is to assess and monitor the world's glaciers primarily using data from optical satellite instruments, such as ASTER (Advanced Spaceborne Thermal Emission and reflection Radiometer). Specifically, GLIMS's objectives are to ascertain the extent and condition of the world's glaciers so that we may understand a variety of Earth surface processes and produce information for resource management and planning. These scientific, management and planning objectives are supported by the monitoring and information production objectives of the United Nations scientific organizations (Kieffer et al. 2000, Bishop et al. 2004, Kargel et al. 2005, Raup et al. 2007b).

A first attempt to obtain an overview of the Earth's glaciers and ice caps was made with the compilation of the World Glacier Inventory by WGMS, which is a database of glacier locations and attributes produced mainly on the basis of aerial photographs and topographic maps. In 1999, the GLIMS project was established to continue this inventorying task based on space technologies and in close cooperation with the National Snow and Ice Data Center (NSIDC) in Boulder, U.S.A. and the WGMS.

GLIMS entails comprehensive satellite multi-spectral and stereo-image acquisition of land ice, use of satellite imaging data to measure inter-annual changes in glacier area, boundaries, and snowline elevation, measurement of glacier ice-velocity fields, and development of a comprehensive digital database to inventory the world's glaciers. This work and the global image archive at the EROS Data Center in Sioux Falls, U.S.A., is useful for a variety of scientific and planning applications. The GLIMS glacier database and GLIMS web site are developed and maintained by the NSIDC. Beside the main GLIMS applications of glacier mapping and monitoring, ASTER proved also to be very suitable for assessing glacier hazards and managing related disasters. ASTER emergency imaging was for instance activated in the case of the rock-ice avalanche at Kolka/Karmadon, North Ossetia, Caucasus, on 20 September 2002. The avalanche deposits of about 10^8 m^3 dammed the rivers flowing trough the Karmadon area to the north. The temporary lakes and the ice dam changed significantly over time (Figure 15.6) (Kääb et al. 2003a).

GLIMS primarily utilizes multispectral satellite images from the Landsat TM and ETM+ series, and from the ASTER sensor. Landsat TM and ETM+ data represent a well-established and robust data source for glacier inventorying and monitoring from space (Chapter 8). Capabilities of the ASTER sensor, available since 2000 onboard the NASA Terra spacecraft, include 3 bands in VNIR (visible and near infrared) with 15 m resolution, 6 bands in the SWIR (short-wave infrared) with 30 m, 5 bands in the TIR (thermal infrared) with 90 m, and a 15 m resolution NIR along-track stereo-band looking backwards from nadir. Of special interest for glaciological studies are the high spatial resolution in VNIR and the stereoscopic and pointing capabilities of ASTER. With topography being a crucial parameter for the understanding of high-mountain phenomena and processes, such as glacier volume changes, DEMs generated from the ASTER along-track stereo bands are especially helpful. Figure 15.7 presents the topography of the Tasman glacier and its environments in New Zealand computed using the backward-looking and nadir-looking bands. Mount Cook, the highest peak of New Zealand, is to the middle left of the depicted section of terrain (Kääb 2005). The nominal ASTER lifetime was designed to be six years, i.e. until 2006, but it has been functioning somewhat longer, but by 2008, in particular the ASTER SWIR bands were deteriorating significantly. The vast archive of ASTER data is useful for worldwide glacier mapping and already provides an important data source for time series of glacier change between 2000 and the present, and an invaluable baseline dataset for comparisons with data from other similar multi-spectral satellite missions.

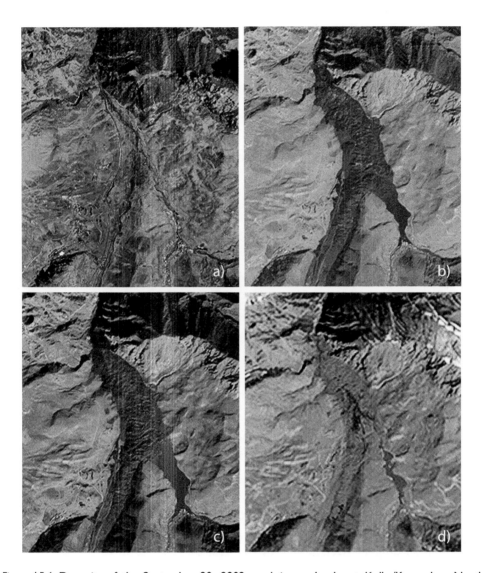

Figure 15.6 Deposits of the September 20, 2002, rock-ice avalanche at Kolka/Karmadon, North Ossetia, Caucasus, represented by ASTER scenes covering 3.8 × 4.5 km. Avalanche direction was from the image bottom to the top. a) June 22, 2001; b) September 27, 2002; c) October 13, 2002; d) April 09, 2004 (after Kääb et al. 2003a). (See colour plate section).

Currently the GLIMS database at NSIDC holds nearly 100 000 glacier outlines in North and South America, Europe, Asia and Antarctica, and is quickly growing (Raup et al. 2007a). NSIDC has developed easy-to-use search and extraction tools which allow download of GLIMS data in a variety of geo-information formats (www.glims.org). GLIMS and ASTER science results are summarized by a number of

Figure 15.7 The topography of Tasman glacier and its environments in New Zealand were computed from ASTER satellite image acquired on April 29, 2000. North is to the top and depicted terrain section is about 25 km by 25 km. (See colour plate section).

publications (Kääb et al. 2003b, Bishop et al. 2004, Kargel et al. 2005, Raup et al. 2007b) and examples given in Figures 15.3–15.7.

REFERENCES

Bishop, M.P., J.A. Olsenholler, J.F. Shroder, R.G. Barry, B. Raup, A.B.G. Bush, L. Copland, J.L. Dwyer, A.G. Fountain, W. Haeberli, A. Kääb, F. Paul, D.K. Hall, J.S. Kargel, B.F. Molnia, D.C. Trabant and R. Wessels (2004). Global land ice measurements from space (GLIMS): remote sensing and GIS investigations of the Earth's Cryosphere. *Geocarto International* 19(2), 57–84.

Haeberli, W. (1998). Historical evolution and operational aspects of worldwide glacier monitoring. In: W. Haeberli, M. Hoelzle and S. Suter (eds.), *Into the Second Century of World Glacier Monitoring: Prospects and Strategies.* pp. 35–51, UNESCO Publishing, Paris.

Haeberli, W., J. Cihlar and R. Barry (2000). Glacier monitoring within the Global Climate Observing System – a contribution to the Fritz Müller memorial. *Annals of Glaciology* 31, 241–246.

Haeberli, W., M. Maisch and F. Paul (2002). Mountain glaciers in global climate-related observation networks. *World Meteorological Organization Bulletin* 51(1), 18–25.

Kääb, A., F. Paul, M. Maisch, M. Hoelzle and W. Haeberli (2002). The new remote sensing derived Swiss glacier inventory: II. First results. *Annals of Glaciology* 34, 362–366.

Kääb, A., R. Wessels, W. Haeberli, C. Huggel, J. Kargel and S.J.S. Khalsa (2003a). Rapid ASTER imaging facilitates timely assessment of glacier hazards and disasters. *EOS, Transactions, American Geophysical Union* 84(13), 117–121.

Kääb, A., C. Huggel, F. Paul, R. Wessels, B. Raup, H. Kieffer and J. Kargel (2003b). Glacier monitoring from ASTER imagery: accuracy and applications. *EARSel eProceedings* 2, 43–53.

Kääb, A. (2005). Remote Sensing of Mountain Glaciers and Permafrost Creep. *Schriftenreihe Physische Geographie. Glaziologie und Geomorphodynamik* 48. University of Zurich, 264 p.

Kääb, A., B. Lefauconnier and K. Melvold (2006). Flow field of Kronebreen, Svalbard, using repeated Landsat7 and ASTER data. *Annals of Glaciology* 42, 7–13.

Kääb, A. (2008). Glacier volume changes using ASTER satellite stereo and ICESat GLAS laser altimetry. A test study on Edgeøya, Eastern Svalbard. *IEEE Transactions on Geoscience and Remote Sensing* 46(10), 2823–2830.

Kargel, J.S., M.J. Abrams, M.P. Bishop, A. Bush, G. Hamilton, H. Jiskoot, A. Kääb, H.H. Kieffer, E.M. Lee, F. Paul, F. Rau, B. Raup, J.F. Shroder, D. Soltesz, D. Stainforth, L. Stearns and R. Wessels (2005). Multispectral imaging contributions to global land ice measurements from space. *Remote Sensing of Environment* 99(1–2), 187–219.

Kieffer, H.H., J.S. Kargel, R. Barry, R. Bindschadler, M. Bishop, D. MacKinnon, A. Ohmura, B. Raup, M. Antoninetti, J. Bamber, M. Braun, I. Brown, D. Cohen, L. Copland, J. DueHagen, R.V. Engeset, B. Fitzharris, K. Fujita, W. Haeberli, J.O. Hagen, D. Hall, M. Hoelzle, M. Johansson, A. Kaab, M. Koenig, V. Konovalov, M. Maisch, F. Paul, F. Rau, N. Reeh, E. Rignot, A. Rivera, M. de Ruyter de Wildt, T. Scambos, J. Schaper, G. Scharfen, J. Shroder, O. Solomina, D. Thompson, K. van der Veen, T. Wohlleben and N. Young (2000). New eyes in the sky measure glaciers and ice sheets. *EOS, Transactions, American Geophysical Union* 81(24), 265, 270–271.

Korona, J., E. Berthier, M. Bernard, F. Rémy and E. Thouvenot (2008). SPIRIT SPOT 5 stereoscopic survey of polar ice: reference images and topographies during the fourth International Polar Year (2007—2009). *ISPRS Journal of Photogrammetry and Remote Sensing* 64(2) 204–212.

Paul, F., A. Kääb, M. Maisch, T. Kellenberger and W. Haeberli (2002). The new remote-sensing-derived Swiss glacier inventory: I. Methods. *Annals of Glaciology* 34, 355–361.

Paul, F., A. Kääb, H. Rott, A. Shepherd, T. Strozzi and E. Volden (2009). GlobGlacier: A new ESA project to map the world's glaciers and ice caps from space. *EARSeL eProceedings* 8(1), 11–26.

Pellikka, P. and K. Kajuutti (eds.) (2005). Remote sensing of glacier area, topography and changes on Svartisen and Hintereisferner – Scientific results of the OMEGA project. *Turku university department of geography publications B6*. Digipaino, Turku, 83 p.

Raup, B., A. Racoviteanu, S.J.S. Khalsa, C. Helm, R. Armstrong and Y. Arnaud (2007a). The GLIMS geospatial glacier database: A new tool for studying glacier change. *Global and Planetary Change* 56(1–2), 101–110.

Raup, B., A. Kääb, J.S. Kargel, M.P. Bishop, G. Hamilton, E. Lee, F. Paul, F. Rau, D. Soltesz, S.J.S. Khalsa, M. Beedle and C. Helm (2007b). Remote sensing and GIS technology in the Global Land Ice Measurements from Space (GLIMS) project. *Computers & Geosciences* 33, 104–125.

Rignot, E. and P. Kanagaratnam (2006). Changes in the velocity structure of the Greenland ice sheet. *Science* 311(5763), 986–990.

WGMS (2005). *Fluctuations of Glaciers 1995–2000* (Vol. VIII). Haeberli, W., Zemp, M., Frauenfelder, R., Hoelzle, M. and Kääb, A. (eds.), IUGG (CCS)/UNEP/UNESCO, World Glacier Monitoring Service, Zurich, 288 p.

Williams, R.S. (Jr.) and D.K. Hall (1998). Use of remote-sensing techniques. In: Haeberli, W., M. Hoelzle and S. Suter (eds.), *Into the Second Century of Worldwide Glacier Monitoring: Prospects and Strategies*. pp. 97–111, UNESCO Publishing, Paris.

WMO (1997). *Global Hierarchical Observing Strategy (GHOST)*. World Meteorological Organisation, 862 p.

Zemp, M., M. Hoelzle and W. Haeberli (2008). Six decades of glacier mass balance observations – a review of the worldwide monitoring network. *Annals of Glaciology* 50(50), 101–111.

Zwally, H.J., B. Schutz, W. Abdalati, J. Abshire, C. Bentley, A. Brenner, J. Bufton, J. Dezio, D. Hancock, D. Harding, T. Herring, B. Minster, K. Quinn, S. Palm, J. Spinhirne and R. Thomas (2002). ICESat's laser measurements of polar ice, atmosphere, ocean, and land. *Journal of Geodynamics* 34(3–4), 405–445.

Chapter 16

Conclusions

W. Gareth Rees
Scott Polar Research Institute, University of Cambridge, England

The significance of glaciers is disproportionately greater than their size. Although they contain less than one percent of the total volume of ice on Earth, they are currently believed to be making the major cryospheric contribution to the global rise in sea-level. They are in general more sensitive to variations in climate, and swifter to respond, than the larger ice caps and ice sheets. The evidence that we have suggests that many of the world's glaciers are indeed responding to a warmer climate, by thinning and retreating. However, *in situ* measurements are not easy to obtain: most glaciers are located in remote places that are difficult to reach. The desirability of remote sensing methods to study glaciers is clear, and has been recognised for many years.

This book arose from the OMEGA project, which is described in detail in the Preface. The project had multiple aims, centred around the development of an operational system, based on the use of remote sensing data, for monitoring glaciers. New data were collected from the two test glaciers – Hintereisferner in Austria, and Svatisen ice caps in Norway – and these datasets will be of great value in establishing baselines for future monitoring. However, a result of at least equal importance to have emerged from the project is that it has given us a very detailed understanding of the state of the art in remote sensing of glaciers: not just what can be achieved now, but where technical advances might take us in the future. The project was timely, not just from the point of view of the increasing international concern about climate change, but also because of recent advances in the technology of remote sensing. The balance between field-based glaciology and remote sensing methods has been altering steadily over the last few decades, and has now reached the point where remote sensing methods are generally preferable to *in situ* measurements (Chapter 2), although fieldwork will probably never become completely unnecessary. Superimposed on this steady technological progress, however, there have also been a number of more abrupt shifts when new techniques have come to exceed some threshold of usefulness. In particular, our ability to measure the surface topography of glaciers – to produce a digital elevation model (DEM) – has recently been profoundly advanced by the development of extremely accurate airborne laser scanning systems. Vertical accuracies of 0.1 m, with horizontal sampling intervals of the order of 1 m, are now routinely achievable, as discussed in Chapter 10. This has in turn led to the need to consider more carefully the accuracy characteristics of DEMs, and their use in determining topographic change over time, topics that are considered in Chapters 13 and 14 respectively. The dramatic improvement in the capabilities of laser scanning systems has come hand in hand with the availability of more accurate positioning data from the global positioning system (GPS) and similar.

The great value of remote sensing methods for studying glaciers has, as was noted earlier, been recognised for many years. Photogrammetry was applied from the late 19th century onwards, initially from a terrestrial perspective and latterly from aircraft (Chapters 4, 6 and 7). Terrestrial photogrammetry, discussed in Chapter 6, is particularly suitable for studying rapid changes over small areas. Vertical aerial photography, discussed in Chapter 7, is a mature technology, well established as a means of mapping and monitoring glaciers, including their topography and dynamics. Photographs represent an efficient way of storing information, and datasets extend back over decades. The greatest problem with aerial photography is the lack of radiometric contrast, and the consequent possibility of undetected error in stereophotogrammetric reconstruction of glacier topography. Other problems are geometric and spectral distortions, such as brightness variations caused by light falloff and bidirectional effects. These problems are not shared by laser scanning, which is now displacing aerial photography, although the value of photographic archives remains great.

Optical remote sensing (in which is included the use of near-infrared and thermal infrared) using spaceborne imagery provides some of the same advantages as aerial photography, although generally at a lower spatial resolution. Since suitable satellite-based sensors have been in operation since the 1970s, long time series of data are potentially available. Although stereo imaging is possible, there are better solutions for determining glacier topography and the most important use of optical imagery is probably for mapping areal extent and surface features, for example in determining the position of the snowline. This type of imagery has proved very versatile, but there are some difficulties associated with it. Like aerial photography, the lack of radiometric contrast from a snow-covered surface can impede attempts at mapping. The imagery can also only be acquired during cloud-free daylight conditions, which means that it can often be difficult to obtain a useable image at a particular point during the balance year. This particular problem is avoided by the use of synthetic aperture radar (SAR) imagery, discussed in Chapter 9, and SAR has received a great deal of attention in the last couple of decades. InSAR also gives topographic information. However, the goal of identifying a reliable proxy for determining mass balance from SAR imagery remains elusive. The use of radar coherence, discussed in Chapter 5, is promising but has not yet received much attention. Chapter 11 discusses the theory and application of ground penetrating radar, which is the main technique for investigating the internal structure and bedrock configuration of glaciers. Other techniques include acoustic and seismic methods. These, and other non-electromagnetic techniques, are discussed in Chapter 5. A particularly interesting idea is the use of gravity surveying from space. This has been demonstrated to work well for large ice masses but does not currently have sufficient spatial resolution to be useful for glaciers.

The current situation is that a combination of spaceborne and airborne data collection technologies, supplemented by *in situ* calibration and validation activities, are able to answer most of the questions that we have relating to the extent and time-evolution of glaciers. For the remotest areas, airborne data collection and field work may not be practicable, so the focus of global monitoring of glaciers remains on space-borne techniques. These are described in Chapter 15. The best-established of these is the Global Land Ice Measurements From Space (GLIMS) project, which has been operational since around 1998, but the European Space Agency's GlobGlacier project, initiated in 2007, has a similar aim. Both of these monitoring projects contribute to the World Glacier Monitoring Service (WGMS) within a coordinated hierarchical global observation strategy, as discussed in Chapter 15. It is very likely that such coordinated approaches to the monitoring of the world's glaciers will continue and grow: Remote sensing has already comprehensively demonstrated its scientific value and cost-effectiveness in the study of glaciers.

What does the future hold for remote sensing of glaciers? Remote sensing in general is a tricky field in which to make reliable predictions, but a number of trends are reasonably clear. Spaceborne observing systems have demonstrated a remarkable degree of continuity over the last few decades, despite the potentially rather short life-time of individual satellites and their sensors. One of the best known examples of this is the Landsat programme, begun in 1972. The operational Landsat satellites are now Landsats 5 and 7, operating since 1984 and 1999 respectively, although the ETM+ instrument on board Landsat 7 has been functioning with reduced capability since 2003. However, other types of spaceborne observation have also been continuously available for many years, for example SAR imagery since the launch of ERS-1 in 1991 and coarse resolution (ca. 1 km) optical and infrared imagery since 1978. Although the highest-resolution data continue to be rather expensive, there has been a welcome trend towards the availability at no or little cost of the coarser resolution data. This has had a tremendously beneficial effect on the research community. At the same time, Moore's law (the observation that computer power doubles roughly every two years) continues, astonishingly, to hold good. This has vastly increased the potential for the collection, storage, transmission and processing of data.

New spaceborne sensors are likely to be deployed. As already implied, to make predictions in this field is to risk looking foolish in only a few years, but one can perhaps suggest that the most promising possibilities from the glaciologist's persepctive are likely to be more diverse (in terms of frequency and/or polarization) radar systems and spaceborne laser profilers with higher sampling rates than are presently possible. Space-borne gravity mapping remains a tantalising prospect for mass-balance measurements. For airborne sensors the possibilities are if anything even greater. The future doubt-less offers steady improvement in the technical characteristics of the existing types of airborne instrument (and one certainly has the impression that the revolution brought about by high resolution laser scanning has only just begun), but there are many types of instrument, of potentially great value in the study of glaciers, that have found only very limited application so far. Just one possibility is that provided by passive microwave radiometry. Finally, a hugely significant trend is the global spread of digital communications, which enables automatic data retrieval from stand-alone data moni-toring and recording systems using the Internet, telecommunication technologies, and other communication technologies introducing possibilities for remote monitoring that we are only just beginning to explore and exploit. It is an exciting time to be studying glaciers.

Copyrights for figures

Authors

Olaf Eisen has a background in geophysics and became involved in glaciology during his studies in Fairbanks, Alaska, U.S.A. After an excursion to the sea ice for his M.Sc. at the University of Karlsruhe, Germany, he pursued the application of geophysical methods on glaciers and ice sheets ever since, with a focus on radar, at the Alfred-Wegener-Institute for Polar and Marine Research in Bremerhaven, Germany, and the VAW at the ETH Zurich. As a fellow of the DFG Emmy Noether-Programme his research group's current interests range from mass-balance estimates of glaciers and ice sheets to the deduction of ice dynamic properties and the distribution of physical properties within ice masses from geophysical measurements.

Olaf Eisen, Alfred Wegener Institute, Columbusstrasse, 27568 Bremerhaven, Germany. olaf.eisen@awi.de

Thomas Geist is a geographer and obtained his Ph.D. degree from the University of Innsbruck for his studies on the application of airborne laser scanning technology in glacier research. He is currently working as an expert in Earth observation at the Austrian Research Promotion Agency *(FFG)* in Vienna and as a lecturer in remote sensing at the University of Innsbruck.

Thomas Geist, FFG – Austrian Research Promotion Agency, Sensengasse 1, 1090 Wien, Austria. thomas.geist@ffg.at

Hardy B. Granberg obtained his Ph.D. degree from McGill University, Montreal. He is a research professor at the Departement of Applied Geomatics and a member of the Centre d'applications et de recherches en télédétection (Centre for remote sensing research and applications, CARTEL) at the Université de Sherbrooke, Canada. His main professional interests are snow and climate, their field measurement, modelling, and remote sensing.

Hardy B. Granberg, Département de géomatique appliqué, Faculté des lettres et sciences humaines, Université de Sherbrooke, Sherbrooke (Québec) J1K 2R1, Canada. hardy.granberg@USherbrooke.ca

Henrik Haggrén is a professor at the Department of Surveying of Helsinki University of Technology in Espoo, Finland. His former research activities have included various applications in photogrammetric mapping and measuring, close range photogramme-try, stereoscopic imaging, videogrammetry and autonomous 3-D measuring systems. Within the OMEGA project he was in charge of the investigation to resurvey old photographs of Hochjochferner, Austria.

Henrik Haggrén, Institute of Photogrammetry and Remote Sensing, Department of Surveying, Helsinki University of Technology, Otakaari 1 F, FIN 02150 Espoo, Finland. henrik.haggren@hut.fi

Johan Hendriks has an M.Sc. degree in physical geography from the University of Utrecht, the Netherlands. In his early career he focused mainly on GIS modelling of arctic rivers and the development of remote sensing techniques. During his Ph.D. studies at the University of Helsinki, he worked on the development and integration of remote sensing applications for change detection in glacial areas. Currently he is the Attaché for Science and Technology to Finland at the Embassy of the Kingdom of the Netherlands.

Johan Hendriks, Department of Geosciences and Geography, University of Helsinki, P.O. Box. 64, FIN-00014 Helsinki, Finland. johan.hendriks@helsinki.fi

Kjell Arild Høgda received a Ph.D. degree in applied physics (remote sensing) from the University of Tromsø, Norway, in 1994. Since 2003 he has been the Research Director of the Earth Observation Group at Norut Tromsø. He has worked with a broad spectrum of satellite remote sensing applications, including snow and ice monitoring, vegetation classification, pollution impact studies, phenology for climate change impact studies, and SAR interferometry for geohazard monitoring.

Kjell Arild Høgda, Norut Tromsø, P.O. Box 6434, N-9294 Tromsø, Norway. kjella@norut.no

Olli Jokinen is a senior research scientist and a docent in remote sensing (especially methods for area-based matching and accuracy estimation) in the Department of Surveying at Helsinki University of Technology. In his doctoral thesis (2000, Helsinki University of Technology), he developed new methods for matching and modelling of multiple 3-D images including improvements in registration algorithms and a new approach for self-calibration of a light striping system. His post-doctoral research concentrated on detection of distortions in multiple DEMs after correcting differences in georeferencing by surface matching. He also derived error bounds for the change in elevation and volume between two elevation surfaces of a glacier.

Olli Jokinen, Institute of Photogrammetry and Remote Sensing, Department of Surveying, Helsinki University of Technology, Otakaari 1 F, FIN 02150 Espoo, Finland. olli.jokinen@hut.fi

Andreas Kääb is a professor of remote sensing at the Department of Geosciences, University of Oslo. His research focus is on the development of terrestrial, airborne and spaceborne remote sensing methods and their application to glaciers, permafrost and related hazards. His study regions include the European and New Zealand Alps, Scandinavia, the Arctic, Central Asia and the Himalayas, North and South America, and the Caucasus.

Andreas Kääb, Department of Geosciences, University of Oslo, Postbox 1047 Blindern, 0316 Oslo, Norway. kaeaeb@geo.uio.no

Kari Kajuutti is a research scientist at the Department of Geography of the University of Turku, Finland. His former research activities have included glacier investigations in Scandinavia and in the U.S.A. by field measurements and remote sensing. In 2001–2004 he worked in the OMEGA project specializing in the usage of terrestrial photography in glacier topography studies.

Kari Kajuutti, Department of Geography, University of Turku, FIN-20014 Turku, Finland. kari.kajuutti@utu.fi

Michael Kuhn is the head of the Institute of Meteorology and Geophysics of the University of Innsbruck, Austria. Kuhn studied meteorology and physics, specializing in snow, ice and climate change. He has been carrying out meteorological and glaciological work in the Alps since 1962. In 1964 he was involved in glaciological and oceanographic studies in the Arctic Ocean, ice islands Arlis II and III, and in 1966–68, he overwintered at Plateau Station in Antarctica making energy balance measurements. Kuhn practised further Antarctic field work in 1969, 1970, 1974 and 1978. Presently, Kuhn is engaged in measuring and modelling the reaction of Alpine snow, ice and water to climatic changes.

Michael Kuhn, Institute of Meteorology and Geophysics, University of Innsbruck, Innrain 52, A-6020 Innsbruck, Austria. michael.kuhn@uibk.ac.at

Tom Rune Lauknes received his M.Sc. degree in applied physics from University of Tromsø, Norway, in 2004. He is currently a researcher in the Earth Observation Group at Norut Tromsø where he is also working on his Ph.D. degree. His research includes development of an interferometric synthetic aperture radar (InSAR) processing system to detect ground displacement related to geological hazards such as landslides and rockslides. He was a visiting scientist with the radar interferometry group at Stanford University during the academic year 2007/2008.

Tom Rune Lauknes, Norut Tromsø, P.O. Box 6434, N-9294 Tromsø, Norway. tomrune@norut.no

Matti Leppäranta, born in 1950, is a specialist in snow and ice geophysics. He worked as a research physicist in the Finnish Institute of Marine Research in 1974-1991 and obtained his Ph.D. degree in 1981 (sea ice dynamics). Since 1992 he has been a professor (physics of the hydrosphere) in the Department of Physics, University of Helsinki, where he is the leader of snow and ice team of 10 scientists and students. He has published two books and 65 scientific articles in peer reviewed scientific periodicals.

Matti Leppäranta, Department of Physics, University of Helsinki, P.O. Box 64 (Gustaf Hällströmin katu 2a), FI-00014 Helsinki, Finland. matti.lepparanta@helsinki.fi

Christoph Mayer is working as a glaciologist and surveyor at the Commission for Glaciology of the Bavarian Academy of Sciences and Humanities in Munich, Germany. His former research activities included investigations of glaciers in the polar regions by field measurements and remote sensing. Since 2004, he has devoted his energy to surveying the variations of glaciers in the Alps and Central Asia.

Christoph Meyer, Commission for Glaciology, Bavarian Academy of Sciences and Humanities, Alfons-Goppel Str. 11, D-80539 Munich, Germany. christoph.mayer@lrz.badw-muenchen.de

Francisco Navarro graduated in Geophysics at Universidad Complutense de Madrid, spent two years at University of California Los Angeles, and obtained his Ph.D. from Universidad Politécnica de Madrid in 1991. He is Professor Titular of Applied Mathematics and director of the Group of Numerical Simulation in Science and Engineering. His research activity has focused on glaciology, particularly on the modelling of glacier dynamics and glaciological applications of ground-penetrating radar. He has participated in 7 Antarctic and 4 Arctic fieldwork campaigns, including over-wintering at the South Pole Station and holding the position of Manager of the Spanish Antarctic Station in 2006-2007 campaign. He is the Spanish delegate to the Standing Scientific Group on Physical Sciences of the Scientific Committee on Antarctic Research (SCAR), President of the Cryospheric Sciences Section of the Spanish Committee of Geodesy and Geophysics, and an elective member of the Council of the International Glaciological Society.

Francisco Navarro, Departamento de Matemática Aplicada ETSI de Telecomunicación, Technical University of Madrid, Av. Complutense, 30 (Ciudad Universitaria), 28040 Madrid, Spain. fnv@mat.upm.es

Frank Paul studied Meteorology at the University of Hamburg, concentrating on remote sensing of the glaciers of the Alps, after which he worked at the Max Planck Institute of Meteorology in Hamburg. He completed his Ph.D. studies on the Swiss glacier inventory 2000 at the Department of Geography of the University of Zurich and then worked as a researcher in several projects dealing with glaciers, remote sensing and GIS. He is currently a principal investigator of the ESA project GlobGlacier.

Frank Paul, Department of Geography, University of Zurich, Winterthurerstrasse 190, CH-8057 Zurich, Switzerland. fpaul@geo.unizh.ch

Tuija Pitkänen has worked as a research scientist at the Institute of Photogrammetry and Remote Sensing of the Helsinki University of Technology, Finland. Her former research activities have included e.g. photogrammetric investigations of the Nisqually glacier, U.S.A.

Rune Storvold has a Ph.D. degree in atmospheric physics from the University of Alaska, Fairbanks, and is Senior Scientist in the Earth Observation Group at Norut Tromsø. Research interest focuses on the development of remote sensing techniques based on unmanned aerial vehicles and satellites using active and passive optical and microwave sensors with main focus on snow and ice in Arctic regions.

Rune Storvold, Norut Tromsø, P.O. Box 6434, N-9294 Tromsø, Norway. rune.storvold@norut.no

Johann Stötter is a professor of geography and head of the Institute of Geography at the University of Innsbruck. He received his academic education at the University of Munich, Germany where he studied geography, cartography and landscape ecology. The main focus of his research interest lies in mountain regions, global change issues, natural hazard and risk research, integrative geography and the application of laser scanning techniques for environmental research.

Johann Stötter, Institute of Geography, University of Innsbruck, Innrain 52, A-6020 Innsbruck, Austria. hans.stoetter@uibk.ac.at

Reviewers

Neil Arnold, *Scott Polar Research Institute, University of Cambridge, England*
Ludwig N. Braun, *Commission for Glaciology, Bavarian Academy of Sciences and Humanities, Germany*
Poul Christofferssen, *Scott Polar Research Institute, University of Cambridge, England*
Bea Csatho, *University of Buffalo, United States*
Julian Dowdeswell, *Scott Polar Research Institute, University of Cambridge, England*
Konrad Eder, *Technical University of Munich, Germany*
Adrian Fox, *British Antarctic Survey, England*
Henrik Haggrén, *Helsinki University of Technology, Finland*
Janne Heiskanen, *University of Helsinki, Finland*
Juha Hyyppä, *Finnish Geodetic Institute, Finland*
Kjell Arild Høgda, *Norut Tromsø, Norway*
Olli Jokinen, *Helsinki University of Technology, Finland*
Kari Kajuutti, *University of Turku, Finland*
Viktor Kaufmann, *Graz University of Technology, Austria*
Kirsi Karila, *Finnish Geodetic Institute, Finland*
Michael Kuhn, *University of Innsbruck, Austria*
Andreas Kääb, *University of Oslo, Norway*
Javier Lapazaran, *Technical University of Madrid, Spain*
Matti Leppäranta, *University of Helsinki, Finland*
Juha Oksanen, *Finnish Geodetic Institute, Finland*
Frank Paul, *University of Zurich, Switzerland*
Petri Pellikka, *University of Helsinki, Finland*
Bruce Raub, *National Snow and Ice Data Center, University of Colorado, United States*
Gareth Rees, *Scott Polar Research Institute, University of Cambridge, England*
Vesa Roivas, *Pöyry Environment, Finland*
Julienne Stroeve, *National Snow and Ice Data Center, University of Colorado, United States*
Tazio Strozzi, *Gamma Remote Sensing, Switzerland*

Subject index

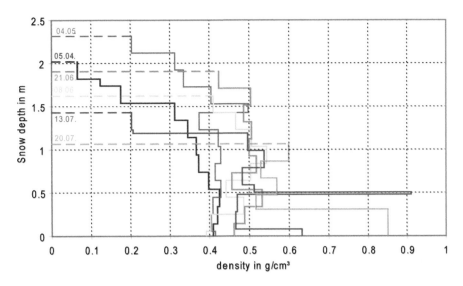

Figure 2.5 Density profiles in the spring to summer snow pack on Hintereisferner (Matzi 2004).

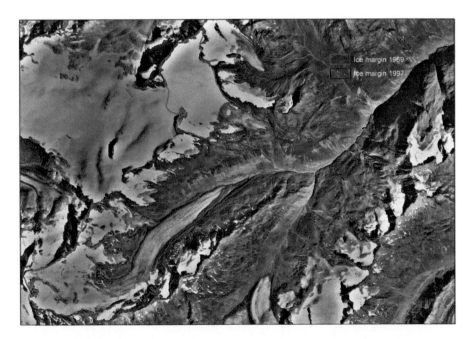

Figure 2.10 Orthophotograph of Hintereisferner, 10°45′ E, 46°8′ N, September 1997.

Figure 2.11 Shaded relief image of the digital elevation model of Hintereisferner, corresponding to Figure 2.10.

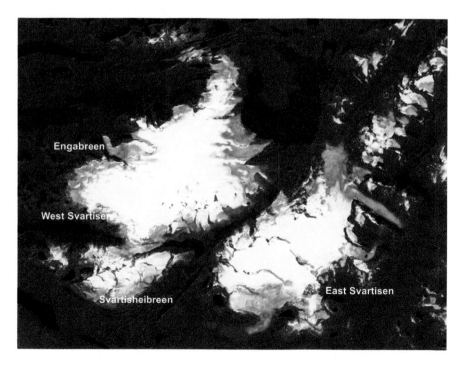

Figure 3.1 Svartisen ice caps in Norway with Engabreen glacier outlet located on the left. Landsat ETM+ satellite image with topographic correction, September 7, 1999 (Heiskanen 2002). The white areas on the glacier are snow surfaces, while grey areas are firn and blue areas bare (blue) ice surfaces.

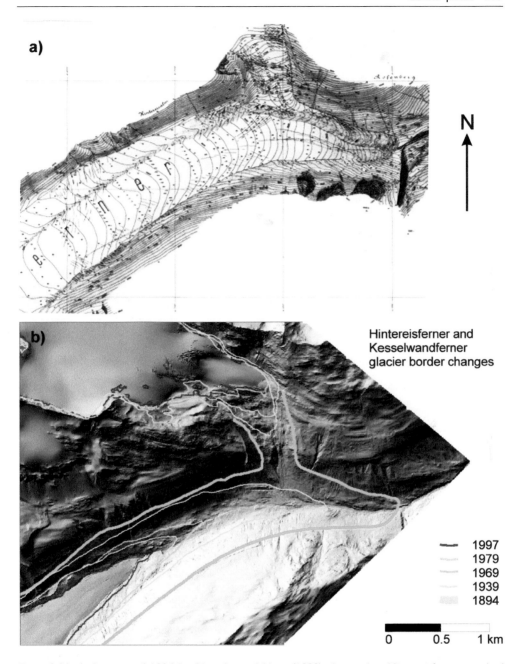

Figure 3.11 a) the map of 1894 by Blümcke and Hess (1899) shows that Hintereisferner reached an elevation of 2275 m in 1894 and thinned 200 m in a century in the location of current snout; b) the position of the snout of Hintereisferner and Kesselwandferner since 1894 digitized from maps various glaciological maps (1894, 1939, 1969, 1979, 1997) and overlaid on a shaded relief model derived from airborne laser scanner data from 2002. The connection between the two glaciers was rapidly lost between 1894 and 1939. Figure material provided kindly by Institute of Meteorology and Geophysics, University of Innsbruck.

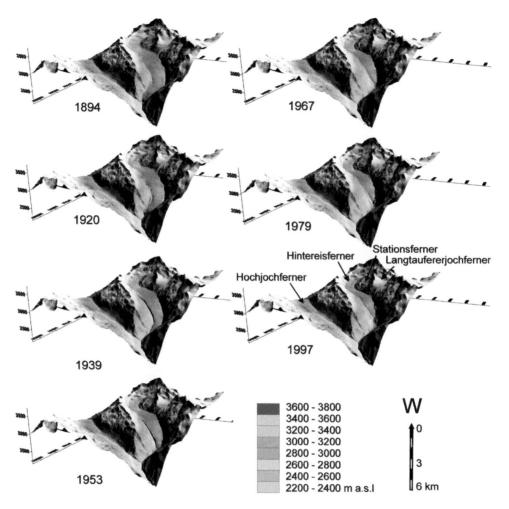

Figure 3.12 The changes of Hintereisferner length and topography between 1894 and 1997 digitized from maps by Norbert Span. The status for Hochjochferner is from 1997 map. Lang-taufererjochferner lost its connection to Hintereisferner in 2000. The figure material provided kindly by Institute of Meteorology and Geophysics, University of Innsbruck.

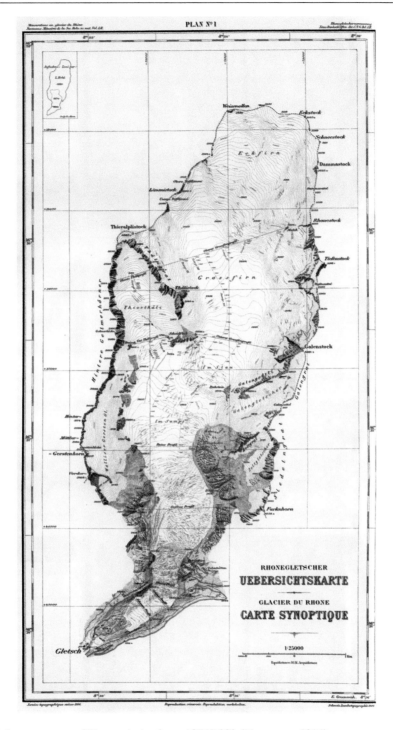

Figure 4.5 Accurate map of Rhone glacier from 1874/1882 (Mercanton 1916).

Figure 5.5 A photograph of Lake Suvivesi taken from the top of Basen nunatak in Dronning Maud Land, Antarctica. A low moraine ridge is seen in the middle of the lake, and Plogen nunatak shows up on the left at the distance of 20 km. The dark area in the foreground is bare ground. Photograph by Matti Leppäranta.

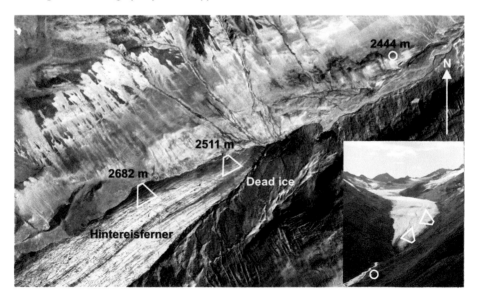

Figure 6.3 The snout of Hintereisferner, shown on a digital camera image mosaic constructed by ALTM images (Parviainen 2006), presents the locations for the acquisition of the terrestrial photographs (triangles) and also the fixed geodetic point on the glacier-free foreground (circle) to which the measurements were tied. An inset photograph shows Hintereisferner in 2003.

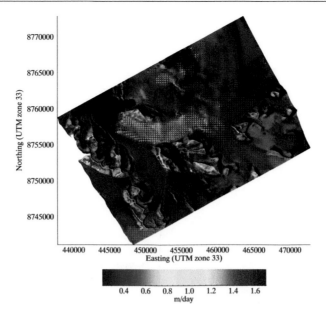

Figure 9.5 Feature tracking glacier velocity from Kronebreen on Svalbard. Data from ERS-1/2 3-day Tandem mission March 8, 1994 and March 11, 1994. Image processed by Yngvar Larsen, Norut Tromsø.

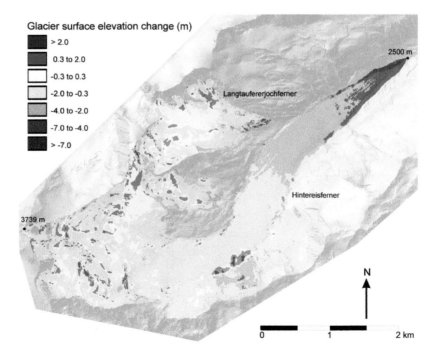

Figure 10.3 Surface elevation change of Hintereisferner for the glaciological year 2001/2002, derived from airborne laser scanner data of October 10, 2001 and September 18, 2002. Some areas increased in elevation due to avalanches like in lower right, for example, but in general glacier was thinning. The overall elevation change of Hintereisferner was −1.3 m.

Figure 10.5 3D-visualisation of Hintereisferner showing general trend of cumulative decrease in elevation between the first and last laser scanning data acquisitions of the OMEGA project; October 10, 2001 and September 26, 2003. Glacier was thinning many metres more in the lower parts of the glacier than in the accumulation areas.

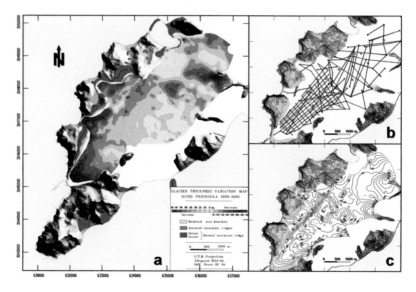

Figure 11.7 a) map of ice thickness changes during 1956–2000 of Johnsons-Hurd glacier, Livingston Island, Antarctica (modified from Molina et al. 2007). It was constructed by subtracting the digital elevation models for both years, retrieved from aerial photographs and satellite images, respectively. Moraines on the proglacial areas are displayed to show earlier Holocene glacier margin positions. The volume change was -0.108 ± 0.048 km^3, which represents a $-10.0 \pm 4.5\%$ change from the 1956 total volume of 1.076 ± 0.055 km^3 computed from the radar-measured ice thickness data; b) network of radar profiles, in which the small circles denote points of CMP measurements; c) resulting ice thickness map.

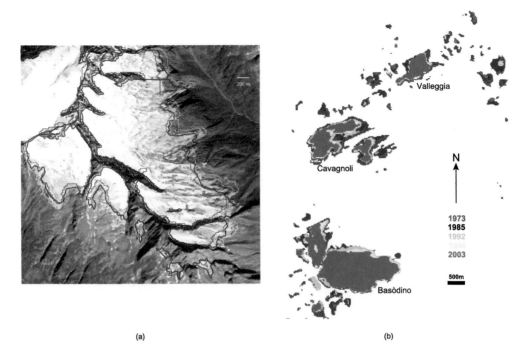

(a)   (b)

Figure 12.4 a) digital overlay of glacier outlines from 1973 (red), 1985 (yellow), 1992 (green) and 1998 (blue) in the Göschener Alp Region, Switzerland. The 1973 outlines are from the digitized glacier inventory, all others are obtained from Landsat TM; b) colour coded glacier areas for five individual years on a pixel-by-pixel basis with advancing Basòdino, disintegrating Cavagnoli and stable Valleggia glaciers. Several snow patches were present in the 1985 image (dark grey).

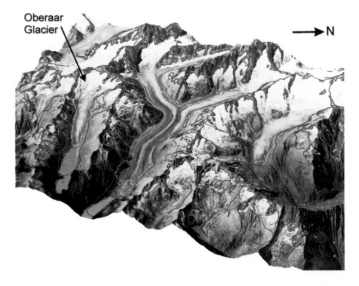

Figure 12.5 Synthetic oblique perspective view of the region to the west of the Grimsel Pass with a fused satellite image (Landsat TM and IRS-1C) draped over a DEM and glacier outlines from 1850 (red), 1973 (blue) and 1998 (green). DEM25 reproduced with permission from swisstopo (BA091556).

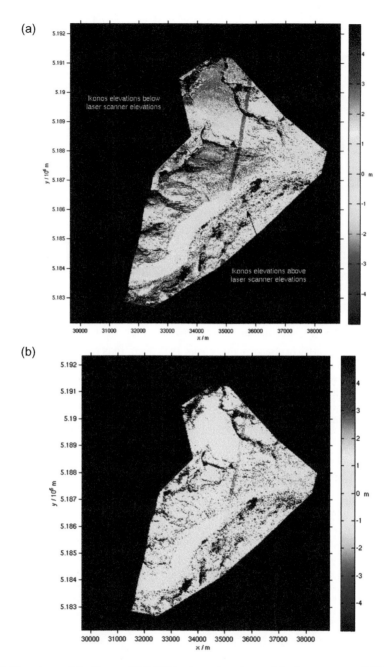

Figure 13.3 Elevations of the Ikonos RSG DEM minus elevations of the laser scanner DEM a) according to the original georeferencing and b) after surface matching. Differences in georeferencing have been corrected and distortions can be analyzed. The coordinate system is M28.

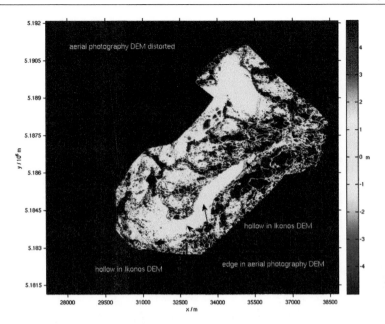

Figure 13.4 Elevations of the aerial photography DEM minus elevations of the Ikonos LPS DEM. There are two longitudinal hollows in the Ikonos LPS DEM. The artificial edges and systematic shape distortion appear in the aerial photography DEM. The coordinate system is M28.

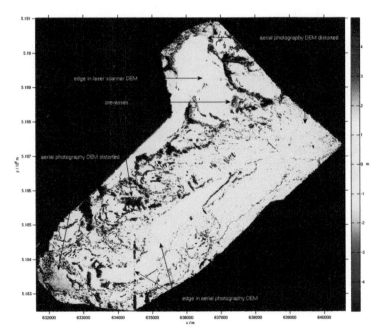

Figure 13.6 Elevations of the laser scanner DEM of 10 m grid spacing minus elevations of the aerial photography DEM. The artificial edges and differences by crevasses have been corrected in the refined laser scanner DEM of 5 m grid spacing. There appear artificial edges and a shape distortion in the aerial photography DEM. The coordinate system is WGS84/UTM32.

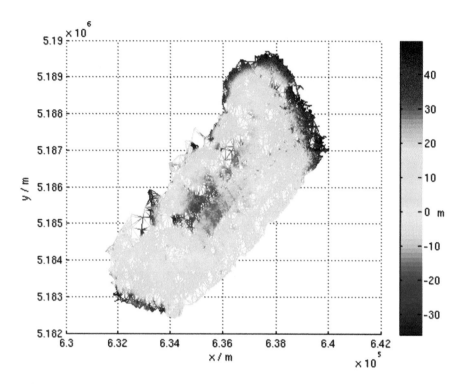

Figure 13.7 Elevations of the laser scanner DEM minus elevations of the digital camera DEM. The digital camera DEM is distorted as the elevations decrease towards borders. The coordinate system is WGS84/UTM32.

(a)

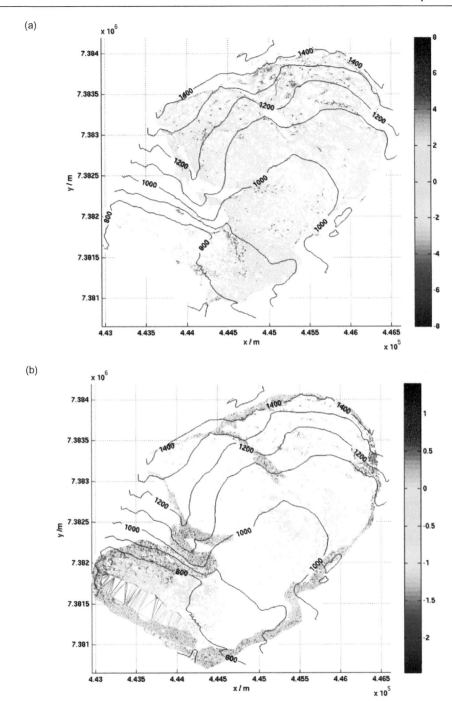

(b)

Figure 14.1 Differences in elevation between the laser scanner and aerial photography DEMs at points where the changes a) exceed the error bounds b) are within the error bounds. Differences in elevation larger than 8 m in absolute value occurring at few points have not been plotted in Figure 14.1a to better distinguish spatial variations of smaller differences. White areas contain no data. The contour lines are according to the aerial photography DEM. Svartisheibreen, Norway, August 25–September 24, 2001.

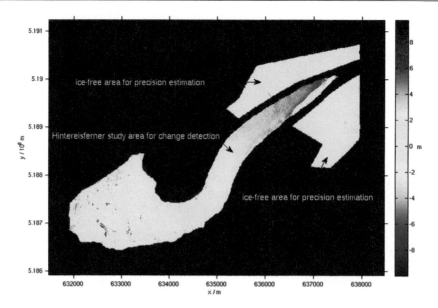

Figure 14.3 Changes in elevation at least occurred in Hintereisferner between August 19, 2002, and August 12, 2003, according to the error bounds and differences in elevation between the laser scanner DEMs in the ice-free test areas for precision estimation. Changes in elevation larger than 10 m in absolute value occurring at few points in the study area have not been plotted in the image to better distinguish spatial variations of smaller differences.

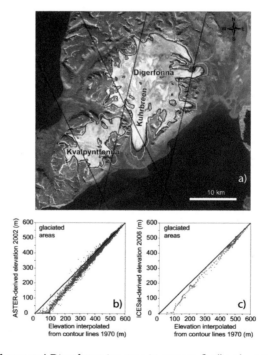

Figure 15.5 a) Kvalpyntfonna and Digerfonna ice caps in eastern Svalbard presented in ASTER satellite image from 2002; b) elevation changes between topographic map of 1970 and DEM derived from the ASTER stereo images; c) elevation changes between topographic map of 1970 and ICESat GLAS LiDAR DEM from 2006. The lines in ASTER image indicate the ICESat ground tracks used (after Kääb 2008).

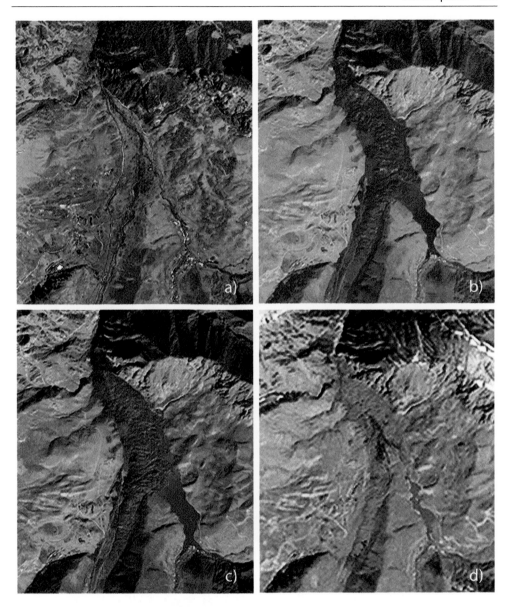

Figure 15.6 Deposits of the September 20, 2002, rock-ice avalanche at Kolka/Karmadon, North Ossetia, Caucasus, represented by ASTER scenes covering 3.8 × 4.5 km. Avalanche direction was from the image bottom to the top. a) June 22, 2001; b) September 27, 2002; c) October 13, 2002; d) April 09, 2004 (after Kääb et al. 2003a).

Figure 15.7 The topography of Tasman glacier and its environments in New Zealand were computed from ASTER satellite image acquired on April 29, 2000. North is to the top and depicted terrain section is about 25 km by 25 km.

T - #0184 - 171019 - C16 - 246/174/16 - PB - 9780367384647